Lecture Notes in Computer Science 9795

Commenced Publication in 1973
Founding and Former Series Editors:
Gerhard Goos, Juris Hartmanis, and Jan van Leeuwen

More information about this series at http://www.springer.com/series/7407

Hien T. Nguyen · Vaclav Snasel (Eds.)

Computational Social Networks

5th International Conference, CSoNet 2016
Ho Chi Minh City, Vietnam, August 2–4, 2016
Proceedings

Editors
Hien T. Nguyen
Ton Duc Thang University
Ho Chi Minh City
Vietnam

Vaclav Snasel
VSB-Technical University of Ostrava
Ostrava
Czech Republic

ISSN 0302-9743   ISSN 1611-3349   (electronic)
Lecture Notes in Computer Science
ISBN 978-3-319-42344-9   ISBN 978-3-319-42345-6   (eBook)
DOI 10.1007/978-3-319-42345-6

Library of Congress Control Number: 2016944351

LNCS Sublibrary: SL1 – Theoretical Computer Science and General Issues

Printed on acid-free paper

This Springer imprint is published by Springer Nature
The registered company is Springer International Publishing AG Switzerland

Preface

The International Conference on Computational Social Network (CSoNet) provides a premier interdisciplinary forum bringing together researchers and practitioners from all fields of social networks. The objective of this conference is to advance and promote the theoretical foundation, mathematical aspects, and applications of social computing. This conference series started in 2013 in Hangzhou, China, under the title of the Workshop on Computational Social Networks (CSoNet 2013) and was co-located with COCOON 2013. The second edition was co-located with COCOON 2014 in August, 2014, at Atlanta, USA, while the third edition was co-located with COCOA 2014 in December, 2014, at Maui, Hawaii, USA. The name of the conference series changed to its current title during the fourth conference held in August, 2015, at Beijing, China. The fifth CSoNet (CSoNet 2016) was held during August 2–4, 2016, in Ho Chi Minh City, Vietnam.

The conference welcomed all submissions focusing on common principles, algorithms, and tools that govern social network structures/topologies, functionalities, social interactions, security and privacy, network behaviors, information diffusions and influence, and social recommendation systems that are applicable to all types of social networks and social media.

The organizers received 79 submissions. Each submission was reviewed by at least two external reviewers or Program Committee members. Finally, a total of 30 papers were accepted as regular papers for presentation at CSoNet 2016 and publication in the proceedings.

We would like to express our appreciation to the numerous reviewers and special session chairs whose efforts enabled us to achieve a high scientific standard for the proceedings. We cordially thank the members of the Technical Program Committee and Steering Committee for their support and cooperation in this publication. We would like to thank Alfred Hofmann, Anna Kramer, and their colleagues at Springer for meticulously supporting us in the timely production of this volume. Moreover, the conference could not have happened without the commitment of the Faculty of Information Technology - Ton Duc Thang University, who helped in many ways to assemble and run the conference. Last but not least, our special thanks go to all the authors who submitted papers and all the participants for their contributions to the success of this event.

June 2016

Hien T. Nguyen
Vaclav Snasel

Organization

Steering Committee

My T. Thai   University of Florida, USA (Chair)
Zhi-Li Zhang   University of Minnesota, USA
Weili Wu   University of Texas—Dallas, USA

Program Committee Co-chairs

Hien T. Nguyen   Ton Duc Thang University, Vietnam
Vaclav Snasel   VSB-Technical University of Ostrava, Czech Republic

Publicity Co-chairs

William Liu   Auckland University of Technology, New Zealand
Jason J. Jung   Chung-Ang University, South Korea
Sanghyuk Lee   Xi'an Jiaotong-Liverpool University, China

Technical Program Committee

Abhijin Adiga	Virginia Tech, USA
Konstantin Avrachenkov	Inria Sophia Antipolis, France
Vladimir Boginski	University of Florida, USA
Tru Cao	Ho Chi Minh City University of Technology, Vietnam
Hocine Cherifi	Université de Bourgogne, France
Luca Chiaraviglio	University of Rome La Sapienza, Italy
Trong Hai Duong	International University, VNU-HCMC, Vietnam
Preetam Ghosh	Virginia Commonwealth University, USA
Van-Nam Huynh	Japan Advanced Institute of Science and Technology, Japan
Jason J. Jung	Chung-Ang University, South Korea
Vasileios Karyotis	National Technical University of Athens, Greece
Donghyun Kim	North Carolina Central University, USA
Jiamou Liu	The University of Auckland, New Zealand
William Liu	Auckland University of Technology, New Zealand
Anh-Cuong Le	Ton Duc Thang University, Vietnam
Parma Nand	Auckland University of Technology, New Zealand
Hien T. Nguyen	Ton Duc Thang University, Vietnam
Panos Pardalos	University of Florida, USA
Hai Phan	New Jersey Institue of Technology, USA
Jaroslav Pokorny	Charles University in Prague, Czech Republic

Tho T. Quan	Ho Chi Minh City University of Technology, Vietnam
Maxim Shcherbakov	Volgograd State Technical University, Russia
David Sundaram	The University of Auckland, New Zealand
Xijin Tang	CAS Academy of Mathematics and Systems Science, China
Mario Ventresca	Purdue University, USA
Li Wang	Taiyuan University of Technology, China
Yu Wang	University of North Carolina at Charlotte, USA
Fay Zhong	California State University East Bay, USA

Contents

Shortest Paths on Evolving Graphs

Yiming Zou[1], Gang Zeng[2], Yuyi Wang[3], Xingwu Liu[2(✉)], Xiaoming Sun[2],
Jialin Zhang[2], and Qiang Li[2]

[1] School of Informatics and Computing, Indiana University Bloomington,
Bloomington, IN, USA
yizou@iu.edu

[2] Institute of Computing Technology, Chinese Academy of Sciences, Beijing, China
{zenggang,liuxingwu,sunxiaoming,zhangjialin,liqiang01}@ict.ac.cn

[3] Distributed Computing Group, ETH Zürich, Zürich, Switzerland
yuyiwang920@gmail.com

Abstract. We consider the shortest path problem in evolving graphs
with restricted access, i.e., the changes are unknown and can be probed
only by limited queries. The goal is to maintain a shortest path between
a given pair of nodes. We propose a heuristic algorithm that takes into
account time-dependent edge reliability and reduces the problem to find
an edge-weighted shortest path. Our algorithm leads to higher precision
and recall than those of the existing method introduced in [5] on both
real-life data and synthetic data, while the error is negligible.

Keywords: Evolving graphs · Heuristic algorithm · Shortest path

1 Introduction

Network structure plays key roles in cyber space, physical world, and human
society. One common feature of various real-life networks is that they are evolving
all the time, e.g., the follow/unfollow action in Twitter, the crash failure in
the sensor network, the traffic accident in the traffic network. Besides, many
interesting networks are very big and changing so rapidly that it becomes difficult
to handle all the change information in time. Take the Web as an example: it
is impossible for any agent to exactly learn about the topological changes in
the last hour, due to the sheering size of Web. Another example is that a third
party vendor cannot capture the exact changes of Facebook networks, because
the changes can be probed only through a rate-limited API. This dynamic and
ignorant nature poses challenges to traditional data analyzing techniques where
inputs are given initially and are fixed during computing. The challenges remain
even if more and more interactive computing styles are emerging such as online,
streaming, incremental algorithms, etc., since all of them basically assume that
the changes are fully known.

The work is partially supported by National Natural Science Foundation of China
(61222202, 61433014, 61502449) and the China National Program for support of
Top-notch Young Professionals.

© Springer International Publishing Switzerland 2016
H.T. Nguyen and V. Snasel (Eds.): CSoNet 2016, LNCS 9795, pp. 1–13, 2016.
DOI: 10.1007/978-3-319-42345-6_1

In this context, Anagnostopoulos et al. [5] introduced an algorithmic framework which probes changes and maintains a specific feature of an evolving graph while this graph changes constantly. In concrete, they assume that, at each time step, a randomly selected edge is removed from the graph, a new edge is added between a random pair of non-adjacent nodes, and the algorithm, unaware of the positions of the changes, is allowed to query one arbitrary node for its neighbors. They studied the connectivity problem in this framework and proposed an algorithm which outputs a valid path connecting two given nodes at every step with probability at least $1 - \frac{\log n}{n}$, where n is the order of this graph. However, the basic assumption of uniform randomness is difficult, if not impossible, to verify for many real-life networks such as Math coauthorship, the Internet at the router level, and the Web [2]. Furthermore, the performance of their algorithm is guaranteed only for dense graphs with average degree at least $\Omega(\ln n)$, but numerous networks in real life are relatively sparse. For instance, Twitter network [11], DBpedia producer network [6], and Amazon buyers network [24] all have average degree less than five in spite of their big size. These limitations compromise the practical relevance of the algorithm in [5].

Instead of making such assumptions, we investigate *real-life* evolving networks. Besides, we go one step further to study the shortest path problem, because of its important role in research and its numerous applications such as GPS navigation in traffic networks and target influence in social networks. Our goal is to find a shortest path (if it exists) between two given nodes S and T in networks at every step, under the condition that changes of the network is unknown and can be probed only through few queries. In fact, shortest path problems are among the most actively-studied topics in dynamic graphs [12, 17, 20, 23], where network changes are fully known, contrary to our setting.

Our contributions in this paper are threefold:

- we find that it is advantageous to exploit historic query results, even though they were far from updated.
- we propose a novel heuristic algorithm which fully uses historic query results by weighting the edges in a time-dependent way and reducing the problem to the edge-weighted shortest path problem.
- we experimentally study this algorithm, and the results show that our algorithm outperforms the previous method (which almost ignores historic information) in terms of recall and precision, while its error is negligible.

Related work. In their seminal paper in 2009 [4], Anagnostopoulos et al. studied the sorting problem on dynamic data, initializing the new algorithmic paradigm where limited-accessible data evolves indefinitely. The paradigm was first applied to evolving graphs in 2012 [5], focusing on the connectivity problem and the minimum spanning tree problem. The work on the connectivity problem has inspired the dynamic shortest path problem, which is the topic of the present paper. Actually, the algorithm in [5] can be slightly modified to solve this problem, though the performance is not good enough on real-life networks (see Sect. 4). Two other problems have also been studied in similar settings: Bahmani et al. [7] designed an algorithm to approximate the PageRank; Zhuang et al. [25] considered the influence maximization problem in dynamic social networks.

The dynamic graph model, where the changes are fully known, has been actively studied. The tasks mainly belong to two categories. One is to efficiently answer queries for some properties of a changing graph [9,13]. The other is to analyze a stream of graphs, subject to limited resources (storage, for example) [18,22]. Both of them assume that the changes of the graphs can be completely observed, fundamentally different from our evolving graph model.

It is worth noting that the evolving data model in this paper is essentially different from the noisy information model [1,15] whose main difficulty is caused by misleading information. In our model, the query results are reliable, while the challenge comes from the restricted access to the data and the key is to design informative query strategies.

In the algorithm community, there are many other models dealing with dynamic and uncertain data, from various points of view. However, none of them captures the two crucial aspects of our evolving scenario: (i) the underlying data keeps changing, and (ii) the data is limited-accessible. For example, local algorithms [8,16] try to capture a certain property with a limited number of queries, but the underlying graphs are typically static; online algorithms [3] know all the data up to now, though the data comes over time and is processed without knowledge about the future data.

2   Model

We follow the framework defined by Anagnostopoulos et al. [5] used for algorithms on evolving graphs. The time is discretized into steps numbered by sequential non-negative integers. An evolving graph is a graph whose edges change over time, modeled by an infinite stream of graphs $G^t = (V, E^t), t \geq 0$, where V is the set of vertices and E^t is the set of edges at time t. For any $t > 0$, $\alpha_t = \frac{|E^t \setminus E^{t-1}|}{|E^{t-1}|}$ is called the evolution rate at time t. The graphs are unknown except that it can be probed by querying local structures. Specifically, at time t, for any node v, a query of v returns all the neighbors of v in G^t. The query operation is highly restricted, in the sense that the number of queries at one time step cannot exceed a prescribed constant β which is called the probing rate in this paper. Note that [5] assumes that the graph evolves in a uniformly random manner, but we don't make this assumption since real-life evolving networks are mainly considered. This paper just deals with undirected graphs, but the algorithm can also be applied to directed graphs, up to minor modifications.

Then we formulate the shortest path problem on the evolving graphs. Given an evolving graph $G^t(t \geq 0)$ with a specified pair of nodes S and T, the objective is to design an algorithm which runs forever and at any time t, produces a shortest (S, T)-path in G^t. When S and T are disconnected at time t, the algorithm should claim *disconnected*. We say that an output path is valid, if and only if it is indeed a path in the current graph.

Obviously, when β is small, it is impossible to design an algorithm exactly solving this problem, so approximation is the only choice. To evaluate the performance of the algorithm, we define three metrics: recall, precision, and error. Formally, *recall* is the proportion of the cases that a valid path is output among the cases that S and T are connected, *precision* is the proportion of the cases that the output path is valid among the cases that a path is output, and *error* is the average value of $(\frac{\text{length of the output path}}{\text{distance between } S \text{ and } T} - 1)$ among the cases that a valid path is output. Note that error is defined only when a valid path is found, because the path length makes sense just in this case.

The motivation of these measures is as follows. Essentially, the objective of the shortest path problem is twofold: validity which means finding a path that does exist, and optimality which means the path should be as short as possible. Contrary to the case of static graphs, the two aspects may not be fulfilled simultaneously in dynamic graphs. Hence they are measured separately: validity is captured by recall and precision, and optimality by error.

3   Algorithm

This section describes two algorithms (Growing-Ball and DynSP) for maintaining a shortest (S, T)-path on an evolving graph modeled by a stream of graphs. Each algorithm boils down into two interacting parts: probing strategy that decides which nodes to probe, and computing strategy which maintains the shortest (S, T)-path based on the probed information up to now. Growing-Ball serves as a baseline, and is a natural adaption of the connectivity algorithm in [5]. DynSP is specially designed for the shortest path problem, and is the main contribution of this paper. Both algorithms are parameterized by a positive integer β, the probing rate. Recall that β, usually small compared with the size of the graphs, is mainly determined by the accessability of the underlying graph stream. It would be desirable that an algorithm performs well even when β is small.

3.1   Growing-Ball

Now we describe the algorithm Growing-Ball, whose basic idea comes from [5]. It was proposed to find an (S, T)-path in a dense evolving graph, and is slightly adapted in the present paper to find a *shortest* (S, T)-path.

Basically, Growing-Ball proceeds phase by phase. At the beginning of each phase, it initializes two singleton balls $B_S = \{S\}$ and $B_T = \{T\}$. Then it goes round by round, alternatively growing the balls B_S and B_T. Until B_S and B_T meet or one of them cannot grow any more, the current phase ends and the next phase is started. In the first case, an (S, T)-path P is naturally obtained and will be output throughout the next phase. In the second case, the next phase will always claim "disconnected".

Now take a close look at a round of growing the ball B_S (likewise for B_T). If all vertices of B_S are labeled as *visited*, B_S cannot grow any more and this round ends. Otherwise, among all the vertices of B_S that are labeled as *unvisited*, choose one (say v) that is the closest to S. Mark v as visited, and query v for all its neighbors in the current graph. Then grow B_S with the neighbors that were not in B_S, and mark them as unvisited. Note that the algorithm has parameter β, meaning that at each time step, B_S and B_T altogether grow for β rounds.

Algorithm 1. Dynamic Shortest Path (DynSP)

Input: S, T, β
Output: A path P or *disconnected*, at every step
    initialize a vertex-labeled graph $G = (V, \emptyset, \ell)$ with labels $\ell[v] = 0$ for all $v \in V$;
    initialize balls $B_0 = \{S\}, B_1 = \{T\}$ and mark S, T as *unvisited*;
    while True **do**
        for $i = 1 : \beta$ **do**
            set $j = i \bmod 2$;
            in B_j, choose an unvisited node v closest to the center and mark v as *visited*;
            query v for all its neighbors in the present stream graph;
            add all of v's news neighbors to B_j and mark them as *unvisited*;
            update v's neighborhood in G according to the query result;
            $\ell[v] = 0$ and $\ell[u] = \ell[u] + 1$ for all nodes $u \neq v$ in G;
            if $B_0 \cap B_1 \neq \emptyset$, **then**
                set queues $B_0 = \{S\}, B_1 = \{T\}$ and mark S, T as *unvisited*;
            end if
        end for   /** f in the next line is the weighting function **/
        weight each edge $e = \{u, v\}$ in G with $w_e = f(\min\{\ell[u], \ell[v]\})$;
        compute an edge-weighted shortest (S, T)-path P in G;
        if P does not exist in G **then**
            Output 'disconnected';
        else
            Output P without edge-weights;
        end if
    end while

3.2 DynSP

By [5], the algorithm Growing-Ball has perfect recall and precision on randomly-evolving dense graphs with high probability. This is mainly due to the fact that the path found in a phase will remain valid in the next phase with high probability. The fact remains true if the graph changes slowly and uniformly, but it may not be the case for real-life networks which may change dramatically and irregularly.

A natural idea to improve Growing-Ball is to recompute a shortest path at every step, rather than directly output the path found in the last phase. Since the recently-probed information is well incorporated, it is more likely that the recomputed result conforms with the ground truth. However, there is a dilemma. On the one hand, historic query results must be used to recompute the path, because query results at the current time step usually don't lead to an (S, T)-path (when the balls don't meet in β rounds). On the other hand, historic query results should be avoided as far as possible, since it might be out-of-date.

The point of our solution is to measure the reliability of historic query results in terms of the elapse time and accordingly make a trade-off between length and reliability of the paths. The algorithm, called DynSP, is shown in Algorithm 1.

Basically, DynSP follows the probing strategy of Growing-Ball and uses a *novel computing strategy*. It maintains a vertex-labeled graph G to incorporate all the information probed up to now, weights the edges of G in an elapsed-time-dependent way, and reduces the original shortest path problem to weighted shortest path problem on G.

Specifically, DynSP carries out three tasks at every time t. First, alternatively grow balls B_S and B_T for totally β rounds, update the topology of the graph G according to the query results, and update the label of each vertex of G. Intuitively, the label of a vertex indicates how long it has been probed since the last query of the vertex. Second, weight each edge of G with a function (called the weighting function) of the elapsed time since the last observation of the edge. The weight of an edge is supposed to capture the risk that the edge gets invalid. It is conceivable that in general the risk increases with the elapsed time, which is actually the smaller label of the endpoints of the edge. Third and last, compute a shortest (S, T)-path in the edge-weighted graph G and output it (with the weights ignored). If S and T are not connected in G, output "disconnected".

As to the weighting function, it determines the trade-off between length and validity of the path. Intuitively, the bigger the weighting function is, the less likely highly-risky edges are used in the path. It seems that any non-negative, non-decreasing function f can be used as the weighting function. In Sect. 4, we will try five candidates: $f(x) = 1$, $f(x) = \ln x$, $f(x) = x$, $f(x) = x^2$, $f(x) = x^3$.

Note that DynSP does not degenerate into Growing-Ball, even if $f(x) = 1$. Actually, DynSP is uses a novel computing strategy (the blue lines in Algorithm 1) which is essentially different from Growing-Ball.

4   Experiments

In this section, we first introduce the data sets that will be used, then describe the setup of the experiments, and finally present the experimental results.

4.1   Data Sets

We use synthetic dynamic networks of various sizes, the human contact network in an American high school on January 14th of 2010, the coauthor network of scientific papers from DBLP, and the network of German Wikipedia articles.

Synthetic Networks. We generate the synthetic dynamic networks by the unweighted evolving graph model in [5]. Specifically, the vertices and the number of edges are fixed. The initial graph G^0 is sampled from the Erdős-Rényi model $G_{n,m}$. At every step $t > 0$, randomly remove αm edges from G^{t-1}, and randomly add αm edges to G^{t-1}, resulting in G^t. We generate two dynamic networks, namely Syn1K and Syn1M, with $m = n = 10^3$ and $m = n = 10^6$, respectively. Both networks have evolution rate $\alpha = 20\%$, and evolve for $T = 1000$ time steps for Syn1K, $T = 100$ for Syn1M.

Contact Network. This data set is borrowed from [21]. It records the 8-hour close-proximity interactions among 789 volunteers from an American high school. The 8 h is discretized into 1440 steps. A contact network is defined at each step with the volunteers as vertices and with edges indicating the occurrences of interactions during that interval. Among the networks, only 1296 consecutive ones will be used in the experiment. The evolution rate is 8–12%.

Coauthor Network. This data is shared with us by the authors of [25]. It records the publication information of 862,770 authors during 31 years. A network is constructed for each year with the authors as vertices and with edges indicating co-authorship in the past or the next two years. In this way, we get an evolving network with 31 time steps. The evolution rate is 11–20%.

Wiki Network. It is available at http://konect.uni-koblenz.de/networks/ and provided by [19]. It includes the editing history of 1.5 million Germany Wikipedia articles. The history is uniformly discretized into 150 time steps. For each step, an undirected graph is constructed with articles as vertices and with edges indicating links in that time interval. We choose to produce undirected graph so as to be consistent with the other data sets. Altogether, we get an evolving network with 150 time steps. The evolution rate is 0.7–6%.

Table 1. Summary of the data sets

Data sets	Average n	Average m	Evolution rate	Time steps	Connectivity	Duration
Syn1K	10^3	10^3	20%	1000	80.1%	2.0
Syn1M	10^6	10^6	20%	100	79.8%	3.1
Contact	789	1.6×10^3	8–12%	1296	92.4%	4.8
Coauthor	862,770	1.2×10^6	11–20%	31	74.8%	5.6
Wiki	1.5×10^6	1.1×10^6	0.7-6%	150	77.3%	7.2

The data sets are summarized in Table 1, where connectivity means the expected fraction of steps in which a randomly-chosen pair is connected and duration means the expected period of time in which a random-chosen pair remains connected once it gets connected.

4.2 Experimental Setup

Since the algorithms are designed to run infinitely, what's really interesting is their performance in steady states. Hence, in the experiments we assume that the algorithms fully know the initial underlying graph G^0. However, G^0 can be unknown in general. The assumption of knowing G^0 is just to guarantee that the simulations quickly reach steady states.

The experiments have one parameter β, which prescribes how many queries can be made at a time step. Conceivably, fixing an algorithm and a data set, the bigger β is, the better the algorithm performs. To further explore the effect of β on performance, we use multiple values of β on each data set. Namely, choose $\beta = 40, 60, 80, ..., 160$ for Syn1K and $\beta = 40, 60, 80, ..., 260$ for the contact network. For the other data sets, choose $\beta = 100, 200, ..., 600$; such β are relatively small, considering the high evolution rate and big size of each data set.

Now we talk about the weighting function f in DynSP. It is hard to figure out a universally optimal weighting function, especially when real-life networks are considered. So, we test five candidates on each data set, namely $f(x) = 1, \ln x, x, x^2$, and x^3. These candidates are chosen because they are very simple, positive, and non-decreasing.

As to the choice of S and T, for each data set, we uniformly and randomly choose 1000 pairs of vertices and run the algorithms with each pair as input. For each algorithm, the average performance on the 1000 inputs reasonably approximates the algorithm's performance on the data set.

4.3   Results

As mentioned in Sect. 2, the algorithms are evaluated in terms of recall, precision, and error. The results will be presented in four aspects. Due to the space limitation, only partial results will be illustrated.

Effect of β on the Performance. In this part, all the figures of DynSP are based on the weighting function $f(x) = x$. However, all the results remain valid for the other weighting function candidates.

The first observation is that on all the data sets, both DynSP and Growing-Ball have better performance (higher recall and precision and smaller error) when β is bigger, as shown in Figs. 1 and 2. This is reasonable, since when β is bigger, more queries can be made in one step, which means more information of the underlying graph can be obtained.

We also observe that on all the data sets, recall and precision quickly approach 1 and the error rapidly approaches 0 when β increases. It is a surprise that extremely good performance on all the large data sets (Figs. 1 and 2) can be achieved when β is as low as 600, even though tens of thousands of edges change at every time step. This is desirable, indicating that our algorithm DynSP works perfectly with few queries even though the networks change dramatically.

Performance Comparison of the Algorithms. We compare the performance of Growing-Ball with that of DynSP, in terms of the experimental results. Throughout this part, the weighting function in DynSP is $f(x) = x$, but all the results remain true for the other weighting functions. Again from Figs. 1 and 2, we make the following observations.

First, DynSP always outperforms Growing-Ball in both recall and precision under the same β. Take the coauthor network as an example. When $\beta = 100$,

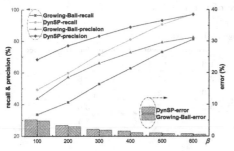 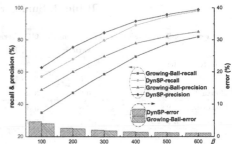

Fig. 1. Growing-Ball and DynSP: performance vs β on Coauthor network

Fig. 2. Growing-Ball and DynSP: performance vs β on Wiki network

recall of DynSP is 0.49, 50 % higher than that of Growing-Ball which is 0.33; precision of DynSP is also 50 % higher than that of Growing-Ball. When $\beta = 600$, recall and precision of DynSP are as high as 0.97, but those of Growing-Ball are no more than 0.84.

The better performance of DynSP is mainly due to the fact that at every time step, DynSP uses the latest probed information to update the path or connectivity. It is hence possible to capture the changes that have occurred recently, and to produce outputs that conform with the ground truth. On the contrary, Growing-Ball always outputs the path that was found in the last finished phase, so the outputs are risky of getting invalid.

Second, when parameterized by the same β, the error of DynSP is no smaller than that of Growing-Ball, but the difference is at most 0.02. It means that the increase in error is negligible. Such small increase in error deserves, since the precision and recall are substantially improved. Hence our trade-off between path length and validity is justified.

The higher error of DynSP may be due to the following reason. Growing-Ball searches for a path locally (only in the union of the two balls), but DynSP searches globally in the graph G. In extreme cases where local paths between S, T consist of highly-risky edges, DynSP may find a bypassing path with lower risk. This helps to improve recall and precision, at the cost that the bypassing path may be a little too long, compared with the true shortest path.

The paths computed by Growing-Ball and DynSP are further analyzed in Table 2, where GT and GB mean ground truth shortest paths and the paths output by Growing-Ball respectively, while *average length* and *length variance* stand for the average length of the paths and the variance of the length respectively. It indicates that Growing-Ball and DynSP almost have the same average length and variance which are very close to those of the ground truth.

DynSP with Different Weighting Functions. Now we check the performance of DynSP when different weighting functions f are used. For each $f(x) = 1, \ln x, x, x^2, x^3$, we run DynSP on every data set. Part of the results including coauthor network, Wiki network are shown in Figs. 3 and 4, with

Table 2. Summary of the results

Data sets	GT average length	GT length variance	GB average length	GB length variance	DynSP average length	DynSP length variance
Syn1K	3.1	1.3	3.3	1.3	3.3	1.3
Syn1M	5.7	2.6	5.7	2.7	5.8	2.8
Contact	2.8	0.9	3.0	1.0	3.1	0.9
Coauthor	4.9	2.8	5.3	3.2	5.3	3.0
Wiki	5.2	3.1	5.4	3.3	5.4	3.4

$\beta = 400, 400$, respectively. Each figure shows how recall, precision, and error vary with the weighting functions, on the corresponding data set. The results with other β are not illustrated here due to space limitation, but they also coincide with the following observation.

From the figures, we can observe that all the performance metrics reach the maximum at $f(x) = x^2$. Namely, when f changes from 1 through to x^2, precision, recall, and error all increase; when f changes from x^2 to x^3, this trend does not continue: precision, recall, and error all decrease (though just slightly). This phenomenon happens in all the experiments, suggesting that this may hold in general. It means that if one wants to choose f for a trade-off between precision, recall, and error, he/she just needs to try one side of x^2.

It is reasonable that all the metrics are relatively small when f is either too small or too big. Intuitively, when f is too small, the difference between new edges and old edges disappears, so it is likely that out-of-dated edges are used in the output path; when f is too big, all old edges will not be used, possibly leading to incorrect decision on a path or connectivity. However, it is a surprise that $f(x) = x^2$ is a universal peak in the experiments, and we have no idea yet to formally prove it.

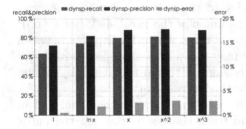

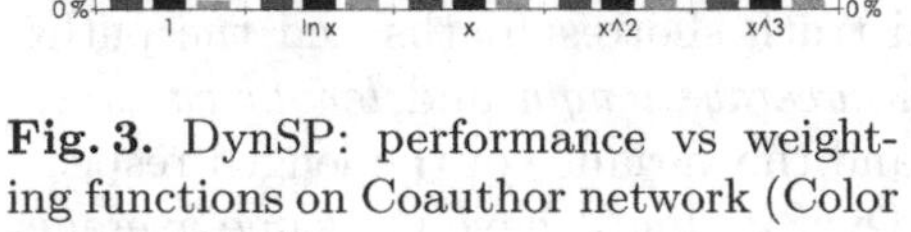

Fig. 3. DynSP: performance vs weighting functions on Coauthor network (Color figure online)

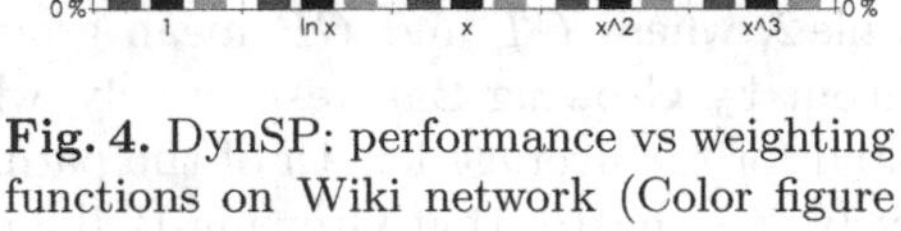

Fig. 4. DynSP: performance vs weighting functions on Wiki network (Color figure online)

Asymptotic Performance of DynSP. Since DynSP is designed to deal with graphs that evolve infinitely, an important issue is whether the performance decays with time. We have done experiments on all data sets except the coauthor

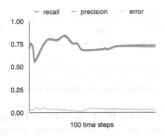
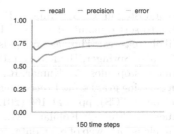

Fig. 5. DynSP: asymptotic performance on Syn1M (Color figure online)

Fig. 6. DynSP: asymptotic performance on Wiki network (Color figure online)

network due to the small size of its time step. The weighting function is $f(x) = x$. At every time step t during an experiment, we record the recall, precision, and error during the period from the beginning till t. By this means, a curve can be naturally obtained for each metrics. Figures 5 and 6 in the following show the curves of the three metrics on Syn1M and Wiki network, with $\beta = 400, 400$, respectively. These figures indicate that the performance of DynSP does not decay with time. This remains true for other β and weighting functions.

5 Conclusion

We study the dynamic shortest path problem, where the underlying graph keeps evolving but the changes are unknown. The only way to learn about the changes is through local queries. The goal is to maintain the shortest path between a given pair of nodes. We propose the DynSP algorithm to solve this problem. Experiments on synthetic and real-life data show that DynSP has high recall and precision with negligible error. This work sheds light on handling dynamic real-life big data that evolves dramatically and irregularly and can be accessed only in a highly-restricted manner.

An interesting future direction is to explore why the weighting function x^2 is so special for DynSP. This may lead to a general guideline on choosing weighting functions in practice. Another future direction is to find a method to determine the probing rate β when the algorithm only has limited access to the network evolution. A more ambitious task is to adaptively tune β during computing, rather than fix it at the beginning.

References

1. Ajtai, M., Feldman, V., Hassidim, A., Nelson, J.: Sorting and selection with imprecise comparisons. ACM Trans. Algorithms **12**, 1–19 (2015)
2. Albert, R.: Statistical mechanics of complex networks (2001)
3. Albers, S.: Online algorithms: a survey. Math. Programm. **97**(1–2), 3–26 (2003)

4. Anagnostopoulos, A., Kumar, R., Mahdian, M., Upfal, E.: Sort me if you can: how to sort dynamic data. In: Albers, S., Marchetti-Spaccamela, A., Matias, Y., Nikoletseas, S., Thomas, W. (eds.) ICALP 2009, Part II. LNCS, vol. 5556, pp. 339–350. Springer, Heidelberg (2009)
5. Anagnostopoulos, A., Kumar, R., Mahdian, M., Upfal, E., Vandin, F.: Algorithms on evolving graphs. In: Proceedings of the 3rd Innovations in Theoretical Computer Science (ITCS), pp. 149–160 (2012)
6. Auer, S., Bizer, C., Kobilarov, G., Lehmann, J., Cyganiak, R., Ives, Z.G.: DBpedia: a nucleus for a web of open data. In: Aberer, K., et al. (eds.) ASWC 2007 and ISWC 2007. LNCS, vol. 4825, pp. 722–735. Springer, Heidelberg (2007)
7. Bahmani, B., Kumar, R., Mahdian, M., Upfal, E.: Pagerank on an evolving graph. In: Proceedings of KDD 2012, pp. 24–32 (2012)
8. Bressan, M., Peserico, E., Pretto, L.: Approximating pagerank locally with sublinear query complexity. ArXiv preprint (2014)
9. Casteigts, A., Flocchini, P., Quattrociocchi, W., Santoro, N.: Time-varying graphs and dynamic networks. Int. J. Parallel Emergent Distrib. Syst. 27(5), 387–408 (2012)
10. Chung, F., Lu, L.: The diameter of sparse random graphs. Adv. Appl. Math. 26(4), 257–279 (2001)
11. De Choudhury, M., Lin, Y.-R., Sundaram, H., Candan, K.S., Xie, L., Kelliher, A.: How does the data sampling strategy impact the discovery of information diffusion in social media? In: Proceedings of ICWSM 2010, pp. 34–41 (2010)
12. Demetrescu, C., Italiano, G.F.: Algorithmic techniques for maintaining shortest routes in dynamic networks. Electr. Notes Theor. Comput. Sci. 171(1), 3–15 (2007)
13. Eppstein, D., Galil, Z., Italiano, G.F.: Dynamic graph algorithms. In: Atallah, M.J. (ed.) Algorithms and Theoretical Computing Handbook. CRC Press, Boca Raton (1999)
14. Erdős, P., Rényi, A.: On the evolution of random graphs. Publ. Math. Inst. Hungar. Acad. Sci. 5, 17–61 (1960)
15. Feige, U., Raghavan, P., Peleg, D., Upfal, E.: Computing with noisy information. SIAM J. Comput. 23(5), 1001–1018 (1994)
16. Fujiwara, Y., Nakatsuji, M., Shiokawa, H., Mishima, T., Onizuka, M.: Fast and exact top-k algorithm for pagerank. In: Proceedings of the 27th AAAI Conference on Artificial Intelligence, pp. 1106–1112 (2013)
17. Huo, W., Tsotras, V.J.: Efficient temporal shortest path queries on evolving social graphs. In: Proceedings of the 26th International Conference on Scientific and Statistical Database Management (SSDBM) (2014). Article No. 38
18. Muthukrishnan, S.: Data streams: algorithms and applications. Found. Trends Theoret. Comput. Sci. 1(2), 117–236 (2005)
19. Preusse, J., Kunegis, J., Thimm, M., Gottron, T., Staab, S.: Structural dynamics of knowledge networks. In: Proceedings of ICWSM 2013 (2013)
20. Ren, C.: Algorithms for evolving graph analysis. Ph.D. thesis, The University of Hong Kong (2014)
21. Salathé, M., Kazandjieva, M., Lee, J.W., Levis, P., Feldman, M.W., Jones, J.H.: A high-resolution human contact network for infectious disease transmission. Proc. Nat. Acad. Sci. 107(51), 22020–22025 (2010)
22. Sarma, A.D., Gollapudi, C., Panigrahy, R.: Estimating pagerank on graph streams. J. ACM 58(3), 13 (2011)
23. Xuan, B.B., Ferreira, A., Jarry, A.: Computing shortest, fastest, and foremost journeys in dynamic networks. Int. J. Found. Comput. Sci. 14(2), 267–285 (2003)

24. Yang, J., Leskovec, J.: Defining and evaluating network communities based on ground-truth. In: Proceedings of 2012 ACM SIGKDD Workshop on Mining Data Semantics, pp. 3:1–3:8. ACM, New York (2012). Article No. 3

25. Zhuang, H., Sun, Y., Tang, J., Zhang, J., Sun, X.: Influence maximization in dynamic social networks. In: Proceedings of the 13th IEEE International Conference on Data Mining (ICDM), pp. 1313–1318. IEEE (2013)

Analysis of a Reciprocal Network Using Google+: Structural Properties and Evolution

Braulio Dumba[✉], Golshan Golnari, and Zhi-Li Zhang

University of Minnesota, Twin Cities, MN, USA
{braulio,golnari,zhzhang}@cs.umn.edu

Abstract. Many online social networks such as Twitter, Google+, Flickr and Youtube are directed in nature, and have been shown to exhibit a nontrivial amount of reciprocity. Reciprocity is defined as the ratio of the number of reciprocal edges to the total number of edges in the network, and has been well studied in the literature. However, little attention is given to understand the connectivity or network form by the reciprocal edges themselves (reciprocal network), its structural properties, and how it evolves over time. In this paper, we bridge this gap by presenting a comprehensive measurement-based characterization of the connectivity among reciprocal edges in Google+ and their evolution over time, with the goal to gain insight into the structural properties of the reciprocal network. Our analysis shows that the reciprocal network of Google+ reveals some important user behavior patterns, which reflect how the social network was being adopted over time.

Keywords: Reciprocal Network · Google+ · Evolution · Reciprocity

1 Introduction

Many online social networks are fundamentally *directed*: they consist of both *reciprocal* edges, i.e., edges that have already been linked back, and *parasocial* edges, i.e., edges have not been or is not linked back [1]. Reciprocity is defined as the ratio of the number of reciprocal edges to the total number of edges in the network. It has been shown that major online social networks (OSN) that are directed in nature, such as Twitter, Google+, Flickr and Youtube, all exhibit a nontrivial amount of reciprocity: for example, the global reciprocity of Flickr [2], Youtube [2], Twitter [3] and Google+ [4] have been empirically measured to be 0.62, 0.79, 0.22 and 0.32, respectively. Reciprocity has been widely studied in the literature. For example, it has been used to compare and classify different directed networks, e.g., reciprocal or anti-reciprocal networks [5]. The authors in [1] investigate the factors that influence parasocial edges to become reciprocal ones. The problem of maximum achievable reciprocity in directed networks is formulated and studied in [6], with the goal to understand how bi-degree sequences (or resources or "social bandwidth") of users determines the reciprocity observed in real directed networks. The authors in [7] propose schemes to extract meaningful sub-communities from dense networks by considering the roles of users and

© Springer International Publishing Switzerland 2016
H.T. Nguyen and V. Snasel (Eds.): CSoNet 2016, LNCS 9795, pp. 14–26, 2016.
DOI: 10.1007/978-3-319-42345-6_2

their respective connections (reciprocal versus non-reciprocal ties). The authors in [8] examine the evolution of reciprocity and speculate that its evolution is affected by the hybrid nature of Google+, whereas the authors in [9] conduct a similar study and conclude that Google+ users reciprocated only a small fraction of their edges: this was often done by very low degree users with no or little activity.

Reciprocal edges represent the most stable type of connections or relations in directed network – they reflect strong ties between nodes or users [10–12], such as (mutual) friendships in an online social network or "following" each other in a social media network like Twitter. Connectivity among reciprocal edges can thus potentially reveal more information about users in such networks. For example, a clique formed by reciprocal edges suggest users involved are mutual friends or share common interests. More generally, it is believed that nontrivial patterns in the *reciprocal network* – the bidirectional subgraph (see Fig. 1) of a directed graph could reveal possible mechanism of social, biological or different nature that systematically acts as organizing principles shaping the observed network topology [5]. Moreover, understanding the dynamic structural properties of the reciprocal network can provide us with additional information to characterize or compare directed networks that go beyond the classic reciprocity metric, a single static value currently used in many studies. However, little attention has been paid in the literature to understand the connectivity between reciprocal edges – the reciprocal network – and how it evolves over time.

In this paper we perform a comprehensive measurement-based characterization of the connectivity and evolution of reciprocal edges in Google+ (thereafter referred to as $G+$ in short), in order to shed some light on the structural properties of $G+$'s reciprocal network. We are particularly interested in understanding how the reciprocal network of G+ evolves over time as new users (nodes) join the social network, and how reciprocal edges are created, e.g., whether they are formed mostly among extant nodes already in the system or by new nodes joining the network. For this, we employ a unique massive dataset collected in a previous study [9]. We start by providing a brief overview of G+ and a description of our dataset in Sect. 2. We then present our methodology to extract the reciprocal network of G+ using Breadth-First-Search (BFS), together with some notations in Sect. 3. In Sect. 4.1, we discuss a few key aggregate properties of the reciprocal network including the growth of the numbers of nodes and edges over time, the in-degree, out-degree, and reciprocal or mutual degree distributions. We then analyze the evolution of the reciprocal network in terms of its density, and categorize the nodes joining the reciprocal network based on the (observed) time they joined the network in Sect. 4.2, and study the types of connections they make (reciprocal edges) in Sect. 4.3. Finally we discuss the implications of our findings and we conclude the paper in Sect. 5. We summarize the major findings of our study as follows:

– We find that the density of G+ – which reflects the overall *degree of social connections* among G+ users – decreases as the network evolves from its second to third year of existence. This finding differs from the observations reported in [8], where it found that G+ social density fluctuates in an increase-decrease fashion in three phases, but it reaches a steady increase in the last phase during its first year of existence.

– Furthermore, we observe that both the density and reciprocity metrics of G+'s reciprocal network also decrease over time. Our analysis reveals that these are due to the fact that the new users joining G+ later tend to be less "social" as they make fewer connections in general. In particular, (i) the number of users creating at least one reciprocal edge is decreasing as the network evolves; (ii) the new users joining the reciprocal network are creating fewer edges than the users in the previous generation.

– We show that if a user does not create a reciprocal edge when he/she joins G+, there is a lower chance that he/she will create one later. In addition, users who already have reciprocal connections with some users tend to create more reciprocal connections with additional users.

To the best our knowledge, our study is the first study on the properties and evolution of a "reciprocal network" extracted from a *directed* social graph.

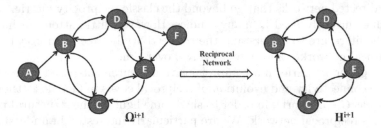

Fig. 1. Illustration of the reciprocal network (H^{i+1}) of a directed graph (Ω^{i+1}). Specifically, (B,C), (C,B), (B,D), (D,B), (D,E), (E,D), (C,E), (E,C) are reciprocal edges; (A,B), (C,A), (D,F), (F,E) are parasocial edges. The reciprocity of Ω^{i+1} is $8/12 = 0.67$

2   Google+ Overview and Dataset

In this section, we briefly describe key features of the Google+ service and a summary of our dataset.

Platform Description: Google has launched in June 2011 its own social networking service called Google+ (G+). The platform was announced as a new generation of social network. Previous works on the literature [8,9] claim that G+ cannot be classified as particularly asymmetric (Twitter-like), but it is also not as symmetric (Facebook-like) because G+ features have some similarity to both Facebook and Twitter. Therefore, they labelled G+ as a hybrid online social network [8]. Similar to Twitter (and different from Facebook) the relationships in

G+ are unidirectional. In graph-theoretical terms, if user[1] x follows user y this relationship can be represented as a directed social edge (x, y); if user y also has a directed social edge (y, x), the relationship x, y is called symmetric [13]. Similar to Facebook, each user has a stream, where any activity performed by the user appears (like the Facebook wall). For more informations about the features of G+ the reader is referred to [14, 15].

Dataset: We obtained our dataset from an earlier study on G+ [9], so no proprietary rights can be claimed. The dataset is a collection of 12 directed graphs of the social links of the users[2] in $G+$, collected from August, 2012 to June, 2013. We used BFS to extract the Largest Weakly Connected Component (LWCC) from all of our snapshots of G+. We label these set of LWCCs as subgraphs Ω^i (for $i = 1, ..., 12$). Since LWCC users form the most important component of G+ network [9], we extract the reciprocal network of G+ from the Ω^i subgraphs (see Sect. 3). However, for consistency in our analysis, we removed from the subgraphs $\Omega^{i=1,...,11}$ those nodes that do not appear in our last snapshot at Ω^{12}. Table 1 summarizes the main characteristics of the extracted Ω^i.

Table 1. Main characteristics of G+ dataset

ID	# nodes	# edges	Start-date	Duration
Ω_1	66,237,724	1,291,890,737	24-Aug-12	17
Ω_2	69,454,116	1,345,797,560	10-Sept-12	11
Ω_3	71,308,308	1,376,350,508	21-Sept-12	13
Ω_4	73,146,149	1,406,353,479	04-Oct-12	15
Ω_5	76,438,791	1,442,504,499	19-Oct-12	14
Ω_6	84,789,166	1,633,199,823	02-Nov-12	35
Ω_7	90,004,753	1,716,223,015	07-Dec-12	40
Ω_8	101,931,411	1,893,641,818	16-Jan-13	40
Ω_9	114,216,757	2,078,888,623	25-Feb-13	35
Ω_{10}	125,773,639	2,253,413,103	01-Apr-13	25
Ω_{11}	132,983,313	2,356,107,044	26-Apr-13	55
Ω_{12}	145,478,563	2,548,275,802	20-Jun-13	N/A

3 Methodology and Basic Notations

In this section, we describe our methodology to extract the reciprocal network of G+. To derive the reciprocal network of G+, we proceed as follows: we extract

---

[1] In this paper we use the terms "user" and "node" interchangeable.

[2] G+ assigns each user a 21-digit integer ID, where the highest order digit is always 1 (e.g., 100000000006155622736).

the subgraphs composed of nodes with at least one reciprocal edge for each of the snapshots of Ω^i. We label these new subgraphs G^i (for $i = 1, 2, ..., 12$). By comparing the set of nodes and edges in each of the sugbraphs G^i, we observe that a very small percentage of nodes depart G^i as it evolves (*unfollowing behaviour* [16]). Therefore, for consistency in our analysis, we removed from the subgraphs $G^{i=1,...,11}$ those nodes that don't appear in our last snapshot at G^{12}. We label these new set of subgraphs L^i (for $i = 1, 2, ..., 12$). However, L^i is not a connected subgraph. Hence, we use BFS to extract the Largest Weakly Connected Component (LWCC) for each of the snapshots of $L^{i=1,...,12}$. We label these extracted LWCCs as subgraphs H^i (for $i = 1, 2, ..., 12$).

In this paper, we consider subgraph H^i as the "reciprocal network" of G+[3]. In the next sections, we will focus our analysis on the structural properties and evolution of H^i. To achieve this, we extract subgraphs H^i_j composed of the set of users that join the network at snapshot i and j represents this subgraph at specific snapshots ($j => i$).

Let ΔH^{i+1} denote the subgraph composed with the set of nodes that join subgraph H^i_j at snapshot $j = i + 1$. Then, we define the following relationship (see Fig. 2):

$$H^{i+1} = H^i \cup \Delta H^{i+1} \tag{1}$$

In the following sections, we use subgraphs ΔH^{i+1}, H^i_j and (1) to analyse the reciprocal network of G+. For clarity of notation, we sometimes drop the superscript i and subscript j from the above notations, unless we are referring to specific snapshots or subgraphs.

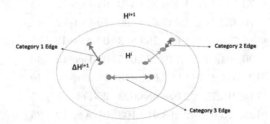

Fig. 2. Illustration of the relationship between subgraphs ΔH^{i+1}, H^{i+1}, H^i and the categories of the edges in subgraph H^i (for i=1,...,12)

4   Reciprocal Network Characteristics and Its Evolution

In this section, we present a comprehensive characterization of the connectivity and evolution of the reciprocal edges in G+, in order to shed an insightful light on the structural properties of the reciprocal network of G+. To achieve this,

---

[3] It contains more than 90 % of the nodes with at least one reciprocal edge in G+. Hence, our analysis of the dataset is eventually approximate.

we proceed as follows: (a) we provide a brief overview of the structural properties of the reciprocal network; (b) we analyse the evolution of the density of the reciprocal network and (c) we categorize the nodes joining the reciprocal network and their edges respectively.

4.1   Overview of the Reciprocal Network

We start by providing a brief overview of some global structural properties of the reciprocal network of G+, more precisely, the growth of its number of nodes and edges, as well as, its degree distributions:

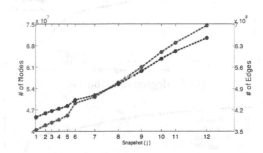

Fig. 3. Growth in the number of nodes and edges in H (Color figure online)

Nodes and Edges: Figure 3 plots the number of nodes (left axis) and edges (right axis) across time. We observe that the number of nodes and edges increase (almost) linearly as H^i evolves. The only exception is between H^i snapshots 5–6 ($19.Oct.12 - 02.Nov.12$), where we observe a significant increase in the number of nodes and edges. The time of this event correlates with the addition of a new G+ feature, on $31.Oct.12$, that allows users to share contents created and stored in Google Drive [17] directly into the G+ stream, as reported in [17]: "share the stuff you create and store in Google Drive, and people will be able to flip through presentations, open PDFs, play videos and more, directly in the G+ stream". Our dataset shows the impact of this event in G+: *it attracts more users to join G+ and many of these users might have already been using Google Drive in the past.*

In-degree, Out-degree and Mutual Degree Distributions: Figure 4 shows the CCDF for mutual degree, in-degree and out-degree for nodes in subgraphs H^i. We can see that these curves have approximately the shape of a Power Law distribution. The CCDF of a Power Law distribution is given by $Cx^{-\alpha}$ and $x, \alpha, C > 0$. By using the tool in [18,19], we estimated the exponent α that best models our distributions. We obtained $\alpha = 2.72$ for mutual degree, $\alpha = 2.41$ for out-degree and $\alpha = 2.03$ for in-degree. We observe that the mutual degree and out-degree distribuition have similar x-axis range and the out-degree curve drops sharply around 5000. We conjecture this is because G+ maintains a policy that allows only some special users to add more than 5000 friends to their circles [4].

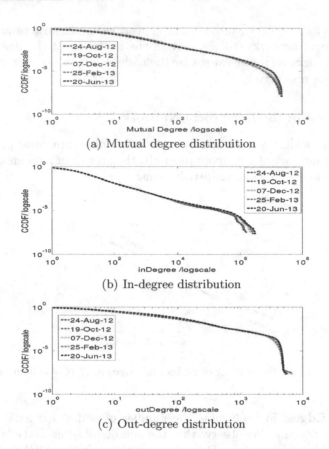

(a) Mutual degree distribuition

(b) In-degree distribution

(c) Out-degree distribution

Fig. 4. Degree distributions for subgraph H^i (Color figure online)

The observed power law trend in the distributions implies that a small fraction of users have disproportionately large number of connections, while most users have a small number of connections - *this is characteristics for many social networks.* We also observe that the shape of the distributions have initially evolved as the number of users with larger degree appeared.

4.2 Density Evolution and Nodes Categories

In this section, we analyze the evolution of the reciprocal network in terms of its density, and categorize the nodes joining the reciprocal network based on the (observed) time they joined the network. Next, we present our analysis:

Density: Figure 5(a) shows the evolution of the density of subgraph H^i, measured as the ratio of links-to-nodes[4]. We observe that as subgraph $H^{i=1,...,12}$ evolves its density decreases. However, if we fix the number of nodes for each

---

[4] We follow the terminology in [22] in order to compare with previous results.

of the snapshots of H^i and analyse their evolution, we observe that the density is increasing (see Fig. 5(a)). From these results, we conclude that the new users (ΔH^{i+1}) joining subgraphs H^i are responsible for the observed decrease in the density. Because these users initially create few connection when they join H^i (*cold start phenomenon*). However, the longer these users stay in the network, they discover more of their friends and consequently they increase their number of connections (edges). From the slopes of the graphs in Fig. 5(a), we observe that the new users are creating fewer links than the new users in the previous generation. Here, we define "previous generation" as the set of new users in the anterior snapshot, for example: the previous generation for new users in ΔH^3 are the users in ΔH^2.

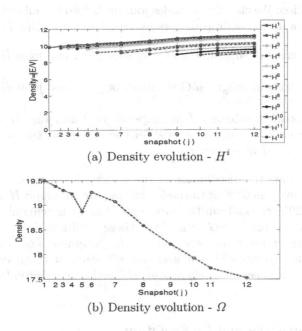

(a) Density evolution - H^i

(b) Density evolution - Ω

Fig. 5. Evolution of the Density for graphs Ω and H (Color figure online)

We also observe that the percentage of total users with at least one reciprocal edges in G+ decreases from 66.7 % to 54.1 % as the network evolves. Consequently, in our analysis, we also observe that the global reciprocity of G+ decreases (almost) linearly from 33.9 % to 25.9 %. From these results, we extract some important points: *(a) the number of users creating at least one reciprocal edge is decreasing as the network evolves and (b) the new users joining the reciprocal network are creating fewer edges than the users in the previous generation. Thus, the new users in G+ are becoming less social.*

Previous studies on social networks show that the social density for Facebook [20] and affiliation networks [21] increases over time. However, it fluctuates on Flickr [22] and is almost constant on email networks [23]. Differently, our dataset shows that the social density of G+ and of its reciprocal network (Fig. 5(a) and (b)) decrease as the network evolves. This is an interesting observation because it contradicts the *densification power law*, which states that real networks tend to densify as they grow [24].

The authors in [8] analysed the evolution of the social density of G+ using a dataset collected in the first year of its existence $(06.Jun.11 - 11.Oct.11)$. They reported that G+ social density fluctuates in an increase-decrease fashion in three phases, but it reaches a steady increase in the last phase [8]. *Differently, our results shows that the social density of G+ is decreasing as the network evolves from its second to third year of existence* – the only exception is between snapshots 5 to 6, due to the events discussed in Sect. 4.1.

Node Categories: We classify the nodes joining H into the following categories (for clarity of notations we drop the superscript i and subscript j):

- Ω: node "x" exists in subgraph Ω at snapshot $j-1$ and joins H at snapshot j
- G: node "x" exists in subgraph G at snapshot $j-1$ and joins H at snapshot j
- L: node "x" exists in subgraph L at snapshot $j-1$ and joins H at snapshot j
- $NewArrival$: node "x" does not exist in the system at snapshot $j-1$ and joins both Ω and H at snapshot j

Figure 6(a) shows the distribution of the nodes joining H by categories. We observe that on average 63 % of the nodes joining the subgraph H are new users in the system, 29 % comes from the subgraph Ω and the remaining percentage comes from either subgraphs G or L. From these results we infer the following: *(a) the majority of users that are joining the reciprocal network of G+ are new users in the system; (b) if a user doesn't create a reciprocal edge when he/she joins G+, it is very unlikely that he/she will ever reciprocate a link in the network.*

4.3   Edge Categories and Its Evolution

In order to understand the connectivity between the nodes in the reciprocal network, we analyse the evolution of the reciprocal edges in H^i. To achieve this, we restrict our analysis[5] to the subgraphs H^1 and H^2. Firstly, we present our edges categories. Secondly, we analyse the evolution of the degree distribution for each edge category:

Edges Categories: We classify the edges created by nodes joining H^i into the following three categories (see Fig. 2 for an illustration):

- Category 1: $e(u, v)$ such that $u \in \Delta H^{i+1}$ and $v \in H^i$
- Category 2: $e(u, v)$ such that $u \in \Delta H^{i+1}$ and $v \in \Delta H^{i+1}$ and $\exists v^* \in H^i$: $e^*(u, v^*)$
- Category 3: $e(u, v)$ such that $u \in H^i$ and $v \in H^i$

---

[5] Similar results are obtained using the other subgraphs ($H^{i=3,\cdots,12}$).

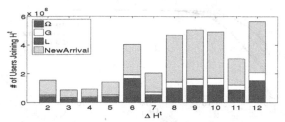

(a) Total number of nodes joining H^i per category

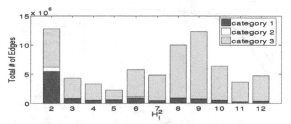

(b) Total number of new edges per category created in H^2_j for each of j snapshots

Fig. 6. Nodes and edges categories for subgraph H (Color figure online)

Figure 6(b) shows the distribution of the edges based on the defined categories. We observe that most of the new edges seen across all snapshots of H^2_j are due to category 3 edges. Furthermore, by looking at the last snapshot of H^i (for $i = 12$), we observe that 69 % of the edges in H^{12}_{12} are between nodes in H^1 only. This result shows that although the density decreases as subgraph H^i evolves, the connectivity of a subset of its nodes is increasing (*densification*) and their connectivity accounts for a huge percentage of the total edges in the system.

Degree Distribution: Figure 7 shows the degree distribution for all categories of edges and how they evolve across time. Figure 7(a) shows the CDF of the degree distribution for category 1 edges. From this figure, we observe that when new nodes (ΔH^2) join H^1_2, initially they create few connections, but the longer they stay in the system the number of connections to nodes already in the system increases significantly (as stated in Sect. 4.2). Furthermore, from our dataset, we observe that 72 % of the nodes in ΔH^2 have only connections (edges) to nodes already in the system (H^1_2).

Figure 7(b) shows the CDF for the degree distribution of category 2 edges. From the results of Fig. 7(a) and (b), we infer that when new nodes (ΔH^2) join H^1_2, they create more connections with the nodes already in the system. Figure 7(c) shows the degree distribution for edges of category 3. We observe that the shape of the degree distribution is decreasing which implies that the network is become more dense (*densification*), as discussed above.

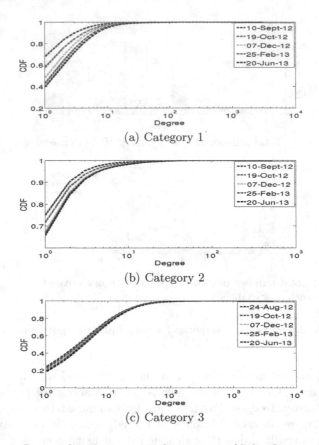

(a) Category 1

(b) Category 2

(c) Category 3

Fig. 7. Degree distribution per edge category (Color figure online)

In summary, our analysis on the categories of nodes and edges, in this section, led to the following key findings: *(a) the majority of users that joins the reciprocal network of G+ are new users in the network and they tend to create reciprocal connections mostly to users who already have reciprocal connections to others; (b) if a user does not create a reciprocal edge when he/she joins G+, there is a lower chance that he/she will create one later.*

5   Implication of Our Results for G+ and Conclusion

In this paper, we present the first study on the properties and evolution of a "reciprocal network", using a very large G+ dataset. Analyzing the connectivity of reciprocal edges is relevant because they are the most stable type of connections in directed network and they represent the strongest ties between nodes: users with large number of mutual edges are less likely to depart from the network and they may form the most relevant community structure[6](the intimacy community [7])

---

[6] We will analyse the community structure in a reciprocal network as future work.

in directed OSN networks. Our analysis show that the reciprocal network of G+ reveals some important patterns of the user's behavior, for example: new users joining G+ are becoming less social as the network involves and they tend to create reciprocal connections mostly to users who already have reciprocal connections to others. Understanding these behaviors is important because they expose insightful information about how the social network is being adopted.

The findings here also provide hints that can help explain why G+ is failing to compete with Twitter and Facebook, as recently reported [25]. Firstly, we observe that although the number of nodes and edges increase as G+ evolves, the density of the network is decreasing. This result supports the claim that some users joined G+ because they need to access some of Google products but they weren't interested in creating connections in the network, in contrast to users in Twitter. Secondly, we observe a decrease in the reciprocity of G+ because the percentage of users with at least a reciprocal edge decrease as the network evolved. Furthermore, the users that are joining the reciprocal network of G+ always create fewer connections than the users in the previous generation. From this result, we infer that many users didn't use G+ to connect and chat with friends, in contrast to users in Facebook[7]. Therefore, since its second year of life, the G+ social network was already showing "signs" that it was failing to compete with others online social network, such as Twitter and Facebook. Many of the studies in the literature about G+ [4,8,9,13] were done using dataset mostly collected on the first year of G+ existence. Thus, they either did not observe or failed to see these signs.

Our work is only a first step towards exploring the connectivity of reciprocal edges in social and other complex networks, or reciprocal networks. There are several interesting directions for future work that we will explore further to uncover the properties of a reciprocal network so as to further understand the structural properties of directed graphs.

Acknowledgments. This research was supported in part by a Raytheon/NSF subcontract 9500012169/CNS-1346688, DTRA grants HDTRA1- 09-1-0050 and HDTRA1-14-1-0040, DoD ARO MURI Award W911NF-12-1-0385, and NSF grants CNS-1117536, CRI-1305237 and CNS-1411636. We thank the authors of [9] for the datasets and the workshop reviewers for helpful comments.

References

1. Gong, N.Z., Xu, W.: Reciprocal versus parasocial relationships in online social networks. Soc. Netw. Anal. Min. **4**(1), 184–197 (2014)
2. Mislove, A., Marcon, M., Gummadi, K.P., Druschel, P., Bhattacharjee, B.: Measurement and analysis of online social networks. In: IMC 2007, pp. 29–42. ACM (2007)
3. Kwak, H., Lee, C., Park, H., Moon, S.: What is twitter, a social network or a news media? In: WWW 2010, pp. 591–600. ACM (2010)

---

[7] The authors in [9] stated similar conclusion.

4. Magno, G., Comarela, G., Saez-Trumper, D., Cha, M., Almeida, V.: New kid on the block: exploring the Google+ social graph. In: IMC 2012, pp. 159–170. ACM (2012)
5. Garlaschelli, D., Loffredo, M.I.: Patterns of link reciprocity in directed networks. Phys. Rev. Lett. **93**, 268–701 (2004)
6. Jiang, B., Zhang, Z.-L., Towsley, D.: Reciprocity in social networks with capacity constraints. In: KDD 2015, pp. 457–466. ACM (2015)
7. Hai, P.H., Shin, H.: Effective clustering of dense and concentrated online communities. In: Asia-Pacific Web Conference (APWEB) 2010, pp. 133–139. IEEE (2010)
8. Gong, N.Z., Xu, W., Huang, L., Mittal, P., Stefanov, E., Sekar, Song, D.: Evolution of the social-attribute networks: measurements, modeling, and implications using Google+. In: IMC 2015, pp. 131–144. ACM (2015)
9. Gonzalez, R., Cuevas, R., Motamedi, R., Rejaie, R., Cuevas, A.: Google+ or Google−? dissecting the evolution of the new OSN in its first year. In: WWW 2013, pp. 483–494. ACM (2013)
10. Wolfe, A.: Social network analysis: methods and applications. Am. Ethnologist **24**(1), 219–220 (1997)
11. Jamali, M., Haffari, G., Ester, M.: Modeling the temporal dynamics of social rating networks using bidirectional effects of social relations and rating patterns. In: WWW 2011, pp. 527–536. ACM (2011)
12. Li, Y., Zhang, Z.-L., Bao, J.: Mutual or unrequited love: identifying stable clusters in social networks with uni- and bi-directional links. In: Bonato, A., Janssen, J. (eds.) WAW 2012. LNCS, vol. 7323, pp. 113–125. Springer, Heidelberg (2012)
13. Schiberg, D., Schneider, F., Schiberg, H., Schmid, S., Uhlig, S., Feldmann, A.: Tracing the birth of an OSN: social graph and profile analysis in Google+. In: WebSci 2012, pp. 265–274. ACM (2012)
14. Google+ Platform. http://www.google.com/intl/en/+/learnmore/
15. Google+. http://en.wikipedia.org/wiki/Google+
16. Kwak, H., Chun, H., Moon, S.: Fragile online relationship: a first look at unfollow dynamics in twitter. In: CHI 2011, pp. 1091–1100. ACM (2011)
17. Google+ New Feature. http://googledrive.blogspot.com/2012/10/share-your-stuff-from-google-drive-to.html
18. Clauset, A., Shalizi, C.R., Newman, M.E.J.: Power-Law Distributions in Empirical Data. SIAM Rev. **51**, 661–703 (2009)
19. Fitting Power Law Distribution. http://tuvalu.santafe.edu/aaronc/powerlaws/
20. Backstrom, L., Boldi, P., Rosa, M., Ugander, J., Vigna, S.: Four degrees of separation. In: WebSci 2012, pp. 33–42. ACM (2012)
21. Leskovec, J., Kleinberg, J., Faloutsos, C.: Graphs over time: densification laws, shrinking diameters and possible explanations. In: KDD 2005, pp. 177–187. ACM (2005)
22. Kumar, R., Novak, J., Tomkins, A.: Structure and evolution of online social networks. In: KDD 2006, pp. 611–617. ACM (2006)
23. Kossinets, G., Watts, D.J.: Empirical analysis of an evolving social network. Sci. **311**, 88–90 (2006)
24. Leskovec, J., Chakrabarti, D., Kleinberg, J., Faloutsos, C., Ghahramani, Z.: Kronecker graphs: an approach to modeling networks. J. Mach. Learn. Res. **11**, 985–1042 (2010)
25. Google Strips Down Google Plus. http://blogs.wsj.com/digits/2015/11/17/google-strips-down-google-plus/

Comparison of Random Walk Based Techniques for Estimating Network Averages

Konstantin Avrachenkov[1], Vivek S. Borkar[2], Arun Kadavankandy[1], and Jithin K. Sreedharan[1(✉)]

[1] Inria Sophia Antipolis, Valbonne, France
{k.avrachenkov,arun.kadavankandy,jithin.sreedharan}@inria.fr
[2] IIT Bombay, Mumbai, India
borkar.vs@gmail.com

Abstract. Function estimation on Online Social Networks (OSN) is an important field of study in complex network analysis. An efficient way to do function estimation on large networks is to use random walks. We can then defer to the extensive theory of Markov chains to do error analysis of these estimators. In this work we compare two existing techniques, Metropolis-Hastings MCMC and Respondent Driven Sampling, that use random walks to do function estimation and compare them with a new reinforcement learning based technique. We provide both theoretical and empirical analyses for the estimators we consider.

Keywords: Social network analysis · Estimation · Random walks on graph

1 Introduction

The analysis of many Online Social Networks (OSN) is severely constrained by a limit on Application Programming Interface (API) request rate. We provide evidence that random walk based methods can explore complex networks with very low computational load. One of the basic questions in complex network analysis is the estimation of averages of network characteristics. For instance, one would like to know how young a given social network is, or how many friends an average network member has, or what proportion of a population supports a given political party. The answers to all the above questions can be mathematically formulated as the solutions to a problem of estimating an average of a function defined on the network nodes.

Specifically, we model an OSN as a connected graph $\mathcal{G}$ with node set $\mathcal{V}$ and edge set $\mathcal{E}$. Suppose we have a function $f : \mathcal{V} \to \mathcal{R}$ defined on the nodes. If the graph is not connected, we can mitigate the situation by considering a modified random walk with jumps as in [2]. Our goal is to propose good estimators for the average of $f(.)$ over $\mathcal{V}$ defined as

$$\mu(\mathcal{G}) = \frac{1}{|\mathcal{V}|} \sum_{v \in \mathcal{V}} f(v). \tag{1}$$

© Springer International Publishing Switzerland 2016
H.T. Nguyen and V. Snasel (Eds.): CSoNet 2016, LNCS 9795, pp. 27–38, 2016.
DOI: 10.1007/978-3-319-42345-6_3

The above formulation is rather general and can be used to address a range of questions. For example to estimate the average age of a network we can take $f(v)$ as an age of node $v \in \mathcal{V}$, and to estimate the number of friends an average network member has we can set $f(v) = d_v$, where d_v is the degree of node v.

In this work, we compare in a systematic manner several random walk based techniques for estimating network averages $\mu(\mathcal{G})$ for a deterministic function f. In addition to familiar techniques in complex network analysis such as Metropolis-Hastings MCMC [6,8,13,15] and Respondent-Driven Sampling (RDS) [9,17,18], we also consider a new technique based on Reinforcement Learning (RL) [1,5]. If a theoretic expression for the limiting variance of Metropolis-Hastings MCMC was already known (see e.g., [6]), the variance and convergence analysis of RDS and RL can be considered as another contribution of the present work.

Metropolis-Hastings MCMC has being applied previously for network sampling (see e.g., [8,10] and references therein). Then, RDS method [9,17,18] has been proposed and it was observed that in many cases RDS practically superior over MH-MCMC. We confirm this observation here using our theoretical derivations. We demonstrate that with a good choice of cooling schedule, the performance of RL is similar to that of RDS but the trajectories of RL have less fluctuations than RDS.

There are also specific methods tailored for certain forms of function $f(v)$. For example, in [7] the authors developed an efficient estimation technique for estimating the average degree. In the extended journal version of our work we plan to perform a more extensive comparison across various methods. Among those methods are Frontier Sampling [14], Snowball Sampling [11] and Walk-Estimate [12], just to name a few.

The paper is organized as follows: in Sect. 3 we describe various random walk techniques and provide error analysis, then, in Sect. 4 we compare all the methods by means of numerical experiments on social networks. Finally, in Sect. 5 we present our main conclusions.

2   Background and Notation

First we introduce some notation and background material that will make the exposition more transparent. A column vector is denoted by bold lower font e.g., $\mathbf{x}$ and its components as x_i. The probability vector $\boldsymbol{\pi}$ is a column vector. A matrix is represented in bold and its components in normal font (e.g., $\mathbf{A}, A_{ij}$.) In addition, $\mathbf{1}$ represents the all-one column vector in $\mathbb{R}^n$.

For a sequence of random variables (rvs), $X_n \xrightarrow{D} X$, denotes convergence in distribution to X [3].

A random walk (RW) is simply a time-homogenous first-order Markov Chain whose state space is $\mathcal{V}$, the set of vertices of the graph and the transition probabilities are given as:

$$p_{ij} = \mathbb{P}\left(X_{t+1} = j | X_t = i\right) = \frac{1}{d_i}$$

if there is a link between i and j, i.e., $(i, j) \in E$, d_i being the degree of node i. Therefore we can think of the random walker as a process that traverses the links of the graph in a random fashion. We can define $\mathbf{P}$ the transition probability matrix (t.p.m) of the Random walk as an $|\mathcal{V}| \times |\mathcal{V}|$ matrix, such that $\mathbf{P}_{ij} = p_{ij}$. Since we consider undirected networks, our random walk is time reversible. When the graph is connected the transition probability matrix $\mathbf{P}$ is irreducible and by Frobenius Perron Theorem there always exists a unique stationary probability vector $\boldsymbol{\pi} \in \mathbb{R}^{1 \times |\mathcal{V}|}$ which solves $\boldsymbol{\pi}\mathbf{P} = \boldsymbol{\pi}$, which is in fact $\pi_i = \frac{d_i}{2|E|}$. Since our state space is finite the Markov chain is also positive recurrent and the quantities such as hitting times, and cover times are finite and well-defined. An important application of Random walks is in estimating various graph functions. The random walk based techniques can be easily implemented via APIs of OSNs and can also be easily distributed.

Let us define the fundamental matrix of a Markov chain given by $\mathbf{Z} := (\mathbf{I} - \mathbf{P} + \mathbf{1}\boldsymbol{\pi}^T)^{-1}$. For two functions $f, g : \mathcal{V} \to \mathbb{R}$, we define $\sigma_{ff}^2 := 2\langle \mathbf{f}, \mathbf{Zf} \rangle_\pi - \langle \mathbf{f}, \mathbf{f} \rangle_\pi - \langle \mathbf{f}, \mathbf{1}\boldsymbol{\pi}^T \mathbf{f} \rangle_\pi$, and $\sigma_{fg}^2 = \langle \mathbf{f}, \mathbf{Zg} \rangle_\pi + \langle \mathbf{g}, \mathbf{Zf} \rangle_\pi - \langle \mathbf{f}, \mathbf{y} \rangle_\pi - \langle \mathbf{f}, \mathbf{1}\boldsymbol{\pi}^T g \rangle_\pi$, where $\langle \mathbf{x}, \mathbf{y} \rangle_\pi = \sum_i x_i y_i \pi_i$, for any two vectors $\mathbf{x}, \mathbf{y} \in \mathbb{R}^{|\mathcal{V}|}$, π being the stationary distribution of the Markov chain. In addition N denotes the number of steps of the random walk. By the Ergodic Theorem for Markov Chains applied to graphs the following is true [6], where f is an arbitrary function defined on the vertex set $\mathcal{V}$.

Theorem 1. *[6] For a RW $\{X_0, X_1, X_2 \dots X_n, \dots\}$ on a connected undirected graph,*

$$\frac{1}{N} \sum_{t=1}^{N} f(X_t) \to \sum_{x \in \mathcal{V}} \pi(x) f(x), \quad N \to \infty,$$

almost surely.

In addition the following central limit theorems also follow for RWs on graphs from the general theory of recurrent Markov chains [13].

Theorem 2. *[13] If f is a function defined on the states of a random walk on graphs, the following CLT holds*

$$\sqrt{N} \left(\frac{1}{N} \sum_{i=1}^{N} f(X_i) - \mathbb{E}_\pi(f) \right) \xrightarrow{D} \mathcal{N}(0, \sigma_{ff}^2)$$

Theorem 3. *[13] If f, g are two functions defined on the states of a random walk, define the vector sequence $\mathbf{z}_t = \begin{bmatrix} f(x_t) \\ g(x_t) \end{bmatrix}$ the following CLT holds*

$$\sqrt{N} \left(\frac{1}{N} \sum_{t=1}^{N} \mathbf{z}_t - \mathbb{E}_\pi(\mathbf{z}_t) \right) \xrightarrow{D} \mathcal{N}(0, \boldsymbol{\Sigma}),$$

where $\boldsymbol{\Sigma}$ is 2×2 matrix such that $\Sigma_{11} = \sigma_{ff}^2$, $\Sigma_{22} = \sigma_{gg}^2$ and $\Sigma_{12} = \Sigma_{21} = \sigma_{fg}^2$.

In the following section we describe some of the most commonly used RW techniques to estimate functions defined on the vertices of a graph. We also give theoretical mean squared error (MSE) for each estimator defined as $MSE = \mathbb{E}[||\hat{\mu}(\mathcal{G}) - \mu(\mathcal{G})|^2]$.

3   Description of the Techniques

In light of the Ergodic theorem of RW, there are several ways to estimate $\mu(\mathcal{G})$ as we describe in the following subsections.

Basic Markov Chain Monte Carlo Technique (MCMC-Technique)

MCMC is an algorithm that modifies the jump probabilities of a given MC to achieve a desired stationary distribution. Notice that by the Ergodic Theorem of MC, if the stationary distribution is uniform, then an estimate formed by averaging the function values over the visited nodes converges asymptotically to $\mu(G)$ as the number of steps tend to infinity. We use the MCMC algorithm to achieve $\pi(x) = 1/|\mathcal{V}|, \forall x \in \mathcal{V}$. Let p_{ij} be the transition probabilities of the original graph. We present here the Metropolis Hastings MCMC (MH-MCMC) algorithm for our specific purpose. When the chain is in state i it chooses the next state j according to transition probability p_{ij}. It then jumps to this state with probability a_{ij} or remains in the current state i with probability $1 - a_{ij}$, where a_{ij} is given as below

$$a_{ij} = \begin{cases} \min\left(\frac{p_{ji}}{p_{ij}}, 1\right) & \text{if } p_{ij} > 0, \\ 1 & \text{if } p_{ij} = 0. \end{cases} \tag{2}$$

Therefore the effective jump probability from state i to state j is $a_{ij}p_{ij}$, when $i \neq j$. It follows that the final chain represents a Markov chain with the following transition matrix $\mathbf{P}^{MH}$

$$P_{ij}^{MH} = \begin{cases} \frac{1}{\max(d_i, d_j)} & \text{if } j \neq i \\ 1 - \sum_{k \neq i} \frac{1}{\max(d_i, d_k)} & \text{if } j = i. \end{cases}$$

This chain can be easily checked to be reversible with stationary distribution $\pi_i = 1/n \; \forall i \in \mathcal{V}$. Therefore the following estimate for $\mu(\mathcal{G})$ using MH-MCMC is asymptotically consistent.

$$\hat{\mu}_{MH}(\mathcal{G}) = \frac{1}{N} \sum_{t=1}^{N} f(X_t).$$

By using the 1D CLT for RW from Theorem 2 we can show the following central limit theorem for MH.

Proposition 1 *(Central Limit Theorem for MH-MCMC). For MCMC with uniform target distribution it holds that*

$$\sqrt{N}\left(\hat{\mu}_{MH}(\mathcal{G}) - \mu(\mathcal{G})\right) \xrightarrow{\mathcal{D}} \mathcal{N}(0, \sigma^2_{MH}),$$

as $N \to \infty$, where $\sigma^2_{MH} = \sigma^2_{ff} = \frac{2}{n}\mathbf{f}^T\mathbf{Z}\mathbf{f} - \frac{1}{n}\mathbf{f}^T\mathbf{f} - \left(\frac{1}{n}\mathbf{f}^T\mathbf{1}\right)^2$

Proof. Follows from Theorem 2 above.   □

Respondent Driven Sampling Technique (RDS-Technique)

This estimator uses the unmodified RW on graphs but applies a correction to the estimator to compensate for the non-uniform stationary distribution.

$$\hat{\mu}_{RDS}^{(N)}(\mathcal{G}) = \frac{\sum_{t=1}^{N} f(X_t)/d(X_t)}{\sum_{t=1}^{N} 1/d(X_t)} := \frac{\sum_{t=1}^{N} f'(X_t)}{\sum_{t=1}^{N} g(X_t)}, \tag{3}$$

where $f'(X_t) := f(X_t)/d(X_t), g(X_t) := 1/d(X_t)$. The following result shows that the RDS estimator is asymptotically consistent and also gives the asymptotic mean squared error.

Proposition 2 *(Asymptotic Distribution of RDS Estimate). The RDS estimate $\hat{\mu}_{RDS}(\mathcal{G})$ satisfies a central limit theorem given below*

$$\sqrt{N}(\hat{\mu}_{RDS}^{(N)}(\mathcal{G}) - \mu(\mathcal{G})) \xrightarrow{\mathcal{D}} \mathcal{N}(0, \sigma^2_{RDS}),$$

where σ^2_{RDS} is given by

$$\sigma^2_{RDS} = d^2_{av}\left(\sigma^2_1 + \sigma^2_2\mu^2(\mathcal{G}) - 2\mu(\mathcal{G})\sigma^2_{12}\right),$$

where $\sigma^2_1 = \frac{1}{|E|}\mathbf{f}^T\mathbf{Z}\mathbf{f}' - \frac{1}{2|E|}\sum_x \frac{f(x)^2}{d(x)} - \left(\frac{1}{2|E|}\mathbf{f}^T\mathbf{1}\right)^2, \sigma^2_2 = \sigma^2_{gg} = \frac{1}{|E|}\mathbf{1}^T\mathbf{Z}\mathbf{g} - \frac{1}{2|E|}\mathbf{g}^T\mathbf{1} - (\frac{1}{d_{av}})^2$ and $\sigma^2_{12} = \frac{1}{2|E|}\mathbf{f}^T\mathbf{Z}\mathbf{g} + \frac{1}{2|E|}\mathbf{1}^T\mathbf{Z}\mathbf{f}' - \frac{1}{2|E|}\mathbf{f}^T\mathbf{g} - \frac{1}{d_{av}}\frac{1}{2|E|}\mathbf{1}^T\mathbf{f}$

Proof. Let $f'(x) := \frac{f(x)}{d(x)}$ and $g(x) := \frac{1}{d(x)}$. Define the vector $\mathbf{z}_t = \begin{bmatrix} f'(x_t) \\ g(x_t) \end{bmatrix}$, and let $\mathbf{z}_N = \sqrt{N}\left(\frac{1}{N}\sum_{t=1}^N \mathbf{z}_t - \mathbb{E}_\pi(\mathbf{z}_t)\right)$. Then by Theorem 3, $\mathbf{z}_N \xrightarrow{\mathcal{D}} \mathcal{N}(0, \boldsymbol{\Sigma})$, where $\boldsymbol{\Sigma}$ is defined as in the theorem. Equivalently, by Skorohod representation theorem [3] in a space $(\Omega, \mathcal{F}, \mathbb{P})$, $\Omega \subset \mathbb{R}^2$, there is an embedding of $\mathbf{z}_N$ s.t. $\mathbf{z}_N \to \mathbf{z}$ almost surely (a.s.), such that $\mathbf{z} \sim \mathcal{N}(0, \boldsymbol{\Sigma})$. Hence the distribution of $\sqrt{N}(\hat{\mu}_{RDS}^{(N)}(\mathcal{G}) - \mu(\mathcal{G}))$ is the same as that of

$$\frac{\sum_{t=1}^N f'(X_t)}{\sum_{t=1}^N g(X_t)} \overset{\mathcal{D}}{=} \frac{\frac{1}{\sqrt{N}}z_1^{(N)} + \mu_{f'}}{\frac{1}{\sqrt{N}}z_2^{(N)} + \mu_g} = \frac{z_1^{(N)} + \sqrt{N}\mu_{f'}}{z_2^{(N)} + \sqrt{N}\mu_g} = \frac{z_1^{(N)} + \sqrt{N}\mu_{f'}}{\sqrt{N}\mu_g(1 + \frac{z_2^{(N)}}{\sqrt{N}\mu_g})}$$

$$= \frac{1}{\sqrt{N}\mu_g}(z_1^{(N)} - \frac{z^{(N)}(1)z_2^{(N)}}{\sqrt{N}\mu_g} + \sqrt{N}\mu_{f'} - \frac{z_2^{(N)}\mu_{f'}}{\mu_g} + \mathcal{O}(\frac{1}{\sqrt{N}}))$$

This gives

$$\sqrt{N} \left(\frac{\sum_{t=1}^{N} f'(X_t)}{\sum_{t=1}^{N} g(X_t)} - \frac{\mu_{f'}}{\mu_g} \right) \xrightarrow{\mathcal{D}} \frac{1}{\mu_g} \left(z_1 - z_2 \frac{\mu_{f'}}{\mu_g} \right),$$

since the term $\mathcal{O}(\frac{1}{\sqrt{N}})$ tend to zero in probability, and using Slutsky's lemma [3]. The result then follows from the fact that $\mathbf{z} \sim \mathcal{N}(0, \boldsymbol{\Sigma})$. □

Reinforcement Learning Technique (RL-Technique)

Consider a connected graph $\mathcal{G}$ with node set $\mathcal{V}$ and edge set $\mathcal{E}$. Let $\mathcal{V}_0 \subset \mathcal{V}$ with $|\mathcal{V}_0| << |\mathcal{V}|$. Consider a simple random walk $\{X_n\}$ on $\mathcal{G}$ with transition probabilities $p(j|i) = 1/d(i)$ if $(i,j) \in \mathcal{E}$ and zero otherwise. Define $Y_n := X_{\tau_n}$ for $\tau_n :=$ successive times to visit $\mathcal{V}_0$. Then $\{(Y_n, \tau_n)\}$ is a semi-Markov process on $\mathcal{V}_0$. In particular, $\{Y_n\}$ is a Markov chain on $\mathcal{V}_0$ with transition matrix (say) $[[p_Y(j|i)]]$. Let $\xi := \min\{n > 0 : X_n \in \mathcal{V}_0\}$ and for a prescribed $f : \mathcal{V} \mapsto \mathcal{R}$, define

$$T_i := E_i[\xi],$$

$$h(i) := E_i \left[\sum_{m=1}^{\xi} f(X_m) \right], \ i \in \mathcal{V}_0.$$

Then the Poisson equation for the semi-Markov process (Y_n, τ_n) is [16]

$$V(i) = h(i) - \beta T_i + \sum_{j \in \mathcal{V}_0} p_Y(j|i) V(j), \ i \in \mathcal{V}_0. \tag{4}$$

Here $\beta :=$ the desired stationary average of f. Let $\{z\}$ be IID uniform on $\mathcal{V}_0$. For each $n \geq 1$, generate an independent copy $\{X_m^n\}$ of $\{X_m\}$ with $X_0^n = z$ for $0 \leq m \leq \xi(n) :=$ the first return time to $\mathcal{V}_0$. A learning algorithm for (4) along the lines of [1] then is

$$V_{n+1}(i) = V_n(i) + a(n)\mathbb{I}\{z = i\} \times$$
$$\left[\left(\sum_{m=1}^{\xi(n)} f(X_m^n) \right) - V_n(i_0)\xi(n) + V_n(X_{\xi(n)}^n) - V_n(i) \right], \tag{5}$$

where $a(n) > 0$ are stepsizes satisfying $\sum_n a(n) = \infty$, $\sum_n a(n)^2 < \infty$. (One good choice is $a(n) = 1/\lceil \frac{n}{N} \rceil$ for $N = 50$ or 100.) Here $\mathbb{I}\{A\}$ denotes indicator function for the set A. Also, i_0 is a prescribed element of $\mathcal{V}_0$. One can use other normalizations in place of $V_n(i_0)$, such as $\frac{1}{|\mathcal{V}_0|} \sum_j V_n(j)$ or $\min_i V_n(i)$, etc. Then this normalizing term $(V_n(i_0)$ in (5)) converges to β as n increases to ∞. This normalizing term forms our estimator $\hat{\mu}_{RL}^{(n)}(\mathcal{G})$ in RL based approach.

The relative value iteration algorithm to solve (4) is

$$V_{n+1}(i) = h(i) - V_n(i_0)T_i + \sum_j p_Y(j|i)V_n(j)$$

and (5) is the stochastic approximation analog of it which replaces conditional expectation w.r.t. transition probabilities with an actual sample and then makes an incremental correction based on it, with a slowly decreasing stepwise that ensures averaging. The latter is a standard aspect of stochastic approximation theory. The smaller the stepwise the less the fluctuations but slower the speed, thus there is a trade-off between the two.

RL methods can be thought of as a cross between a pure deterministic iteration such as the relative value iteration above and pure MCMC, trading off variance against per iterate computation. The gain is significant if the number of neighbours of a node is much smaller than the number of nodes, because we are essentially replacing averaging over the latter by averaging over neighbours. The V-dependent terms can be thought of as control variates to reduce variance.

MSE of RL Estimate

For the RL Estimate the following concentration bound is true [4]:

$$\mathbb{P}\left\{|\hat{\mu}_{RL}^{(N)}(\mathcal{G}) - \mu(\mathcal{G})| \geq \epsilon\right\} \leq K \exp(-k\epsilon^2 N).$$

Thus it follows that MSE is $\mathcal{O}(\frac{1}{\sqrt{N}})$ because

$$\mathbb{E}|\hat{\mu}_{RL}^{(N)}(\mathcal{G}) - \mu(\mathcal{G})|^2 = \int_0^\infty \mathbb{P}\left\{|\hat{\mu}_{RL}^{(N)}(\mathcal{G}) - \mu(\mathcal{G})|^2 \geq \epsilon\right\} d\epsilon$$

$$= \int_0^\infty \mathbb{P}\left\{|\hat{\mu}_{RL}^{(N)}(\mathcal{G}) - \mu(\mathcal{G})| \geq \epsilon^{1/2}\right\} d\epsilon$$

$$\leq \int_0^\infty K \exp(-k\epsilon N)\, d\epsilon = \mathcal{O}\left(\frac{1}{N}\right).$$

4   Numerical Comparison

The algorithms explained in Sect. 3 are compared in this section using simulations on two real-world networks. For the figures given below, the x-axis represents the budget B which is the number of allowed samples, and is the same for all the techniques. We use the normalized root mean squared error (NRMSE) for comparison for a given B and is defined as

$$\text{NRMSE} := \sqrt{\text{MSE}}/\mu(\mathcal{G}), \quad \text{where MSE} = E\left[(\hat{\mu}(\mathcal{G}) - \mu(\mathcal{G}))^2\right].$$

For the RL technique we choose the initial or super-node $\mathcal{V}_0$ by uniformly sampling nodes assuming the size of $\mathcal{V}_0$ is given a priori.

4.1 Les Misérables network

In Les Misérables network, nodes are the characters of the novel and edges are
formed if two characters appear in the same chapter in the novel. The number
of nodes is 77 and number of edges is 254. We have chosen this rather small
network in order to compare all the three methods in terms of theoretical limiting
variance. Here we consider four demonstrative functions: (a) $f(v) = \mathbb{I}\{d(v) > 10\}$
(b) $f(v) = \mathbb{I}\{d(v) < 4\}$ (c) $f(v) = d(v)$, where $\mathbb{I}\{A\}$ is the indicator function for
set A and (d) for calculating $\mu(\mathcal{G})$ as the average clustering coefficient

$$
C := \frac{1}{|\mathcal{V}|} \sum_{v \in \mathcal{V}} c(v), \quad \text{where } c(v) = \begin{cases} t(v)/\binom{d_v}{2} & \text{if } d(v) \geq 2 \\ 0 & \text{otherwise,} \end{cases} \tag{6}
$$

with $t(v)$ as the number of triangles that contain node v. Then $f(v)$ is taken as
$c(v)$ itself.

The average in MSE is calculated from multiple runs of the simulations. The
simulations on Les Misérables network is shown in Fig. 1 with $a(n) = 1/\lceil \frac{n}{10} \rceil$
and the super-node size as 25.

Study of Asymptotic MSE: In order to show the asymptotic MSE expressions derived in Propositions 1 and 2, we plot the sample MSE as MSE × B in
Figs. 2a, b and c. These figures correspond to the three different functions we

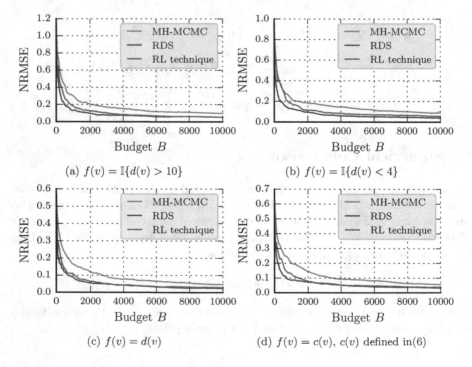

(a) $f(v) = \mathbb{I}\{d(v) > 10\}$ (b) $f(v) = \mathbb{I}\{d(v) < 4\}$

(c) $f(v) = d(v)$ (d) $f(v) = c(v)$, $c(v)$ defined in (6)

Fig. 1. Les Misérables network: NRMSE comparisons (Color figure online)

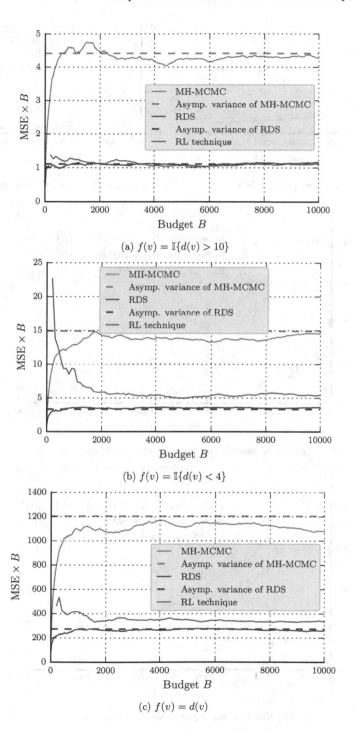

(a) $f(v) = \mathbb{I}\{d(v) > 10\}$

(b) $f(v) = \mathbb{I}\{d(v) < 4\}$

(c) $f(v) = d(v)$

Fig. 2. Les Misérables network: asymptotic MSE comparisons (Color figure online)

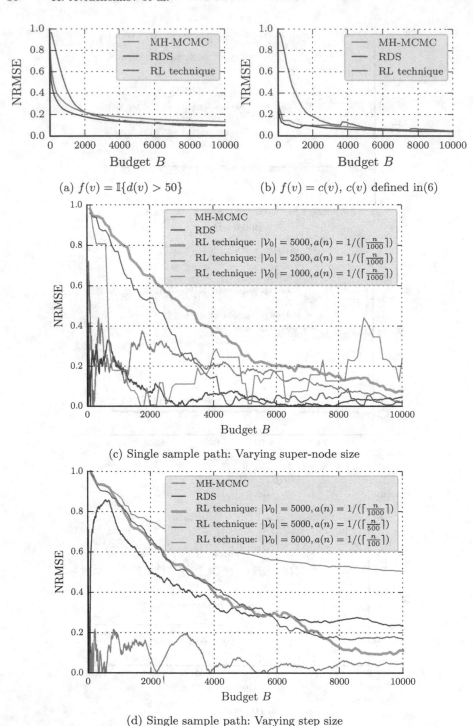

(a) $f(v) = \mathbb{I}\{d(v) > 50\}$ (b) $f(v) = c(v)$, $c(v)$ defined in(6)

(c) Single sample path: Varying super-node size

(d) Single sample path: Varying step size

Fig. 3. Friendster network: (a) and (b) NRMSE comparison, (c) and (d) Single sample path comparison with $f(v) = \mathbb{I}\{d(v) > 50\}$ (Color figure online)

have considered. It can be seen that asymptotic MSE expressions match well with the estimated ones.

4.2   Friendster Network

We consider a larger graph here, a connected subgraph of an online social network called Friendster with 64,600 nodes and 1,246,479 edges. The nodes in Friendster are individuals and edges indicate friendship. We consider the functions (a). $f(v) = \mathbb{I}\{d(v) > 50\}$ and (b). $f(v) = c(v)$ (see (6)) used to estimate the average clustering coefficient. The plot in Fig. 3b shows the results for Friendster graph with super-node size 1000. Here the sequence $a(n)$ is taken as $1/\lceil \frac{n}{25} \rceil$.

Now we concentrate on *single* sample path properties of the algorithms. Hence the numerator of NRMSE becomes absolute error. Figure 3c shows the effect of increasing super-node size while fixing step size $a(n)$ and Fig. 3d shows the effect of changing $a(n)$ when super-node is fixed. In both the cases, the green curve of RL technique shows much stability compared to the other techniques.

4.3   Observations

Some observations from the numerical experiments are as follows:

1. With respect to the limiting variance, RDS always outperforms the other two methods tested. However, with a good choice of parameters the performance of RL is not far from RDS;
2. In the RL technique, we find that the normalizing term $1/|\mathcal{V}_0| \sum_j V_n(j)$ converges much faster than the other two options, $V_t(i_0)$ and $\min_i V_t(i)$;
3. When the size of the super-node decreases, the RL technique requires smaller step size $a(n)$. For instance in case of Les Misérables network, if the super-node size is less than 10, RL technique does not converge with $a(n) = 1/(\lceil \frac{n}{50} \rceil + 1)$ and requires $a(n) = 1/(\lceil \frac{n}{5} \rceil)$;
4. If step size $a(n)$ decreases or the super node size increases, RL fluctuates less but with slower convergence. In general, RL has less fluctuations than MH-MCMC or RDS.

5   Conclusion and Discussion

In this work we studied and compared the performances of various random walk-based techniques for function estimation on OSNs and provide both empirical and theoretical analyses of their performance. We found that in terms of asymptotic mean squared error (MSE), RDS technique outperforms the other methods considered. However, RL technique with small step size displays a more stable sample path in terms of MSE. In the extended version of the paper we plan to test the methods on larger graphs and involve more methods for comparison.

Acknowledgements. This work was supported by CEFIPRA grant no. 5100-IT1 "Monte Carlo and Learning Schemes for Network Analytics," Inria Nokia Bell Labs ADR "Network Science," and by the French Government (National Research Agency, ANR) through the "Investments for the Future" Program reference ANR-11-LABX-0031-01.

References

1. Abounadi, J., Bertsekas, D., Borkar, V.S.: Learning algorithms for markov decision processes with average cost. SIAM J. Control Optim. **40**(3), 681–698 (2001)
2. Avrachenkov, K., Ribeiro, B., Towsley, D.: Improving random walk estimation accuracy with uniform restarts. In: Kumar, R., Sivakumar, D. (eds.) WAW 2010. LNCS, vol. 6516, pp. 98–109. Springer, Heidelberg (2010)
3. Billingsley, P.: Probability and Measure. Wiley, New York (2008)
4. Borkar, V.S.: Stochastic Approximation. Cambridge University Press, Cambridge (2008)
5. Borkar, V.S., Makhijani, R., Sundaresan, R.: Asynchronous gossip for averaging and spectral ranking. IEEE J. Sel. Top. Sig. Process. **8**(4), 703–716 (2014)
6. Brémaud, P.: Markov Chains: Gibbs Fields, Monte Carlo Simulation, and Queue. Springer, New York (2013)
7. Dasgupta, A., Kumar, R., Sarlos, T.: On estimating the average degree. In: Proceedings of the WWW, pp. 795–806 (2014)
8. Gjoka, M., Kurant, M., Butts, C.T., Markopoulou, A.: Walking in facebook: a case study of unbiased sampling of osns. In: Proceedings of the IEEE INFOCOM, pp. 1–9 (2010)
9. Goel, S., Salganik, M.J.: Respondent-driven sampling as Markov chain Monte Carlo. Stat. Med. **28**(17), 2202–2229 (2009)
10. Leskovec, J., Faloutsos, C.: Sampling from large graphs. In: Proceedings of the 12th ACM SIGKDD, pp. 631–636 (2006)
11. Maiya, A.S., Berger-Wolf, T.Y.: Sampling community structure. In: Proceedings of the WWW, pp. 701–710 (2010)
12. Nazi, A., Zhou, Z., Thirumuruganathan, S., Zhang, N., Das, G.: Walk, not wait: faster sampling over online social networks. Proc. VLDB Endowment **8**(6), 678–689 (2015)
13. Nummelin, E.: MC's for MCMC'ists. Int. Stat. Rev. **70**(2), 215–240 (2002)
14. Ribeiro, B., Towsley, D.: Estimating and sampling graphs with multidimensional random walks. In: Proceedings of the 10th ACM SIGCOMM, pp. 390–403 (2010)
15. Robert, C., Casella, G.: Monte Carlo Statistical Methods. Springer Science & Business Media, New York (2013)
16. Ross, S.M.: Applied Probability Models with Optimization Applications. Courier Corporation, Chelmsford (2013)
17. Salganik, M.J., Heckathorn, D.D.: Sampling and estimation in hidden populations using respondent-driven sampling. Sociol. Methodol. **34**(1), 193–240 (2004)
18. Volz, E., Heckathorn, D.D.: Probability based estimation theory for respondent driven sampling. J. Off. Stat. **24**(1), 79 (2008)

Integrating Networks of Equipotent Nodes

Anastasia Moskvina[1] and Jiamou Liu[2(✉)]

[1] Auckland University of Technology, Auckland, New Zealand
anastasia.moskvina@aut.ac.nz
[2] The University of Auckland, Auckland, New Zealand
jiamou.liu@auckland.ac.nz

Abstract. When two social groups merge, members of both groups should socialize effectively into the merged new entity. In other words, interpersonal ties should be established between the groups to give members appropriate access to resource and information. Viewing a social group as a network, we investigate such integration from a computational perspective. In particular, we assume that the networks have equipotent nodes, which refers to the situation when every member has equal privilege. We introduce the network integration problem: Given two networks, set up links between them so that the integrated network has diameter no more than a fixed value. We propose a few heuristics for solving this problem, study their computational complexity and compare their performance using experimental analysis. The results show that our approach is a feasible way to solve the network integration problem by establishing a small number of edges.

1 Introduction

All social groups evolve through time. When two social groups merge, new relations need to be set up. Take, as an example, a merger between two companies. The success of mergers and acquisitions of companies often hinges on whether firms can socialize employees effectively into the merged new entity [1]. Therefore a big challenge faced by the top managers of both companies is how to *integrate* the two companies to ensure coherence and efficient communication. This paper approaches this challenge from a computational perspective. To motivate our formal framework, we make three assumptions: (1) the integration takes place assuming *equipotency* of nodes; (2) creating weak ties between the networks can be encouraged and forced; and (3) structural properties such as distance provide a measure of effective communication and resource accessibility.

The first condition assumes the networks follow *peer-to-peer* relational dynamic, which refers to social structures where information and resources are distributed. In such a social structure, as discussed by Baker in [3], members have no formal authority over each other, and have equal privileges regardless their roles [5]. Examples of such social groups include volunteer organizations, teams of scientists, and companies that embrace a holacracy management style [16]. Baker claims that in order for such a peer-to-peer network to operate efficiently, there must be

© Springer International Publishing Switzerland 2016
H.T. Nguyen and V. Snasel (Eds.): CSoNet 2016, LNCS 9795, pp. 39–50, 2016.
DOI: 10.1007/978-3-319-42345-6_4

clear and open communication; moreover each individual should be aware of the resources available from other nodes.

The second condition arises from the nature of interpersonal relations. Social networks are usually the result of complex interactions among autonomous individuals whose relationships cannot be simply controlled and forced. Ties between people differ by strength; while strong ties denote frequent interactions which form a basis for trust, weak ties plays an important role in information flow. In business networks, although a firm is seldom in control of strong relationships among its employees [15], it can normally prepare the ground for future weak ties: conferences and meetings, group assignments, special promotions etc. can be instruments of bringing people together.

The third condition discusses how the integrated network provides members with appropriate access to resource and information. *Distance* is an important factor of information dissemination in a network [8]: a network with a small diameter means that members are in general close to each other and information could be passed from one person to any others within a small number of steps [17]. This argument has been used to explain how small-world property – the property that any node is reachable from others via only a few hops – becomes a common feature of most real-world social networks [2].

Extending these ideas, we define *network integration* as the process when one or more edges are established across two existing networks in such a way that the integrated network has a bounded diameter Δ. Furthermore, a new edge always costs effort and time to establish and maintain. Thus, we also want to minimize the number of new edges to be created during the integration process. We propose two heuristics to perform network integration. The first is a naive greedy method that iteratively creates edges to minimize the diameter of the resulting network. The second method separately discusses two cases: (1) When Δ is at least the diameter of the original networks, we create edges by considering center and peripheral nodes in the networks. (2) When Δ is smaller than the original diameter of the original networks, we first reduce the distance between nodes in the respective networks and then apply the procedure in case (1). The experiments verify that, our second heuristic significantly outperforms the first, both in terms of running time, and in terms of the output edge set.

The rest of the paper is organized as follows: Sect. 2 presents the formal framework of network integration and shows that it is a computationally hard problem. Section 3 presents a naive greedy heuristic Naive. Section 4 discusses our Integrate algorithm. Section 5 presents experimental results on our algorithms using both generated and real-world data. Section 6 discusses related works before conclusion in Sect. 7.

2 Preliminaries and Problem Setup

We define a *network* as an undirected unweighted connected graph $G = (V, E)$ where V is a set of nodes and E is a set of (undirected) edges on V. We write an edge $\{u, v\}$ as uv. A *path* (of *length k*) is a sequence of nodes $u_0, u_1, \ldots, u_k$ where

$u_i u_{i+1} \in E$ for any $0 \leq i < k$. The *distance* between u and v, denoted by $\text{dist}(u, v)$, is the length of a shortest path from u to v. The *eccentricity* of u is $\text{ecc}(u) = \max_{v \in V} \text{dist}(u, v)$. The *diameter* of the network G is $\text{diam}(G) = \max_{u \in V} \text{ecc}(u)$. The *radius* $\text{rad}(G)$ of G is $\min_{u \in V} \text{ecc}(u)$. For two sets V_1, V_2, we use $V_1 \otimes V_2$ to denote the set of all edges $\{uv \mid u \in V_1, v \in V_2\}$.

Definition 1. *Let $G_1 = (V_1, E_1)$ and $G_2 = (V_2, E_2)$ be two networks. Fix a set of edges $E \subseteq V_1 \otimes V_2$. We define the integrated network of G_1, G_2 with edge set E as the graph $G_1 \oplus_E G_2 = (V_1 \cup V_2, E_1 \cup E_2 \cup E)$.*

When integrating two organizations, each person normally has constraints over who he or she may connect to; this is determined largely by *privilege*, i.e., the type of social inequality created from difference in positions, titles, ranks, etc. [5]. In this paper we focus on the simpler case of social networks with equipotent nodes, and therefore assume all nodes have unbounded and equal privilege. Take an integer $\Delta \geq 1$, we propose the *network integration problem* $\text{NI}_\Delta(G_1, G_2)$:

INPUT. Two networks $G_1 = (V_1, E_1), G_2 = (V_2, E_2)$ where $V_1 \cap V_2 = \varnothing$.

OUTPUT. A set $E \subseteq V_1 \otimes V_2$ such that $\text{diam}(G_1 \oplus_E G_2) \leq \Delta$.

In the rest of the paper we investigate $\text{NI}_\Delta(G_1, G_2)$ on two networks $G_1 = (V_1, E_1)$ and $G_2 = (V_2, E_2)$ where $V_1 \cap V_2 = \varnothing$. The problem naturally depends on the value of Δ. When $\Delta = 1$, it is easy to see that $\text{NI}_\Delta(G_1, G_2)$ has a solution if and only if both networks G_1, G_2 are complete. When $\Delta \geq 2$, since $G_1 \oplus_{V_1 \otimes V_2} G_2$ has diameter 2, $\text{NI}_\Delta(G_1, G_2)$ guarantees to have a solution.

Throughout, we assume $\Delta \geq 2$. We are interested in a solution E to the problem $\text{NI}_\Delta(G_1, G_2)$ that contains the least number of edges; such an E is called an *optimal solution*. The brute-force way of finding optimal solutions for $\text{NI}_\Delta(G_1, G_2)$ examines all possible sets of edges until it finds a required solution set E. This will take time $2^{O(|V_1| \cdot |V_2|)}$. In fact, obtaining optimal solutions is computationally-hard; the following theorem implies that this problem is unlikely to be polynomial-time solvable.

Theorem 2. *The problem of deciding, given two graph G_1, G_2 and an integer $k > 0$, whether $\text{NI}_\Delta(G_1, G_2)$ has a solution set E with cardinality $\leq k$, is hard for $W[2]$, the second level of the W-hierarchy.*

Proof. Let $G = (V, E)$ be a graph. A *distance-r dominating set* of G is a set S of nodes such that for all $u \in V$, there is some $v \in S$ with $\text{dist}(u, v) \leq r$. As shown in [11], finding the smallest distance-r dominating set in G with diameter $r + 1$ is complete for $W[2]$ (for any fixed r). In fact, the $W[2]$-hardness also holds if G is *diametrically uniform*, i.e., if all nodes have the same eccentricity.

Now let $G_1 = (V_1, E_1)$ be a diametrically uniform graph with diameter $\Delta \geq 2$ and let G_2 be a graph that contains only a single node $\{u\}$. For any distance-$(\Delta - 1)$ dominating set $S \subseteq V_1$, the set of edge $S \otimes \{u\}$ is a solution of $\text{NI}_\Delta(G_1, G_2)$. Conversely, suppose $S \subseteq V_1$ is not distance-$(\Delta - 1)$ dominating. Then there is a node $w \in V_1$ that is at distance at least Δ away from any node $v \in S$. This means that $\text{dist}(w, u)$ in the integrated network is at least $\Delta + 1$ and S is not a solution of $\mathsf{NI}_\Delta(G_1, G_2)$. Thus $\mathsf{NI}_\Delta(G_1, G_2)$ has a size-k solution if and only if G_1 has a size-k distance-$(\Delta - 1)$ dominating set.   □

3 The Naive Greedy Method

We thus turn to heuristics for finding small solution sets of $\mathsf{NI}_\Delta(G_1, G_2)$. As a first step we introduce a greedy heuristic that approximates a solution.

Definition 3. *A set of edges $E \subseteq V_1 \otimes V_2$ is called a* naive greedy set *if we can write it as $\{e_1, \ldots, e_\ell\}$ such that for all $1 \le i \le \ell$, and $e' \in (V_1 \otimes V_2) \setminus \{e_1, \ldots, e_{i-1}\}$, $\mathsf{diam}(G_1 \oplus_{\{e_1, \ldots, e_i\}} G_2) \le \mathsf{diam}(G_1 \oplus_{\{e_1, \ldots, e_{i-1}, e'\}} G_2)$. A* naive greedy solution *to $\mathsf{NI}_\Delta(G_1, G_2)$ is a solution that is a naive greedy set.*

As its name suggests, a naive greedy set can be constructed incrementally using a greedy strategy that locally optimizes diameter in the integrated network. Naive greedy solutions to $\mathsf{NI}_\Delta(G_1, G_2)$ are not necessarily optimal, and vice versa:

Example. Let both G_1 and G_2 be paths of length 5, i.e., G_1 contains nodes $a_1, \ldots, a_5$ while G_2 contains nodes $b_1, \ldots, b_5$ with edges $a_i a_{i+1}$, $b_i b_{i+1}$ for any $1 \le i < 5$. Suppose $\Delta = 3$. The only optimal solution E contains four edges, i.e., $E = \{a_1 b_3, a_3 b_1, a_3 b_5, a_5 b_3\}$. However, for any edge $e \in E$, $\mathsf{diam}\left(G_1 \oplus_{\{e\}} G_2\right) = 7$, while $\mathsf{diam}\left(G_1 \oplus_{\{a_3 b_3\}} G_2\right) = 5$. Thus E is not a naive greedy solution, nor will any naive greedy solution be optimal.

Theorem 4. *There exists an algorithm $\mathsf{Naive}_\Delta(G_1, G_2)$ that runs in time $O(n^6)$ and computes a naive greedy solution for $\mathsf{NI}_\Delta(G_1, G_2)$ where $n = |V_1 \cup V_2|$.*

Proof. The algorithm $\mathsf{Naive}_\Delta(G_1, G_2)$ iteratively adds edges $e_1, e_2, \ldots$ to the solution set E. It also computes a matrix $\mathsf{D} \colon (V_1 \cup V_2)^2 \to \mathbb{N}$ that represents the distance between nodes in the current integrated graph. See Procedure 1

   Since $\Delta \ge 2$, the algorithm will terminate. Furthermore, the set of edges created by the algorithm is a naive greedy solution. At each iteration, computing each matrix D_e takes time $O(n^2)$; computing F takes $O(n^2)$. Since there are $O(n^2)$ edges in $(V_1 \otimes V_2) \setminus E_i$, this iteration runs in $O(n^4)$. Since there are at most n^2 iterations, the algorithm takes times $O(n^6)$.   □

We remark that when $\Delta > 2$, the maximum number of edges required is $O(n)$, and hence $\mathsf{Naive}_\Delta(G_1, G_2)$ will take $O(n^5)$. The algorithm $\mathsf{Naive}_\Delta(G_1, G_2)$ is still too inefficient in most practical cases and hence in subsequent sections we discuss more efficient heuristics for $\mathsf{NI}_\Delta(G_1, G_2)$.

4 Efficient Algorithms for $\mathsf{NI}_\Delta(G_1, G_2)$

We separately discuss two cases: (1) when the integrated network's diameter is at least the diameters of the given networks, i.e. $\Delta \ge \max\{\mathsf{diam}(G_1), \mathsf{diam}(G_2)\}$; and (2) when we improve the diameter, i.e. $\Delta < \max\{\mathsf{diam}(G_1), \mathsf{diam}(G_2)\}$.

Procedure 1. $\mathsf{Naive}_\Delta(G_1, G_2)$; Output E

Initialize D (for the disjoint union of G_1, G_2); Set $E := \varnothing$
while $\mathsf{diam}\,(G_1 \oplus_E G_2) > \Delta$ **do**
    for $e := xy \in (V_1 \otimes V_2) \setminus E$ **do**
        for $(u, v) \in (V_1 \cup V_2)^2$ **do** $\triangleright$ define a temporary $D_e : (V_1 \cup V_2)^2 \to \mathbb{N}$
            $D_e(u, v) := \min\{D_i(u, v), D_i(u, x) + D_i(y, v) + 1\}$
        end for
        Set $\mathsf{diam}_e := \max\{D_e(u, v) : (u, v) \in (V_1 \cup V_2)^2\}$.
    end for
    Set $F := \{e \in (V_1 \otimes V_2) \setminus E_i \mid \mathsf{diam}_e \le \mathsf{diam}_{e'} \text{ for all } e' \in (V_1 \otimes V_2) \setminus E_i\}$.
    Pick a random edge $e_i \in F$ and set $D := D_{e_i}$, $E := E \cup \{e_i\}$
end while

4.1 The Case When $\Delta \ge \max\{\mathsf{diam}(G_1), \mathsf{diam}(G_2)\}$

Firstly, when integrating two networks, it makes sense to establish a link between the most central persons in the networks, as they have the closest proximity to other nodes. Furthermore, if x, y are nodes that are furthest apart in the integrated network, they are unlikely to communicate effectively thanks to their shear distance; this, in a certain sense, represents a form of *structural hole* [4]. Hence in integrating these two networks, it makes sense to connect x, y by an edge. Formally, the *center* $C(G)$ of a graph $G = (V, E)$ is the set of all nodes that have the least eccentricity, i.e., $C(G) = \{v \in V \mid \mathsf{ecc}(v) = \mathsf{rad}(G)\}$. A pair of nodes (x, y) in G forms a *peripheral pair*, denoted by $(x, y) \in P(G)$, if $\mathsf{dist}(x, y) = \mathsf{diam}(G)$. Our heuristic first creates an edge between two nodes that are in $C(G_1)$ and $C(G_2)$ respectively, and then iteratively "bridges" peripheral pairs.

Definition 5. *A set $E \subseteq V_1 \otimes V_2$ is called a* center-periphery set *if we can write it as $\{e_0, \ldots, e_\ell\}$ such that (1) $e_0 \in C(G_1) \otimes C(G_2)$; and (2) for all $1 \le i \le \ell$, $e_i \in P\left(G_1 \oplus_{\{e_0, e_1, \ldots, e_{i-1}\}} G_2\right)$. A* center-periphery solution *is a solution to $\mathsf{NI}_\Delta(G_1, G_2)$ that is also a center-periphery set.*

Clearly, if $\Delta > \mathsf{rad}(G_1) + \mathsf{rad}(G_2)$, then for any $uv \in C(G_1) \otimes C(G_2)$, we have $\mathsf{diam}\,(G_1 \oplus_{\{uv\}} G_2) \le \max\{\mathsf{diam}(G_1), \mathsf{diam}(G_2), \mathsf{rad}(G_1) + \mathsf{rad}(G_2) + 1\} \le \Delta$. Thus $\{uv\}$ forms a solution of $\mathsf{NI}_\Delta(G_1, G_2)$. In this case, center-periphery solutions coincide with optimal solutions.

Theorem 6. *There exists an algorithm $\mathsf{CtrPer}_\Delta(G_1, G_2)$ that has $O(n^4)$ running time and computes a center-periphery solution for $\mathsf{NI}_\Delta(G_1, G_2)$ assuming $\Delta \ge \max\{\mathsf{diam}(G_1), \mathsf{diam}(G_2)\}$, where $n = |V_1 \cup V_2|$.*

Proof. The $\mathsf{CtrPer}_\Delta(G_1, G_2)$ algorithm also maintains a matrix $D : (V_1 \cup V_2)^2 \to \mathbb{N}$ such that $D(u, v)$ is the distance between u, v. The eccentricity of each node can be easily extracted from D allowing the algorithm to identify the centers $C(G_1)$ and $C(G_2)$, respectively. The algorithm then iteratively adds edges that connect peripheral pairs in the integrated graph until its diameter becomes at most Δ. See Procedure 2.

Suppose the algorithm creates a set $E \subseteq V_1 \otimes V_2$ and $\mathsf{diam}\,(G_1 \oplus_E G_2) > \Delta$. The algorithm will update the matrix D and then picks (u, v) with the largest $D(u, v)$. By definition of D, (u, v) forms a peripheral pair in $G_1 \oplus_E G_2$. We need to show that uv is a valid edge to add, that is, u, v cannot both lie in one of V_1 and V_2. Indeed, if $\{u, v\} \subseteq V_1$ or $\{u, v\} \subseteq V_2$, then $\mathsf{dist}(u, v) \leq \max\{\mathsf{diam}(G_1), \mathsf{diam}(G_2)\} \leq \Delta < \mathsf{diam}\,(G_1 \oplus_E G_2)$. Thus $uv \in V_1 \otimes V_2$. Now either $E \cup \{uv\}$ is a solution, or $\mathsf{diam}\,(G_1 \oplus_{E \cup \{uv\}} G_2) > \Delta$. In the latter case the algorithm repeats the iteration to find another peripheral pair. Thus the algorithm will terminate and produce a center-periphery solution to $\mathsf{NI}_\Delta(G_1, G_2)$.

It takes $O(n^3)$ to initialize the matrix D using Floyd-Warshall algorithm. At each iteration, the algorithm takes $O(n^2)$ to update D and finds a peripheral pair. Since there are at most n^2 iterations, the algorithm takes time $O(n^4)$. $\square$

Procedure 2. $\mathsf{CtrPer}_\Delta(G_1, G_2)$: $\Delta \geq \max\{\mathsf{diam}(G_1), \mathsf{diam}(G_2)\}$; Output E

Initialize the matrix D so that $D(u, v) = \mathsf{dist}(u, v)$ in the un-integrated graphs
Take a node $u \in C(G_1)$ and a node $v \in C(G_2)$
Set $e := uv$, $E := \{e\}$
while $(\mathsf{diam}\,(G_1 \oplus_E G_2) > \Delta)$ **do**
    for $(x, y) \in (V_1 \cup V_2)^2$ **do** $\triangleright$ define $D' : (V_1 \cup V_2)^2 \to \mathbb{N}$
        $D'(x, y) := \min\{D(x, y), D(x, u) + D(v, y) + 1\}$ where $e = uv$
    end for
    $D := D'$ $\triangleright$ update matrix D
    Pick (u, v) with the largest $D(u, v)$. Set $e := uv$ and $E := E \cup \{e\}$
end while

4.2 The Case When $\Delta < \max\{\mathsf{diam}(G_1), \mathsf{diam}(G_2)\}$

When the diameter bound Δ is less than the diameters of the two component networks G_1, G_2, the goal is to *improve* the connectivity of each original network through integration. In other words, the integration should "bring people closer". In this case $\mathsf{CtrPer}_\Delta(G_1, G_2)$ no longer applies as it is possible for both nodes in a peripheral pair to lie in the same component graph G_1 or G_2, forbidding us to create the edge xy. We therefore need to first decrease the distance between nodes in each G_1 and G_2. Suppose a, b are two people in an organization with large distance. When their organization merges with another organization, a and b can be brought closer if they both know a 'third person' c in the other organization, i.e., the ties ac and bc allows a, b to be only 2 steps away.

Definition 7. *Let $E \subseteq V_1 \otimes V_2$ be a set of edges and $i \in \{1, 2\}$. The diameter of G_i relative to E is the maximum distance between any two nodes in V_i in the integrated network $G_1 \oplus_E G_2$; we denote this by $\mathsf{diam}_E(G_i)$. A set of edges $E \subseteq V_1 \otimes V_2$ is a Δ-bridge if $\mathsf{diam}_E(G_i) \leq \Delta$ for both $i \in \{1, 2\}$.*

Theorem 8. *For any $\Delta \geq 2$, there exists an algorithm* $\mathsf{Bridge}_\Delta(G_1, G_2)$ *that runs in time $O(n^4)$ and computes a Δ-bridge E, where $n = |V_1 \cup V_2|$.*

Proof. The algorithm has two phases. In phase $i \in \{1,2\}$, it makes $\mathsf{diam}_E(G_i) \leq \Delta$. Phase i consists of several iterations; at each iteration, the algorithm takes a pair $(u,v) \in V_i$ with maximum distance and a node $w \in V_{3-i}$, and builds two edges uw and vw. See Procedure 3. Throughout, the algorithm computes and maintains a matrix D: $(V_1 \cup V_2)^2 \to \mathbb{N}$ such that $D(u,v)$ is the current distance between nodes u, v. When a pair of new edges uw, vw are added, the new distance $D'(x,y)$ between any pair of nodes $(x,y) \in V_i^2$ is calculated as follows:

$$D'(x,y) = \min\{D(x,y), D(x,u)+D(w,y)+1, D(x,u)+D(v,y)+2,$$
$$D(x,v)+D(w,y)+1, D(x,v)+D(u,y)+2\} \quad (1)$$

In the worst case, the algorithm adds edges uw, vw for any pair $(u,v) \in V_i^2$ where $i \in \{1,2\}$. Thus the algorithm terminates in at most n^2 iterations. Finding nodes u, v, w and updating the matrix D at each iteration takes time $O(n^2)$. Therefore the total running time is $O(n^4)$. $\qquad\square$

Procedure 3. $\mathsf{Bridge}_\Delta(G_1, G_2)$: $\Delta < \max\{\mathsf{diam}(G_1), \mathsf{diam}(G_2)\}$; Output E

Initialize the matrix D so that $D(u,v) = \mathsf{dist}(u,v)$ in the un-integrated graphs
Initialize $E := \varnothing$
for $i = 1, 2$ **do**   ▷ The two phases
    while $\mathsf{diam}_E(G_i) > \Delta$ **do**
        Take a pair of nodes $u, v \in V_i$ with maximum $D(u,v)$
        Take a node w in V_{3-i}
        $E := E \cup \{uw, wv\}$
        for $(x,y) \in (V_1 \cup V_2)^2$ **do**   ▷ define $D' : (V_1 \cup V_2)^2 \to \mathbb{N}$
            Compute $D'(x,y)$ as in (1)
        end for
        $D := D'$ ▷ update matrix D
    end while
end for

Remark. Suppose the $\mathsf{Bridge}_\Delta(G_1, G_2)$ algorithm adds edges uw, vw. Here w plays the role as a *bridging node* that links u and v. Naturally, the choice of w affects the performance of the algorithm: by carefully choosing the bridging node w, we may reduce the number of new ties that need to be created. Imagine that G_1, G_2 represent two organizations.

1. To allow smooth flow of information between the two organizations and avoid *information gate keepers*, we should have many bridging nodes in G_2.
2. A node with a higher degree means it has better access to resource and information, and thus is a more appropriate bridging nodes.

Therefore, we introduce the following heuristics to $\text{Bridge}_\Delta(G_1, G_2)$: Suppose the algorithm has selected a set E of edges and picked $(u, v) \in V_i$ where $i \in \{1, 2\}$ with the largest $D(u, v)$. To pick a bridging node w:

Heuristic 1. For any node $w \in V_{3-i}$, let $b(w) = |\{v \mid wv \in E\}|$. The chosen bridging node w is taken from $B_i = \{w \in V_{3-i} \mid b(w) \leq b(w') \text{ for all } w' \in V_{3-i}\}$.
Heuristic 2. The chosen bridging node w has the highest degree in B_i.

We now extend $\text{Bridge}_\Delta(G_1, G_2)$ to an algorithm that solves $\text{NI}_\Delta(G_1, G_2)$.

Algorithm 4. $\text{Integrate}_\Delta(G_1, G_2)$; Output E

Run $\text{Bridge}_\Delta(G_1, G_2)$ to obtain a set $E \subseteq V_1 \otimes V_2$
Run $\text{CtrPer}_\Delta(G_1, G_2)$ to add edges to E (instead of building E from scratch)

Theorem 9. *The $\text{Integrate}_\Delta(G_1, G_2)$ algorithm runs in time $O(n^4)$ and computes a solution to $\text{NI}_\Delta(G_1, G_2)$ for any networks G_1, G_2 and $\Delta \geq 2$, where $n = |V_1 \cup V_2|$.*

5   Experiments and Case Studies

To test the algorithms, we generate two types of random graphs: the first (NWS) is Newman-Watts-Strogatz's small-world networks, which have small average path lengths and high clustering coefficients [13]. The second (BA) is Barabasi-Albert's preferential attachment model which produces scale-free graphs whose degree distribution of nodes follows a power law [2]. In Figs. 1 and 2, we integrate two graphs of each type using the $\text{Integrate}_\Delta(G_1, G_2)$ algorithm. The statistics for each graph is shown in the table below. For the NWS graphs, Δ ranges from 6 to 11, while for the BA graphs, Δ ranges from 4 to 7. The figures show how the two networks dissolve into each other with decreasing Δ: when very few edges link the two networks, the network exhibits a clear community structure; this, however, becomes less clear as more edges are created.

	NWS Graph 1	NWS Graph 2	BA Graph 1	BA Graph 2
Number of nodes/edges	50/77	50/78	50/141	50/141
Diameter/radius	7/5	8/5	4/3	4/3

Experiment 1 (Running times). We implement both algorithms and record their running times on 300 generated NWS and BA networks. The results indicate that $\text{Integrate}_\Delta(G_1, G_2)$ outperforms $\text{Naive}_\Delta(G_1, G_2)$ significantly, with the former runs more than 3000 times faster on networks with 1000 nodes to add a single edge in the solution set. Figure 3 plots how much longer (on average) $\text{Naive}_\Delta(G_1, G_2)$ takes

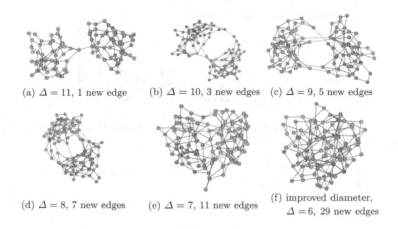

(a) $\Delta = 11$, 1 new edge   (b) $\Delta = 10$, 3 new edges   (c) $\Delta = 9$, 5 new edges

(d) $\Delta = 8$, 7 new edges   (e) $\Delta = 7$, 11 new edges   (f) improved diameter,
$\Delta = 6$, 29 new edges

Fig. 1. Integrating two NWS networks with different Δ

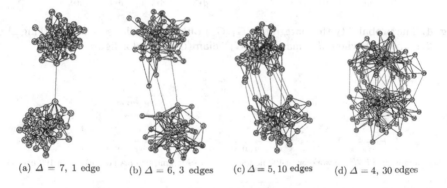

(a) $\Delta = 7$, 1 edge   (b) $\Delta = 6$, 3 edges   (c) $\Delta = 5$, 10 edges   (d) $\Delta = 4$, 30 edges

Fig. 2. Integrating two BA networks with different Δ

to add a single edge to the solution set compared to $\mathsf{Integrate}_\Delta(G_1, G_2)$, against the number of nodes in the networks.

Experiment 2 (Solution size). We compare the output of $\mathsf{Integrate}_\Delta(G_1, G_2)$ against the $\mathsf{Naive}_\Delta(G_1, G_2)$ algorithm. While $\mathsf{Naive}_\Delta(G_1, G_2)$ may output smaller solutions when Δ is large, $\mathsf{Integrate}_\Delta(G_1, G_2)$ is more likely to produce smaller solutions as Δ decreases. Figure 4 plots the percentage of the cases where $\mathsf{Integrate}_\Delta(G_1, G_2)$ returns smaller sets. Note that $\mathsf{Integrate}_\Delta(G_1, G_2)$ almost always returns smaller sets whenever $\Delta < \max\{\mathsf{diam}(G_1), \mathsf{diam}(G_2)\}$. Figure 5 plots the average output size of $\mathsf{Integrate}_\Delta(G_1, G_2)$ and $\mathsf{Naive}_\Delta(G_1, G_2)$, against absolute and relative values of Δ. Here, each graph consists of 100 nodes. Even though $\mathsf{Naive}_\Delta(G_1, G_2)$ may outperform $\mathsf{Integrate}_\Delta(G_1, G_2)$ when Δ is large, the difference is not very significant; as Δ decreases, the advantage of $\mathsf{Integrate}_\Delta(G_1, G_2)$ becomes increasingly significant.

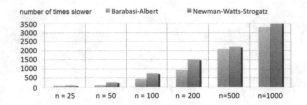

Fig. 3. The number of times Naive$_\Delta(G_1, G_2)$ runs slower than Integrate$_\Delta(G_1, G_2)$ (Color figure online)

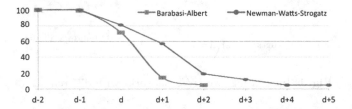

Fig. 4. The probability that Integrate$_\Delta(G_1, G_2)$ outputs smaller sets with varying $\Delta \in \{d - 2, \ldots, d + 5\}$ where $d = \max\{\text{diam}(G_1), \text{diam}(G_2)\}$ (Color figure online)

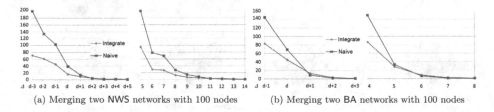

(a) Merging two NWS networks with 100 nodes   (b) Merging two BA networks with 100 nodes

Fig. 5. Comparing the Integrate$_\Delta(G_1, G_2)$ algorithm and the Naive$_\Delta(G_1, G_2)$ algorithm: average number of edges with different parameter Δ (Color figure online)

6   Related Works

This paper studies the integration between two social networks of equipotent nodes. This problem relates to several established topics in network science:

Firstly, *strategic network formation* aims to explain how a network evolves in time [7]. A well known example along this line is on the rise of the Medici Family in the XV century [14], which explains how inter-family ties shape political structures. In a certain sense, the network integration problem can be regarded as network formation between two established networks. However, the network formation models are typically about the transformation within a single network, while this paper initiates the perspective of integrating several different networks.

Secondly, the topic of *interdependent networks* aims to model a complex environment where multiple networks interact and form a type of *network of networks* [6]. The networks in such a complex environment are non-homogeneous, i.e., the networks are of different types. For example, one may be interested in the interdependence between a telecommunication network and a transportation networks and how such interaction affects robustness of the entire infrastructure. The focus here is on robustness of the combined structure: how does a failure in one network affects the other network. Compared to interdependent networks, the problem in this paper involves homogeneous networks and concerns a type of dynamic that 'dissolves' the two networks into one. This is more suitable for the social context discussed in this paper.

A third related area is *link prediction*, which aims to infer potential ties between nodes of a network [9]. Here, most approaches take into account surrounding contexts such as homophily and maximum likelihood.

7   Conclusion and Future Works

This paper amounts to our effort to study integration of social networks from a computational perspective, and is a continuation of our earlier work on network socialization [12], where we study how an individual joins an established network, in order to take an advantageous position in the network.

The simple formulation of the problem means that several natural limitations exist: Firstly, the equipotency assumption restricts us to a special class of social networks. In practice, individuals may have different constraints (e.g. titles, roles, positions, etc.) forbidding certain ties to be created. Hence as a future work we plan to enrich our framework by introducing privileges to nodes and study how networks are integrated with privileged-based constraints on new edges to be forged. Secondly, the paper focuses on optimising the number of new edges between networks, which may not be the most crucial factor when merging social groups. Indeed, every edge is established with certain cost; it may thus be an interesting future work to develop a cost model for the establishment of ties in a social network. Thirdly, the paper concerns with diameter of the integrated network, which is a strong measure on access to resources and information; it may make sense to consider other weaker notions. For example, a more relevant measure of integration may be the distance from any node in one network to any node in the other network, or the average distance between nodes. Lastly, we would like to extend our notion of network integration to more elaborated forms of networks. For example, in [10], a framework of hierarchical networks is defined which incorporates both formal ties in an organization and information ties. This framework allows the definition of a notion of *power* in a hierarchical network. It is then natural to ask how power is affected during integration of two hierarchical networks.

References

1. Aguilera, R., Dencker, J., Yalabik, Z.: Institutions and organizational socialization: integrating employees in cross-border mergers and acquisitions. Int. J. Hum. Resour. Manag. **15**(8), 1357–1372 (2006)

2. Albert, R., Barabasi, A.-L.: Statistical mechanics of complex networks. Rev. Mod. Phys. **74**, 47 (2002)
3. Baker, M.: Peer-to-Peer Leadership: Why the Network Is the Leader. Berrett-Koehler Publishers, San Francisco (2013)
4. Burt, R.: Structural holes and good ideas. Am. J. Soc. **110**(2), 349–399 (2004)
5. Casella, E.: The Archaeology of Plural and Changing Identities: Beyond Identification. Springer, New York (2005)
6. Danziger, M., Bashan, A., Berezin, Y., Shekhtman, L., Havlin, S.: An introduction to interdependent networks. In: Mladenov, V.M., Ivanov, P.C. (eds.) NDES 2014. Communications in Computer and Information Science, vol. 438, pp. 189–202. Springer, Berlin (2014)
7. Jackson, M.: A survey of models of network formation: stability and efficiency. In: Group Formation in Economics, pp. 11–57. Cambridge University Press, Cambridge (2005)
8. Leskovec, J., Kleinberg, J., Faloutsos, C.: Graph evolution: densification and shrinking diameters. ACM TKDD **1**(1), 2 (2007)
9. Liben-Nowell, D., Kleinberg, J.: The link prediction problem for social networks. In: Proceedings of CIKM, pp. 556–559 (2003)
10. Liu, J., Moskvina, A.: Hierarchies, ties and power in organisational networks: model and analysis. In: Proceedings of ASONAM, pp. 202–209 (2015)
11. Lokshtanov, D., Misra, N., Philip, G., Ramanujan, M.S., Saurabh, S.: Hardness of r-DOMINATING SET on graphs of diameter $(r + 1)$. In: Gutin, G., Szeider, S. (eds.) IPEC 2013. LNCS, vol. 8246, pp. 255–267. Springer, Heidelberg (2013)
12. Moskvina, A., Liu, J.: How to build your network? a structural analysis. In: Proceedings of IJCAI 2016, AAAI Press (2016, to appear)
13. Newman, M., Watts, D., Strogatz, S.: Random graph models of social networks. Proc. Natl. Acad. Sci. USA **99**, 2566–2572 (2002)
14. Padgett, J., Ansell, C.: Robust action and the rise of the medici, 1400–1434. Am. J. Sociol. **98**(6), 1259–1319 (1993)
15. Ritter, T., Wilkinson, I., Johnston, W.: Managing in complex business networks. Ind. Mark. Manage. **33**(3), 175–183 (2004)
16. Robertson, B.: Holacracy: The New Management System for a Rapidly Changing World Hardcover. Henry Holt and Co., New York City (2015)
17. Wang, X., Chen, G.: Complex networks: small-world, scale-free and beyond. IEEE Circ. Syst. Mag. **3**(1), 6–20 (2003). First Quarter

Identify Influential Spreaders in Online Social Networks Based on Social Meta Path and PageRank

Vang V. Le[1], Hien T. Nguyen[1(✉)], Vaclav Snasel[2], and Tran Trong Dao[3]

[1] Faculty of Information Technology, Ton Duc Thang University,
Ho Chi Minh City, Vietnam
{levanvang,hien}@tdt.edu.vn
[2] Department of Computer Science, VSB-Technical University of Ostrava,
Ostrava, Czech Republic
vaclav.snasel@vsb.cz
[3] Division of MERLIN, Ton Duc Thang University, Ho Chi Minh City, Vietnam
trantrongdao@tdt.edu.vn

Abstract. Identifying "influential spreader" is finding a subset of individuals in the social network, so that when information injected into this subset, it is spread most broadly to the rest of the network individuals. The determination of the information influence degree of individual plays an important role in online social networking. Once there is a list of individuals who have high influence, the marketers can access these individuals and seek them to impress, bribe or somehow make them spread up the good information for their business as well as their product in marketing campaign. In this paper, according to the idea "Information can be spread between two unconnected users in the network as long as they both check-in at the same location", we proposed an algorithm called SMPRank (Social Meta Path Rank) to identify individuals with the largest influence in complex online social networks. The experimental results show that SMPRank performs better than Weighted LeaderRank because of the ability to determinate more influential spreaders.

Keywords: Influential spreader · LeaderRank · Random walk · PageRank · Social meta path

1 Introduction

Today, the online social network such as Facebook, Twitter becoming a popular channel for transmission of information such as news, brochures, and marketing, ... The booming in the number of OSN users poses a major challenge is how information can be spread to the users in the most effective and optimal way with a fixed cost. One way to do is to find users who have the greatest degree of spread (influential spreaders) and inject information into these people to get the benefit, information from them will be widely spread in online social networks and lead to the most effective marketing result.

© Springer International Publishing Switzerland 2016
H.T. Nguyen and V. Snasel (Eds.): CSoNet 2016, LNCS 9795, pp. 51–61, 2016.
DOI: 10.1007/978-3-319-42345-6_5

Given a network G(V, E) - V is the user set and E is the edge set of G which represents the connections between users in G. X is a subset of V and a function $influence(X)$ is the influence function which maps the seed user set X to the number of users influenced by users in X. Identifying influential spreaders aims at selecting the optimal subset $X^\star$ which contains n seed users to maximize the propagation of information across the networks.

$$X^\star = argmax_{X \subseteq V} influence\,(X)$$
$$|X| = n \tag{1}$$

How to determine efficiently the individuals who have the highest degree of influence in social networks is a major challenge up to the present [4–9,14]. Recently, Lu et al. [10] proposed an algorithm LeaderRank to identify influential spreaders in directed network which is a simple variant of PageRank. The authors said that the connection matrix between individuals (adjacency matrix) in social networks is relatively sparse and they introduced the concept "ground node" (an additional node) and create virtual connections from the ground node to all existing nodes in social networks and set the weight of virtual edge a value of 1. This approach has limited success in shortening the convergence time when running the PageRank algorithm to determine the ranking of the node. However, it has one drawback is whether individuals who have more fans or less fans then receives the same weight value of 1 from the ground node and this slightly estate reasonable. Li et al. [1] proposed the Weighted LeaderRank algorithm, an improvement of standard LeaderRank by allowing nodes with more fans get more scores from the ground node. Weighted LeaderRank is a straightforward and efficient method, however, it is less relevant to real network in which the information diffusion depends not only on the network structure but also the network behavior. In fact, when applying the Weighted LeaderRank to actual dataset (Twitter), the obtained result is not the most influential spreaders.

In this paper, we further improve the Weighted LeaderRank algorithm by applying the definition of social meta path which introduced by Zhan et al. [3]. Our approach, which called SMPRank is the hybrid method of Weighted Leader-Rank method and a part of social meta path. The experiments on the real social network (Twitter) show that the SMPRank can considerably improve the spread-ability of the original Weighted LeaderRank. Our approach is based on the idea:

(1) Typically, information can only spread from a user to another user if and only if they are connected to each other (friends or following). However, our approach assumes that even if there is no direct connection to each other, the information is still able to exchange if they both check-in at the same location (by talking directly).
(2) Even between connected users, the information may be spread stronger between users who often communicate to each other and weaker between users who rarely communicate to each other. For instance, A and B are two followers of C, usually each 10 tweets C writes then A retweets 5 and B retweets 3 mean that information may be spread from C to A stronger than from C to B.

The main contribution of our research is the improvement of the accuracy, our influential spreaders obtained from SMPRank is closer to observed dataset than the result obtained from Weighted LeaderRank. The remaining parts of this paper are organized as follows. We summary the related work in Sect. 2. In Sects. 3, we introduce the proposed SMPRank method. Experiments are given in Sect. 4. Finally, we conclude the paper in Sect. 5.

2   Related Work

Identifying the most influential spreaders in a network is critical for ensuring efficient diffusion of information. For instance, a social media campaign can be optimized by targeting influential individuals who can trigger large cascades of further adoptions. This section presents briefly some related works that illustrate the various possible ways to measure the influence of individuals in the online social network.

Cataldi et al. [12] propose to use the well known PageRank algorithm [11,13] to calculate the influence of individuals throughout the network. The PageRank value of a given node is proportional to the probability of visiting that node in a random walk of the social network, where the set of states of the random walk is the set of nodes. It directly applies the standard random walk process to determine the score of every node. Accordingly, the score of each node in the network will be calculated step by step from t_0 to t_n. At the time t_i, the score of node u will be calculated based on the score of u and the score of u's neighbors in the previous step t_{i-1}. The random walk can be described by an iterative process as formulate (2). In that: $S_u(t_i)$ is the score of node u at the time t_i, $w_{v,u}$ is the weight of connection from v to u, it has value of 1 if existing a connection from v to u and opposite it has value of 0.

$$S_u(t_i) = \sum_{v \in Neighbor(u)} \frac{w_{u,v}}{outdeg(v)} * S_u(t_{i-1}) \tag{2}$$

Recently, Lu et al. [10] proposed an algorithm LeaderRank to identify influential spreaders in directed network which is a simple variant of the algorithm PageRank [2]. To reduce the convergence time of PageRank, it adds an additional node called ground node, by creating many virtual connections from real nodes to ground node it improves the sparseness of original connection matrix. The Fig. 1 demonstrates the LeaderRank by set the value of 1 to all virtual connections from real nodes to ground node and vice versa.

Li et al. [1] proposed the Weighted LeaderRank algorithm, an improvement of standard LeaderRank by allowing nodes with more fans get more scores from the ground node. Instead of setting the value of 1 to all virtual connections, Weighted LeaderRank sets the difference values to difference virtual connections. The virtual connections from ground node to high in-degree real node will get higher weight value compare with the virtual connections from ground node to low in-degree real node. For example, in the Fig. 2, the connection from ground

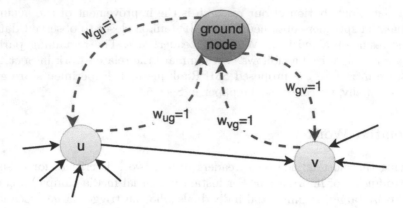

Fig. 1. An example of LeaderRank algorithm

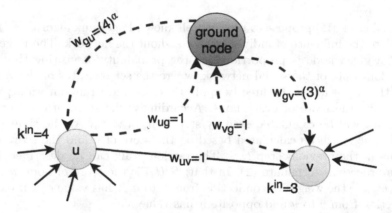

Fig. 2. An example of Weighted LeaderRank algorithm

node g to node u will get higher weight value than the connection from ground node g to node v because u's in-degree is higher than v's in-degree.

The methods we have just described above exist a drawback that they only exploit the structure (topology) of the network, and ignore other important properties, such as nodes' features and the way they interact with other nodes in the network.

Zhan et al. [3] proposed a new model M&M to resolve the Aligned Heterogeneous network Influence maximization (AHI) problem. The explosion of online social networks lead to a person can participate and have multiple accounts on different online social networks. Information can be spread not only on internal network but also it can be exchanged together between difference networks. If a user A participate onto two online social networks X and Y simultaneously, the information A received on the network X can be forwarded to the network Y this means that information can be spread through difference channels: internal and external channel. Through this idea, the author proposed a definition of path, meta path and social meta path.

3 Proposed Model

Typically, information can only spread from user A to user B if and only if A and B are connected to each other (friends or following). However, in our approach we assume that even if there is no direct connection with each other, the information is still able to spreading from A to B (i.e., A and B check-in at the same location on the same event, A is the host of the event and B is the client that attends the event - information will spread from A to B). The Fig. 3 demonstrates the idea of our algorithm, the actual network doesn't have a direct connection from node v to node u but it may exist a hidden connection from v to u (represented by dotted line) through another channel such as v and u check-in the same location on the same event.

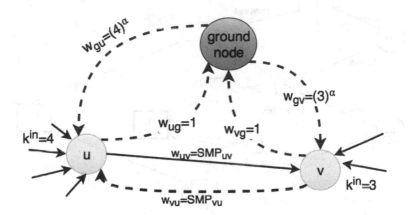

Fig. 3. An example of SMPRank algorithm

In this paper, we will follow the definitions of concepts Social Meta Path proposed in [3]. The Fig. 4 illustrates the schema of Twitter network which we chose to do the experiment. Depend on the network schema, we select 3 social meta paths as below:

(1) Follow

$$MP^1: User \xrightarrow{follow} User$$

(2) Co-location check-in

$$MP^2: User \xrightarrow{write} Tweet \xrightarrow{checkin} Location \xrightarrow{checkin^{-1}} Tweet \xrightarrow{write^{-1}} User$$

(3) Re-tweet

$$MP^3: User \xrightarrow{write} Tweet \xrightarrow{retweet} Tweet \xrightarrow{write^{-1}} User$$

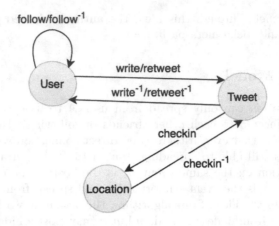

Fig. 4. The network schema of Twitter

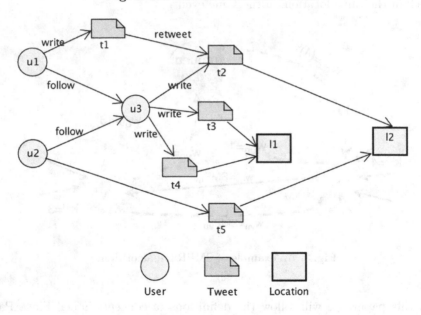

Fig. 5. An example of Twitter network with User, Tweet, and Location

Based on the social media path information, we calculated the value of $\theta^i_{u,v}$ based on Formula (3). In which u, v are vertices of the network, i is in $[1, 3]$ represents the three types of social meta paths selected above. The values of $\theta^i_{u,v}$ represent the power of information transmission from vertex u to vertex v through the ith social meta path channel.

Applying the formula (3) to the example in the Fig. 5 we get the values as shown in Table 1.

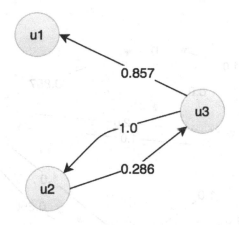

Fig. 6. Re-draw the network base on Table 2

$$\theta_{u,v}^i = \frac{2 * |MP_{u,v}^i|}{|MP_{(u,)}^i| + |MP_{(,v)}^i|} \tag{3}$$

After obtaining the value which represent the power of information transmission from user u to user v in each channel (each social meta path) individually, we will calculate the aggregation weight $w(u,v)$ based on Formula (4). In which α^i is the ratio of each type of meta social path, the greater value of α^i then the information will likely spread greater through the ith social meta path. The value of $w(u,v)$ represents the degree of information that can be transmitted from u to v (u, v is not necessarily to be a friend of each other).

Table 1. The value of $\theta_{u,v}^i$ for example in the Fig. 5

	u_1	u_2	u_3
u_1		$\theta^1 = 0, \theta^2 = 0, \theta^3 = 0$	$\theta^1 = 0, \theta^2 = 0, \theta^3 = 0$
u_2	$\theta^1 = 0, \theta^2 = 0, \theta^3 = 0$		$\theta^1 = 0, \theta^2 = 2, \theta^3 = 0$
u_3	$\theta^1 = 1, \theta^2 = 0, \theta^3 = 1$	$\theta^1 = 1, \theta^2 = 2, \theta^3 = 0$	

In the experimental process, our team selected the optimal value for α^1, α^2, α^3 respectively 5, 1, 1.

$$w(u,v) = \frac{\sum_{i=1}^{3} \alpha^i * \theta_{u,v}^i}{\sum \alpha^i} \tag{4}$$

Applying Formula (4) to the example in the Fig. 5 along with value of $\theta_{(u,v)}^i$ calculated in Table 1 we will calculate the value of $w(u,v)$ as shown in Table 2.

Based on the result in Table 2, we re-draw the network as shown in the Fig. 6. Next step, we apply the algorithm Weighted Rank Leader in the [1] and

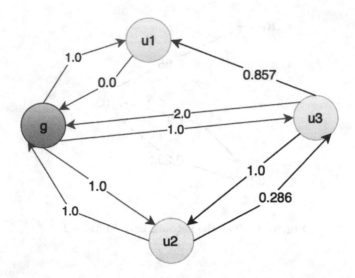

Fig. 7. Re-draw the network after add the ground node

Algorithm 1. Calculate Weight

1: **function** CALCULATEWEIGHT(G, MP) ▷ Where G - Input network, MP -
   Meta-path values
2:     $k \leftarrow 3$
3:     $\alpha \leftarrow [5, 1, 1]$
4:     **for** $u \in V$ **do**
5:         **for** $v \in V$ **do**
6:             $w_{u,v} \leftarrow 0$
7:             **for** $i = 1$ to k **do**
8:                 $\theta^i_{u,v} \leftarrow \frac{2 * MP^i_{u,v}}{MP_{(u,)} + MP_{(,v)}}$
9:                 $w_{u,v} \leftarrow w_{u,v} + \alpha^i * \theta^i_{u,v}$
10:            **end for**
11:        **end for**
12:    **end for**
13: **end function**

Table 2. The value of $w(u, v)$ for example in the Fig. 5

	u_1	u_2	u_3
u_1	0	0	0
u_2	0	0	$2/7 = 0.286$
u_3	$6/7 = 0.857$	$7/7 = 1$	0

Table 3. The value of $w(u,v)$ with ground node for example in the Fig. 5

	g	u_1	u_2	u_3
g	0	1	1	1
u_1	0	0	0	0
u_2	1	0	0	0.286
u_3	2	0.857	1	0

proceed adding a ground node (virtual node) along with the virtual edges which connecting the ground node to existing other nodes (real nodes) in the network. The weight of virtual connections (virtual edges) from real node (u) to ground node (g) and vice versa are calculated according to the principle (5)

$$w(u,g) = 1$$
$$w(u,g) = k_u^{out}. \tag{5}$$

Apply above principle to example in the Fig. 5 along with $w(u,v)$ in Table 2 we will calculate the final weight matrix as shown in Table 3 (Fig. 7).

Finally, after obtaining the weight $w(u,v)$ of all edges in the network (which has an additional ground node virtual node), we proceed to run the PageRank algorithm and obtain the ranking list which represents the ordering of user's influential degree in the network. The users who have higher ranking value will have greater impact in the network.

4   Experiments

To validate the effectiveness of our SMPRank algorithm, we run the experiments on real datasets of Twitter social network. We use and extend the dataset of Jure Leskovec which published on website: http://snap.stanford.edu/data/egonets-Twitter.html The original dataset contains only the users and the connections between users (following relationship). We extended the original dataset by collecting all the tweets of users (each users we collect the maximum 3,200 tweets). For each tweet collected in the previous step, we proceed to gather information such as: number of likes (favorite), number of user retweet, along with the number of followers of their retweet users (Table 4).

Table 4. The statistic of real Twitter dataset

Number of nodes	Number of edges	Number of tweets
76,120	55,458,375	2,941,374

We divide the collected dataset into two parts, the first part contains only tweets written before 30/12/2015 (for running the algorithm), the second part consists of tweets written after 30/12/2015 (for testing the effectiveness of the algorithm). Run the SMPRank and Weighted LeaderRank algorithm on first part dataset we have the output $Rank_{SMP}$ and $Rank_{WL}$

$$Influence(u) = \frac{\sum_{t \in tweets(u)} Infection(t)}{|tweets(u)|}. \qquad (6)$$

We use Formula (6) to calculate the actual influence degree of each user in the network. In which, $Influence(u)$ is the influence rate of user u, $tweets(u)$ is the set of tweets written by the user u, $|tweets(u)|$ is the number of tweets written by the user u, $Infection(t)$ is calculated according to Formula (7).

$$Infection(t) = infect\_rate * |follower(u^t)| + \sum_{t_i \in tweets(t)} infect\_rate * |follower(u^{t_i})| \qquad (7)$$

In Formula (7), t is a tweet, u^t is the user who write the tweet t, $Infection(t)$ is the number of users who saw the tweet t (seen times) which obtained from formula (7), $retweet(t)$ is the set of tweets that are retweeted from tweet t, $infect\_rate$ (in range $[0, 1]$) represents the rate of information diffusion. For instance, $infect\_rate = 0.5$ means that if a user has 10 followers then every tweet written by this user will have 5 followers see that tweet.

Applying Formula (6) to all users on the test data (part 2 of the dataset) we calculated the influence's values of all users, then the actual user's ranking ($Rank_{Actual}$) will be determined based on the strategy: users who have higher influence value will have higher ranking. We compare the SMPRank and Weighted LeaderRank by measuring the Pearson correlation coefficient of each pair ($Rank_{SMP}$, $Rank_{Actual}$) and ($Rank_{WL}$, $Rank_{Actual}$). The empirical data at Table 5 show that the results of SMPRank ranking better than Weighted LeaderRank because of higher correlation coefficient value.

Table 5. The Pearson correlation coefficient comparing between Weighted Leader-Rank, SMPRank and the ground truth ranking.

	Actual ranking
Weighted LeaderRank	0.713
SMP rank	0.852

5   Conclusion

Weighted LeaderRank is an efficient method, however, it calculates user's ranking only based on the network structure and ignores the behavior of users (write tweets, retweet, check-in). In this paper, we further improve the Weighted LeaderRank algorithm by apply the definition of social meta path which introduced by Zhan et al. [3]. Typically, information can only spread from user A to user B if and only if A and B are connected to each other (friends or following). However, our approach assumes that even if there is no direct connection to each other, the information is still able to exchange if they both check-in at the same location (by talking directly). Our approach, which called SMPRank is the hybrid method of Weighted LeaderRank method and social meta path. Experiments on the real social network (Twitter) show that the SMPRank can considerably improve the degree of spreadability of the original Weighted LeaderRank.

References

1. Li, Q., Zhou, T., Lü, L., Chen, D.: Identifying influential spreaders by weighted LeaderRank. Phys. A Stat. Mech. Appl. **404**, 47–55 (2014)
2. Zhang, T., Liang, X.: A novel method of identifying influential nodes in complex networks based on random walks. J. Inf. Comput. Sci. **11**(18), 6735–6740 (2014)
3. Zhan, Q., Zhang, J., Wang, S., Yu, P.S., Xie, J.: Influence maximization across partially aligned heterogenous social networks. In: Cao, T., Lim, E.-P., Zhou, Z.-H., Ho, T.-B., Cheung, D., Motoda, H. (eds.) PAKDD 2015. LNCS, vol. 9077, pp. 58–69. Springer, Heidelberg (2015)
4. Zhou, T., Fu, Z.-Q., Wang, B.-H.: Epidemic dynamics on complex networks. Prog. Nat. Sci. **16**(5), 452–457 (2006)
5. Lü, L., Zhou, T.: Link prediction in complex networks: a survey. Phys. A Stat. Mech. Appl. **390**, 1150–1170 (2011)
6. Lu, L., Chen, D.-B., Zhou, T.: The small world yields the most effective information spreading. New J. Phys. **13**, 123005 (2011)
7. Doerr, B., Fouz, M., Friedrich, T.: Why rumors spread so quickly in social networks. Commun. ACM **55**, 70–75 (2012)
8. Aral, S., Walker, D.: Identifying influential and susceptible members of social networks. Science **337**, 337–341 (2012)
9. Silva, R., Viana, M., Costa, F.: Predicting epidemic outbreak from individual features of the spreaders. J. Stat. Mech. Theor. Exp. **2012**, P07005 (2012)
10. Lu, L., Zhang, Y.-C., Yeung, C.H., Zhou, T.: Leaders in social networks, the delicious case. PLoS One **6**, e21202 (2011)
11. Brin, S., Page, L.: The anatomy of a large-scale hypertextual web search engine. Comput. Netw. ISDN Syst. **30**, 107–117 (1998)
12. Cataldi, M., Di Caro, L., Schifanella, C.: Emerging topic detection on Twitter based on temporal and social terms evaluation. In: Proceedings of the Tenth International Workshop on Multimedia Data Mining, MDMKDD 2010, pp. 4–13 (2010)
13. Page, L., Brin, S., Motwani, R., Winograd, T.: The pagerank citation ranking: bringing order to the web. In: WWW 1998, pp. 161–172 (1998)
14. Kempe, D., Kleinberg, J., Tardos, E.: Maximizing the spread of influence through a social network. In: Proceedings of the Ninth ACM SIGKDD International Conference on Knowledge Discovery and Data Mining, KDD 2003, pp. 137–146. ACM, New York (2003)

Immunization Strategies Based on the Overlapping Nodes in Networks with Community Structure

Debayan Chakraborty[1], Anurag Singh[1(✉)], and Hocine Cherifi[2]

[1] Depatment of Computer Science and Engineering,
National Institute of Technology, New Delhi 110040, Delhi, India
{debayan.chakraborty,anuragsg}@nitdelhi.ac.in
[2] Le2i UMR CNRS 6306, University of Burgundy, Dijon, France
hocine.cherifi@gmail.com

Abstract. Understanding how the network topology affects the spread of an epidemic is a main concern in order to develop efficient immunization strategies. While there is a great deal of work dealing with the macroscopic topological properties of the networks, few studies have been devoted to the influence of the community structure. Furthermore, while in many real-world networks communities may overlap, in these studies non-overlapping community structures are considered. In order to gain insight about the influence of the overlapping nodes in the epidemic process we conduct an empirical evaluation of basic deterministic immunization strategies based on the overlapping nodes. Using the classical SIR model on a real-world network with ground truth overlapping community structure we analyse how immunization based on the membership number of overlapping nodes (which is the number of communities the node belongs to) affect the largest connected component size. Comparison with random immunization strategies designed for networks with non-overlapping community structure show that overlapping nodes play a major role in the epidemic process.

Keywords: Immunization · Diffusion · Complex networks · Overlapping community · Membership number

1 Introduction

The effect of network structure on the spread of diseases is a widely studied topic, and much research has gone into this field [1–8]. The topological feature of network have been used for immunization within network [9–15]. These works have mainly studied the various immunization strategies and their effect on epidemic outbreak within a social network or contact network. The study of networks according to the degree distribution, and further the influence of immunization on degree distribution and targeted attacks has been explored by scholars in recent past [16 18]. But the community-based study of the network has not received much attention. In this level of abstraction, which has

© Springer International Publishing Switzerland 2016
H.T. Nguyen and V. Snasel (Eds.): CSoNet 2016, LNCS 9795, pp. 62–73, 2016.
DOI: 10.1007/978-3-319-42345-6_6

been termed as the mesoscopic level, the concern lies with the properties of the communities. Communities are sets of nodes which show more level of inter-connectivity amongst themselves, than with the rest of the network. We can distinguish two type of community structure in the literature depending on the fact that they share some nodes or not. Non-overlapping communities are stand-alone groups where a node belongs to a single community while in overlapping communities a node can belongs to more than one community. Recent research and analysis of real-world networks have revealed that a significant portion of nodes lies within the overlapping region of two communities [19]. Thus, we look to explore the effect of immunization with these overlapping area of commu-nities on the overall spread of the epidemic within the system. Recently the studies of few researchers have considered community structure in the field of epidemiology or pharmaco-vigilance [20–22]. But, mostly they have taken these sets as stand-alone groups and have again, not explored the communities as they truly are in real world, overlapping sets of shared nodes. Results in the recent literature show that the knowledge of degree distribution and after that, degree distribution based immunization strategies are not sufficient enough to predict viral outbreak or epidemics in general. Further, the behavior shown by an epidemic on networks with varying community structures also show a certain degree of independence amongst themselves [23–25]. Thus, confirming the fact that community structures also play a vital role in the spreading process for epi-demics within the network. So community structure has to be factored into the immunization process. In this level of abstraction, the focus lies on nodes of con-nectivity within two or more communities. In fact, Salathe' and Jones [21] had studied the effect of immunization through these bridge nodes and edges in their paper. However, their community bridge finder model analysed the communities as non-overlapping groups. Further studies have been done by Samukhin et al. [26], who analyzed the Laplacian operator of an uncorrelated random network and diffusion process on the networks. Gupta et al. [27,28] analyse the properties of communities and the effect of their immunization within their paper. They take the community nodes and analyse them on their out-degree, in-degree and difference of two on the communities to which they belong. Their study shows that community-based degree parameter can help in identifying key structural bridge/hub nodes within any given network. Their analysis further consolidated the importance of communities and their effect on the overall immunization strategy once they are taken into account. The major drawback of all these studies are that; they take networks with no underlying overlap within commu-nity structures. Even if there exists a certain amount of overlapping within these networks, they overlook those regions and analyse these areas as independent sets. In this paper, we look at community overlaps and study their immuniza-tion. We analyse the effect of two targeted immunization strategies of nodes within the overlapping regions based on the membership number. We use the classical SIR model of epidemics to analyze the spread of diseases within the network. The Experiment are conducted on a real-world network with ground truth community structure (Pretty Good Privacy). A comparative study with an immunization strategy that is agnostic of the overlapping community structure (random acquaintance [29]) and an immunization strategy derived for non-overlapping community structure (Community Bridge Finder [21]) is performed. The remaining of the paper is arranged as follows: In Sect. 2, we present a short introduction to the classical SIR model following which we present briefly the existing immunization strategies which are agnostic to network structure. In Sect. 3, the overlapping community structure is defined along with the statistic associated with these structures. In Sect. 4 we discuss the experimental results following which we conclude our findings in Sect. 5.

2   Background

2.1   Classical SIR Model

The property of the connection of the individual nodes and the nodes that are in the neighborhood have a direct effect on their ability to propagate information within a system, and their ability to stop the information is also worthy. To characterize the immunization of nodes we first look into the spread of the epidemic within a network. We present the classical **SIR** model which we use to study the general characteristic of diffusion within a system. The model uses rate definitions to define the change of state of each node within the Susceptible, Infected and Recovered states with rates α, λ and β. Whenever infected contacts a susceptible, the susceptible becomes infected at a rate α. Whenever an infected spontaneously changes to a recovered (simulating the random cure of the individual on diffusion), it does so at the rate β. $S(t)$, $I(t)$, $R(t)$ gives the evolution of each set within the network. For example, $S(t)$ gives us the fraction of nodes which are susceptible to infection at time t. The spreading rate $\lambda = \alpha/\beta$ describes the ability of the epidemic to spread within the network. High spreading rate signifies epidemic can spread more quickly within the network.

2.2   Immunization Strategies

Largest Connected Component lcc of a network is that component of the network which contains the most number of nodes within them and each node is reachable from every other node. In a sense, no node is dis-connected from another node within the component. In effect, the size of the largest connected component will tell us the maximum limit to which an epidemic can spread. Starting from the full network $\mathcal{N}$ one can remove the nodes according to an immunization strategy and check for the effect on lcc. As one transforms the network $\mathcal{N}$, the size of the lcc is also subject to change. The transformed network $\mathcal{N}'$ has the largest component of size lcc'. With N being the number of nodes in $\mathcal{N}$ and N' being the number of nodes in $\mathcal{N}'$, one can say that, since $N > N'$, $lcc > lcc'$, $N \in \mathcal{N}$ and $N' \in \mathcal{N}'$. Thus with each transformation of network $\mathcal{N}$ one aims at reducing the size of lcc as much as possible. A good immunization strategy is one, which, with the least number of nodes removed, transforms the network in such a manner, that the lcc' of the transformed network $\mathcal{N}'$ is the least. Here, we present stochastic strategies of immunization. Stochastic models are usually agnostic about the global structure and thus are used here for comparative analysis with the proposed strategy which too uses no prior information about the network.

2.3   Random Acquaintance

Random Acquaintance is one of the stochastic strategies for immunization present in current literature. Random Acquaintance was first introduced by Cohen *et al.* in their paper [29]. It works by picking a random node and then choosing its neighbour at random. A number n is taken before the start of the process and if an acquaintance of the randomly selected node is selected more than or equal to n times, then, it is Immunized. In case where $n = 1$, any acquaintance will be immunized immediately. Without any prior global knowledge of the network, this strategy identifies highly connected individuals.

2.4   Community Bridge Finder CBF

CBF is a random walk based method to find nodes connecting multiple communities. It was presented first by Salathe *et al.* in their work [21]. A random node is selected at the start, and then a random path is followed until a node is reached which is not connected to more than one of the previously visited nodes. The idea is based on the belief that this node which is not connected to more than one of the previously visited nodes, in the random walk is more likely to belong to a different community. This strategy too has no prior information about the network structure.

3   Overlapping Community Structure

Studies have been carried out by scholars for detecting communities within a network. However, till now, there has been no widely accepted definition of community structures in literature. A common notion is to consider those sets/groups of nodes which show high inter-connectivity amongst themselves than with the rest of the network, as communities. These densely connected sets of nodes may thus carry structural hubs within themselves. Overlapping communities are sets of nodes with common nodes shared within them. These common nodes of the sets lie within the overlapping region of the communities.The total number of nodes shared amongst two or more communities, n_{ov} gives us the total number of nodes within the overlapping region in the network.

3.1   Definitions

Community Size: The community size s defines the number of nodes in each community. If $C_1, C_2, C_3....C_z$ signify each of the z communities in a network $\mathcal{N}$ then the size of a community, s is $|C_i|$ for $i \in [1, 2, ...z]$ and it signifies the number of nodes in the community i.

Membership Number: The Membership, m, of each node i, $i \in [1, 2, ...N]$ where N signifies the total number of nodes in the network, defines the number of communities to which any node i belongs to. Thus for m greater than 1, one can state that the node belongs to the overlapping region in the network. A quick analysis of the degree distribution of nodes within the overlapping regions reveals that the degree distribution of the overlapping region also follows a power law characteristic [19].

Membership Function: The membership function $m()$ gives us the membership number of any node i, within the network, provided we know the communities within the network. A finite number of nodes dictate restricted communities present within the network, and thus, the variation of m is also finite. If the largest membership within the network is x, the smallest being 0, one can divide the entire network as a group of disjoint sets where each set contains the nodes whose membership number is equal to the membership number of all other elements of the set. Thus, for all $i \in \mathcal{N}$, $m(i) = m$, where $m \in [1, 2, 3....x]$ and $i \in M_m$, where $M_m \in [M_1, M_2, M_3.....M_x]$ and $M_1 \cap M_2 \cap M_3 \cap ..... \cap M_x = 0$. Further, for all $i, j \in \mathcal{N}$ if $i, j \in M_m$, then $m(i) = m(j) = m$.

Overlap Size: The overlap size s_{ov} is the number of nodes shared between any two communities. $C_1, C_2, C_3....C_z$ signify each of the z communities in a network $\mathcal{N}$. The intersection of two communities C_i, C_j is given by C_{ij} and the size of the overlapping region s_{ov} is defined as $|C_{i,j}|$ which signifies the number of nodes shared by the two communities.

3.2 Immunization of Overlapping Nodes

In this work membership based immunization strategy has been proposed. The membership number metric had been explored by Palla *et al.* [19] and we have studied the effect of immunization based on this metric on the overall diffusion process. We study the effect of immunization on the *lcc*. We have looked into the importance of high membership nodes as well as low membership nodes. As it is shown in [19] the power law nature of m makes it interesting to analyse the effect of membership number based immunization on the *lcc*. A strategy based on membership based immunization is proposed here. If nodes $i, j, k, l, ...$ are arranged in sequence of their membership number and then removal is initialized, two possible strategies emerge.

In our analysis, nodes, $i, j, k, l, ...$ are removed and analysed.

For any nodes, $i, j, k, ...,$

Immunization starting from highest overlap membership to lowest overlap membership ($HLMI$):

i is removed before j and j is removed before k if
$$m(i) > m(j) > m(k), \forall i, j, k \in \mathcal{N}$$

Algorithm 1. HLMI algorithm

> **Input** Graph $G(V, E)$, Membership measure (m), No. of nodes to be immunized
> **Output** Graph with Immunized/Removed Nodes
> Calculate the Membership number m for each node in the graph using membership function $m()$;
> Sort the nodes in decreasing order of their membership values ;
> Remove top n nodes with highest membership values from the Graph G ;
> Return graph G after removal of immunized nodes ;

Algorithm 2. LHMI algorithm

> **Input** Graph $G(V, E)$, Membership measure (m), No. of nodes to be immunized
> **Output** Graph with Immunized/Removed Nodes
> Calculate the Membership number m for each node in the graph using membership function $m()$;
> Sort the nodes in increasing order of their membership values ;
> Remove top n nodes with lowest membership values from the Graph G ;
> Return graph G after removal of immunized nodes ;

Immunization starting from lowest overlap membership to highest overlap membership ($LHMI$):

i is removed before j and j is removed before k if
$$m(i) < m(j) < m(k), \forall i, j, k \in \mathcal{N}$$

The distribution of membership number m and the degree distribution of corresponding set of nodes with equal membership number shows that small membership node sets have a larger number of nodes while large membership node sets have a small number of nodes in them. The outliers of the small membership nodes show that the spread of degree values in each set is vast, and high degree nodes are spread in all the sets of different membership. The small membership number nodes, which are large in number by intuition should perform comparatively better than the immunization removing high membership nodes first since, the small membership set contains more nodes with larger variance in its degree distribution.

4   Experimental Results

4.1   Analysis of the PGP Dataset

We have used the Pretty Good Privacy network [30] dataset for our work. The network consists of email addresses which have signatures associated with them. The groups in this network are the email domains which are present in the dataset as ground-truth communities, where every node explicitly states its full involvement in the community it belongs. The network does show a certain degree of overlap amongst its various groups. The dataset consists of 81036 nodes and 190143 edges, with 17824 groups. Further, the link between two nodes is undirected and un-weighted in nature. The network confirms to the power-law degree distribution followed by scale free networks and contains no cyclic or multiple edges. The ground-truth community structures have been studied and their overall effect on epidemic/diffusion process has been analysed. Our analysis of ground truth data shows us that the size of communities within a network follow a power-law property confirming with the existing literature [19]. Thus, there exists a significant number of small communities and comparatively less number of communities which are quite large in size. The underlying overlapping nature of the communities present in the dataset is explicitly shown within the ground-truth community data. We find the nodes within the overlap regions and study the size of all the overlapping areas within the network. It too showcases power-law characteristics to a certain degree. The membership number m too follows a power-law characteristic. Hence, a larger portion of the nodes within the network shows lower membership to communities. The result for m and the distribution for s_{ov} together give us an interesting insight into the nature of overlap within the network and the participating nodes. This distribution confirms with the existing literature [19] wherein we see the inherent power-law nature of the metrics associated with any node within a community. More nodes were found to exist within the lower membership sets while higher membership set showcased less number of nodes present within them. Also, the number of outliers for lower membership nodes are large in number. One may conclude that the high degree nodes are spread through all memberships within the network. It is thus an important reason to study the immunization from two viewpoints as explained in Sect. 3. The analysis of community metrics mainly s, the size of communities, s_{ov}, the size of the overlap region, membership number m, and the degree distribution of the nodes in the overlap region n_{ov}, are presented in Fig. 1. We show that all the metrics follow the characteristic as given by Palla et al. [19]. The metric $d(n_{ov})$ signifies the degree of the nodes within overlapping region in the network. The function $p(.)$ of any metric, signifies the cumulative distribution of the corresponding metric. It might be concluded that the PGP network dataset and the facebook dataset adheres to all the prerequisites for accepting the PGP and facebook network as a network containing overlapping communities within the network.

4.2 Evaluation of the Immunization Strategies

The comparative results for the CBF and random acquaintance have been studied and have been presented in Fig. 2. We find that the immunization strategy based on removal of nodes in overlap region show a comparatively better performance than the stochastic methods of immunization. We see that the performance of CBF and random acquaintance are acceptable only after fifty percent of the nodes have been removed from the network. This necessitates that half or more of the entire population be vaccinated/immunized to ensure no outbreak. Although it is a good approach still the requirement to immunize half of the population becomes a problem. Considering areas of vast population or even highly populated network of nodes, this strategy presents us with an uphill task

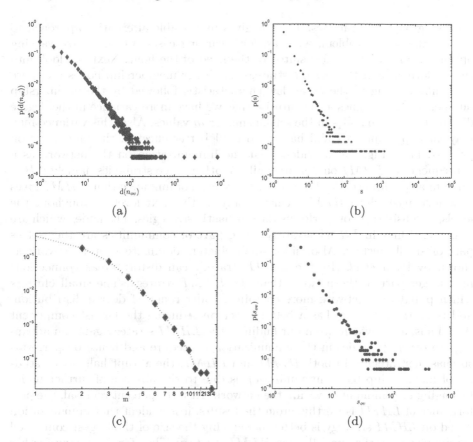

Fig. 1. (a) Shows the degree distribution in the overlapping nodes within the network, **(b)** shows the variation of community sizes in the network, **(c)** the cumulative degree distribution of the membership number m and **(d)** the cumulative distribution of the overlap size. All the above studies were made on the PGP dataset. (Color figure online)

of immunizing a considerable amount of the population. It is a hefty price to pay, but at the same time it is a must since CBF and random acquaintance being stochastic strategies have no knowledge of the entire network. A similar fraction of nodes is required to be immunized if we have an idea about the communities present within the network and follow the $HLMI$ strategy of immunization. It does not give an overall better performance than the stochastic methods. However, its performance is comparable to the one for CBF and random acquaintance and is thus a strategy which may be accepted as at par with the existing stochastic methods. The $HLMI$ strategy, on the other hand, outperforms stochastic based strategies at lower levels of immunization. One point which becomes important if the community knowledge is readily available. It must be understood that the $HLMI$ strategy gives good performance when half of the population is vaccinated. But, the performance at lower levels of immunization at the same time brings to notice that there is a chance of a trade off between the two groups of strategies. Thus, it gives us a viable alternative approach to solving the same problem, whereas, with varying cases, we may take varying options as and when we feel suited to the need of the hour. Next, we look into the performance of the $LHMI$ strategy. As lower membership nodes are more in number, owing to the power-law characteristic followed by the membership number m [19] it comes as no surprise that we have more variance in the degree distribution of the nodes in the sets of smaller m values. Also, this variance may play an important part and have an overall better effect in the immunization process. From Fig. 2 it is evident that the immunization on the network as a consequence of $LHMI$ outperforms all the other three strategies namely CBF, random acquaintance and $HLMI$. At every level of immunization, $LHMI$ gives a performance which $HLMI$ achieves only at the next level. Throughout the levels, it consistently outperforms the stochastic strategies. The nodes which are immunized first in $LHMI$ strategy being part of communities are themselves part of small clusters. Also, since small clusters dominate in any network as shown by Palla *et al.* [19] the $LHMI$ strategy can distort/break connections in a larger part of the network than the $HLMI$ strategy. The small clusters which populate a network more contain a wider range of degree distribution, and thus, their removal has a better effect on reducing the largest component size. Thus, as is evident from our findings, the $LHMI$ strategy gains an advantage to other strategies in the immunization procedure and hence outperforms stochastic strategies. In both $HLMI$ and $LHMI$, the abrupt halt in the studies of membership based immunization is due to the absence of further nodes belonging to communities within the network. At forty percent removal, the performance of $LHMI$ is worthy. From the results, it is evident that immunization based on $LHMI$ strategy is better at reducing the size of the largest connected component than the one following $HLMI$ strategy. Therefore, one may further conclude that the removal of sets with higher variance in the degree distribution of their nodes give a comparatively better result than those where the number of nodes and variability in their degree distribution is less. In Fig. 3 the evolution of Infected (I), Susceptible (S) and Recovered (R) nodes are shown. The evolution in the network after immunization based on our proposed strategy ($LHMI$) is compared with the evolution within the network with no immunization. Initially we start with one infected node within the network and we study the gradual evolution at consequent time evolution. The value of λ, is set to 1. Comparative studies in all these figures show that our proposed method ($LHMI$) is efficient in arresting the fast growth within the system and thus is capable of stopping an epidemic from occurring. The overall performance may be attributed to the stopping of infection spread, which the $LHMI$ algorithm does quite efficiently in the beginning stages of diffusion. Thus our proposed strategy based on membership based immunization is a viable alternative to the stochastic strategies present in the current literature.

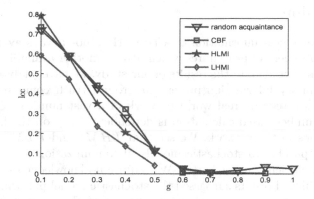

Fig. 2. Fraction immunized (g) and its effect on the largest connected component (lcc)

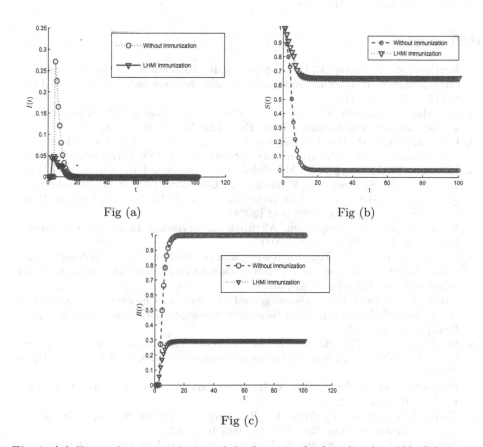

Fig. 3. (a) Shows the time evolution of the fraction of infected nodes, $I(t)$, **(b)** shows the time evolution of the fraction of susceptible nodes, $S(t)$ and **(c)** shows the time evolution for the fraction of recovered nodes $R(t)$ within the network (Color figure online)

5   Conclusion

The global topological information of a network is not always available to us. Thus, the requirement of procedures which utilize another available information of communities is needed. In the results of our study, we have analysed the effect of local community information(present in ground truth communities) based immunization strategy on real world network of a vast number of nodes. The membership number based calculation is dependent solely on the knowledge of the communities in the network. We see that $LHMI$ and $HLMI$ give results which are comparable to stochastic models of immunization and work on par with the same efficiency if not better. We require no knowledge of the network, and yet the achieved results surpassed the stochastic model performances which need at least some local connection information of the studied nodes. Thus, we find that community information may be effectively utilized for developing efficient immunization strategies.

References

1. Barthélemy, M., Barrat, A., Pastor-Satorras, R., Vespignani, A.: Velocity and hierarchical spread of epidemic outbreaks in scale-free networks. Phys. Rev. Lett. **92**(17), 178701 (2004)
2. Boccaletti, S., Latora, V., Moreno, Y., Chavez, M., Hwang, D.U.: Complex networks: structure and dynamics. Phys. Rep. **424**(4), 175–308 (2006)
3. Gong, K., Tang, M., Hui, P.M., Zhang, H.F., Younghae, D., Lai, Y.C.: An efficient immunization strategy for community networks. PloS ONE **8**(12), e83489 (2013)
4. Halloran, M.E., Ferguson, N.M., Eubank, S., Longini, I.M., Cummings, D.A., Lewis, B., Xu, S., Fraser, C., Vullikanti, A., Germann, T.C., et al.: Modeling targeted layered containment of an influenza pandemic in the united states. Proc. Nat. Acad. Sci. **105**(12), 4639–4644 (2008)
5. Pastor-Satorras, R., Vespignani, A.: Epidemic spreading in scale-free networks. Phys. Rev. Lett. **86**(14), 3200 (2001)
6. Singh, A., Singh, Y.N.: Rumor spreading and inoculation of nodes in complex networks. In: Proceedings of the 21st International Conference Companion on World Wide Web, pp. 675–678. ACM (2012)
7. Singh, A., Singh, Y.N.: Nonlinear spread of rumor and inoculation strategies in the nodes with degree dependent tie stregth in complex networks. Acta Phys. Pol., B **44**(1), 5–28 (2013)
8. Singh, A., Singh, Y.N.: Rumor dynamics with inoculations for correlated scale free networks. In: 2013 National Conference on Communications (NCC), pp. 1–5. IEEE (2013)
9. Pastor-Satorras, R., Vespignani, A.: Immunization of complex networks. Phys. Rev. E **65**(3), 036104 (2002)
10. Gallos, L.K., Liljeros, F., Argyrakis, P., Bunde, A., Havlin, S.: Improving immunization strategies. Phys. Rev. E **75**(4), 045104 (2007)
11. Tanaka, G., Urabe, C., Aihara, K.: Random and targeted interventions for epidemic control in metapopulation models. Sci. Rep. **4**, 5522 (2014)
12. Glasser, J., Taneri, D., Feng, Z., Chuang, J.H., Tüll, P., Thompson, W., McCauley, M.M., Alexander, J.: Evaluation of targeted influenza vaccination strategies via population modeling. PloS ONE **5**(9), e12777 (2010)

13. Madar, N., Kalisky, T., Cohen, R., ben Avraham, D., Havlin, S.: Immunization and epidemic dynamics in complex networks. Eur. Phys. J. B **38**(2), 269–276 (2004)
14. Christakis, N.A., Fowler, J.H.: Social network sensors for early detection of contagious outbreaks. PloS ONE **5**(9), e12948 (2010)
15. Krieger, K.: Focus: vaccinate thy neighbor. Physics **12**, 23 (2003)
16. Cohen, R., Erez, K., Ben-Avraham, D., Havlin, S.: Breakdown of the internet under intentional attack. Phys. Rev. Lett. **86**(16), 3682 (2001)
17. Callaway, D.S., Newman, M.E., Strogatz, S.H., Watts, D.J.: Network robustness and fragility: percolation on random graphs. Phys. Rev. Lett. **85**(25), 5468 (2000)
18. Albert, R., Jeong, H., Barabási, A.L.: Attack and error tolerance of complex networks. Nature **406**(6794), 378–382 (2000)
19. Palla, G., Derényi, I., Farkas, I., Vicsek, T.: Uncovering the overlapping community structure of complex networks in nature and society. Nature **435**(7043), 814–818 (2005)
20. Zhang, H., Guan, Z.H., Li, T., Zhang, X.H., Zhang, D.X.: A stochastic sir epidemic on scale-free network with community structure. Physica A **392**(4), 974–981 (2013)
21. Salathé, M., Jones, J.H.: Dynamics and control of diseases in networks with community structure. PLoS Comput. Biol. **6**(4), e1000736 (2010)
22. Becker, N.G., Utev, S.: The effect of community structure on the immunity coverage required to prevent epidemics. Math. Biosci. **147**(1), 23–39 (1998)
23. Shang, J., Liu, L., Li, X., Xie, F., Wu, C.: Epidemic spreading on complex networks with overlapping and non-overlapping community structure. Physica A **419**, 171–182 (2015)
24. Chen, J., Zhang, H., Guan, Z.H., Li, T.: Epidemic spreading on networks with overlapping community structure. Physica A **391**(4), 1848–1854 (2012)
25. Shang, J., Liu, L., Xie, F., Wu, C.: How overlapping community structure affects epidemic spreading in complex networks. In: 2014 IEEE 38th International on Computer Software and Applications Conference Workshops (COMPSACW), pp. 240–245. IEEE (2014)
26. Samukhin, A., Dorogovtsev, S., Mendes, J.: Laplacian spectra of, and random walks on, complex networks: are scale-free architectures really important? Phys. Rev. E **77**(3), 036115 (2008)
27. Gupta, N., Singh, A., Cherifi, H.: Community-based immunization strategies for epidemic control. In: 2015 7th International Conference on Communication Systems and Networks (COMSNETS), pp. 1–6. IEEE (2015)
28. Gupta, N., Singh, A., Cherifi, H.: Centrality measures for networks with community structure. Physica A **452**, 46–59 (2016)
29. Cohen, R., Havlin, S., Ben-Avraham, D.: Efficient immunization strategies for computer networks and populations. Phys. Rev. Lett. **91**(24), 247901 (2003)
30. Hric, D., Darst, R.K., Fortunato, S.: Community detection in networks: structural communities versus ground truth. Phys. Rev. E **90**(6), 062805 (2014)

Improving Node Similarity for Discovering Community Structure in Complex Networks

Phuong N.H. Pham[1], Hien T. Nguyen[2(✉)], and Vaclav Snasel[3]

[1] Faculty of Information Technology, University of Food Industry,
Ho Chi Minh City, Vietnam
phuongpnh@cntp.edu.vn
[2] Faculty of Information Technology, Ton Duc Thang University,
Ho Chi Minh City, Vietnam
hien@tdt.edu.vn
[3] Faculty of Electrical Engineering and Computer Science,
VSB-Technical University of Ostrava, Ostrava, Czech Republic
vaclav.snasel@vsb.cz

Abstract. Community detection is to detect groups consisting of densely connected nodes, and having sparse connections between them. Many researchers indicate that detecting community structures in complex networks can extract plenty of useful information, such as the structural features, network properties, and dynamic characteristics of the community. Several community detection methods introduced different similarity measures between nodes, and their performance can be improved. In this paper, we propose a community detection method based on an improvement of node similarities. Our method initializes a level for each node and assigns nodes into a community based on similarity between nodes. Then it selects core communities and expands those communities by layers. Finally, we merge communities and choose the best community in the network. The experimental results show that our method achieves state-of-the-art performance.

Keywords: Community detection · Similarity measure · Complex networks

1 Introduction

In recent times, many academic works have been published on complex networks in attempts to determine their structural features [1,3,20]. A network is viewed as a graph with vertices connected through edges. Communities are subgraphs of a network consisting of densely connected nodes, while links between the subgraphs are sparse [15]. Generally, a complex network has many nodes and a huge number of connections among nodes, such as the Internet, World Wide Web, citation networks, co-citation networks, biological, and metabolic networks, computer networks and social networks [18]. A community normally contains nodes having similar properties; for example, in web services, a community is a

© Springer International Publishing Switzerland 2016
H.T. Nguyen and V. Snasel (Eds.): CSoNet 2016, LNCS 9795, pp. 74–85, 2016.
DOI: 10.1007/978-3-319-42345-6_7

group of websites sharing the same topics or a group of users sharing common interests in a social network; in contrast a citation is between two authors of or between two research papers in a co-citation network. In metabolic networks, communities may be related to functional modules such as cycles and pathways or a group of proteins functioning in a similar way within a cell in the protein-protein interaction networks [7,10]. Finding the similarity of communities in complex networks is an issue because doing so can help determines structures and changes in the network based on properties of each element in the network.

One of the most important issues that researchers have recently considered in finding communities in complex networks involves structures and properties of each element in communities. Moreover, based on structures and properties, researchers can find useful information about communities in complex networks. To date, many methods have been developed to detect community structure in networks. Each has advantages and disadvantages. Two of the classical algorithms are Kernighan-Lin [17], which introduced a greedy algorithm to optimize the value of the edges within a community, and the spectral bisection method [9], which is based on the eigenvector of the Laplacian matrix of the graph. In 2002, Girvan and Newman proposed an algorithm based on the iterative removal of edges with high betweenness scores that appears to identify the community of networks with some sensitivity [26]; they used the concept of modularity to measure the quality of algorithms for community detection. Newman [24] proposed a fast algorithm for detecting community structure based on the idea of modularity. Furthermore, Clauset et al. [6] introduced an algorithm based on hierarchical clustering to detect community structure in very large networks. Some algorithms for detecting communities are based on node similarity with no need to know in advance the quantity of communities or its measure.

In this paper, we propose a method that improves similarity between nodes for community detection. It initializes a level for each node and assigns nodes to a community based on their similarity. Then it selects the core communities, and expands those communities by layers. Finally, we merge communities and choose the best community.

The structure of this paper is organized as follows. The general view about community detection and related works are introduced in Sects. 1 and 2, respectively. Section 3 presents our proposed method. Section 4 presents experiments and results. Finally, we draw a conclusion in the last section.

2 Related Work

In this section, we present related works about community detection in complex networks. Detecting community structures in complex networks is considered a challenging task. Many methods of detecting communities in complex networks are introduced, normally including two traditional methods called graph partitioning and hierarchical clustering. Those included in graph partitioning are the Kernighan-Lin algorithm [17], Radicchi et al. [29], and normalized cut by Shi and Malik [31]. The drawbacks of these methods include that: the quantity of communities and their size of clusters must be known and the repetition of graph division is not reliable. Those included in hierarchical clustering can be classified in two categories called agglomerative algorithms and divisive algorithms such as finding communities based on random walks [28] and edge betweenness by Newman [23]. Many algorithms detect communities in complex networks by maximizing modularity such as Blondel *et al.* [2], Good *et al.* [14], Dinh and Thai [8]. However, according to Fortunato and Barthelemy [11], the drawback of the algorithms based on modularity is that they fail to detect communities whose size of clusters is smaller than the inherent size based on the network edges. Moreover, Clauset *et al.* [6] can usually obtain a large value of modularity while the accuracy is not necessarily high. Chen *et al.* [4] proposed an agglomerative clustering algorithm using the max-min modularity quality measure. The algorithm considers both topology of the network and provided domain knowledge.

Pan *et al.* [27] presented a community detection method based on node similarity with a running time lower than the other methods and the computational complexity of $O(nk)$ where k is the mean node degree of the whole network. The CDHC [34] proposed an algorithm for community detection based on hierarchical clustering which aims to extend modularity on the basic of modularity. [33] presented an algorithm for detecting communities based on edge structure and node attributes. The algorithm can detect overlapping communities in networks.

3   Proposed Method

In this section, we present how we use the Normalized Google Distance (NGD) to measure similarity between nodes and between communities. Then, we present in detail our algorithm for detecting communities based on node similarity. Furthermore, we also analyze the complexity of our algorithm.

3.1   Measurement Quantity

3.1.1   Node Similarity

Similar measurement is a quantity used to measure the closeness between two nodes or two objects. Much work has introduced such as Jaccard Index [16], Zhou-Lu-Zhang Index [36], Sorensen Index [32] and so on. In our research, similar measurement for two nodes in a network is based on Normalized Google Distance (NGD) proposed by Cilibrasi and Vitanyi [5] as follows.

$$SIM_{x,y} = 1 - \frac{max(log|X|, log|Y|) - log|X \cap Y|}{N - min(log|X|, log|Y|)}, \qquad (1)$$

where X and Y are sets of adjacent nodes of node x and y, respectively. N is the total number of nodes in the network. Based on the idea of similarity of nodes, we proposed a formula for calculating the similarity of communities by using a set of adjacent node as presented in Eq. 2.

3.1.2 Community Similarity

This is the second quantity we use to measure the similarity between communities. Based on NGD and the idea of connection of adjacent nodes, the quantity to measure the similarity between two communities is as follows.

$$CSIM_{a,b} = 1 - \frac{max(log|A|, log|B|) - log|A \cap B|}{M_c - min(log|A|, log|B|)}, \tag{2}$$

where $M_c = \Sigma_{c \in C} \mid C \mid$, C is a set of all communities, a and b are the two communities, A is a set of nodes adjacent with community a, B is a set of nodes adjacent with community b. In this formula, a node adjacent a community is if and only if it can connect to at least one node in that community.

3.2 Detecting Community Structure

Our algorithm includes four phases, described below:

Phase1. Initializing communities: All nodes in the network are placed in descending order by degree. Then we mark the level for all nodes by marking nodes with an equal degree in the same level. Moreover, we combine nodes with high similarity in the same level to create communities.

Phase2. Finding core communities (CCs): We sort communities in descending order by total degree. After that, we select first K communities and mark as core communities, in which K is the parameter.

Phase3. Classifying communities: We mark core communities at layer 0. Then we choose the communities adjacent to the core communities without belonging to any layer, so we mark these communities in layer 1. In general, communities that are adjacent to communities in layer p and do not belong to any layer are marked in layer $p+1$, until all communities has to be assigned the layer.

Phase4. Merging communities: For each community c in layer $p(p>0, p = 1 \dots n - 1 layer)$, we find a community h, which has maximal similarity with community c, in layer $p - 1$ then we merge community c into the core community that contains community h. It repeats this process until the last layer.

4 Experiments

In this section, we evaluated our proposed algorithm with several famous real-world dataset such as Zachary's Karate network, American College Football network, Dolphin Social network and Books about US politics (PolBooks). Moreover, we also applied our method with computer-generated networks. Finally, we compared the results of our algorithm with those of state-of-the-art methods. The experimental results showed that our method has better efficiency than others in terms of F-Measure and NMI. Note that we implemented the algorithms by Python programming language on a PC with Core-i3 2.2 GHz processor and 4 GB memory.

4.1   Datasets

Zachary's Karate Club [35] is a classical data set used to test many community detection algorithms. Zachary's karate club is a friendship network with 34 members of a US karate club over two years. The members of the karate club are divided into two groups: instructor (node 0) and administrator (node 33). In Fig. 1 we show a network structure extracted from the karate club. In our work, our algorithms found exactly two communities of Zachary's karate club with a period of 0.33 s and a value of modularity of 0.371.

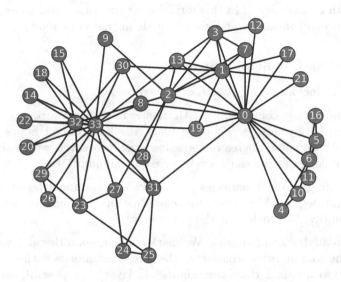

Fig. 1. Community detection in the Zachary's karate club network using our method

The American College Football network [12] presents the schedule of games between American college football teams in a single season. This dataset includes 115 teams and is divided into 12 groups with intraconference games being more frequent than interconference games. In this data, nodes in the graph represent teams and each edge represents regular season games between the two teams having a connection. Our method detected correctly 10 groups with modularity of partition of 0.574.

The Dolphins network [22] shows the associations between 62 dolphins living in Doubtful Sound in New Zealand with edges representing social relations between individuals by Lusseau from seven years of field studies of the dolphins. The dolphin network is divided into two groups. Figure 2 shows the community structure detected by our algorithm with modularity of 0.378.

PolBooks[1] is a network of books about US politics published around the time of the 2004 presidential election and sold by the online bookseller Amazon.com.

---

[1] http://www-personal.umich.edu/~mejn/netdata/.

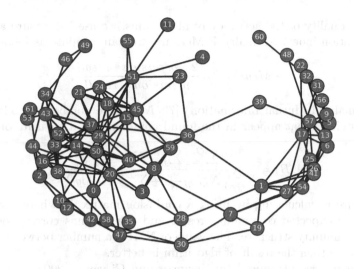

Fig. 2. Community detection in the Dolphin network using our method

Edges between books represent frequent co-purchasing of books by the same buyers.

In addition, to evaluate the performance of the community detection in the large network, we are also implementing a dataset with a large network such as Jazz musicians, PowerGrid and Internet. The experimental results show that our method has better efficiency with higher modularity than other algorithms.

Jazz musicians [13] is a dataset that includes 198 bands that performed between 1912 and 1940 with bands connected if they have a musician in common. The database lists the musicians that played in each band without distinguishing which musicians played at different times.

PowerGrid (see footnote 1) is an undirected, unweighted network representing the topology of the Western States Power Grid of the United States and data compiled by Watts and Strogatz, which includes 4,941 nodes and 6,594 edges.

Internet (see footnote 1) is a symmetrized snapshot of the structure of the Internet at the level of autonomous systems that was created by Newman in 2006.

4.2   Evaluation

We use the following measures to evaluate the performance of our method presented in this paper. Given a community set C produced by an algorithm and the ground truth community set S. The precision and recall are defined as:

$$precision = \frac{|C \cap S|}{|C|}, \tag{3}$$

$$recall = \frac{|C \cap S|}{|S|}. \tag{4}$$

To test the quality of the accuracy of algorithms, we use F-Measure and NMI as an evaluation index. Normally, F-Measure [21] can compute as follows.

$$F - Measure = \frac{2 * precision * recall}{precision + recall},$$ (5)

NMI (Normalized Mutual Information) [7]: N_{ij} is the number of nodes in the real community i that appear in the found community j, a measure of NMI as follows.

$$I(A, B) = \frac{-2 \sum_{j=1}^{C_A} \sum_{j=1}^{C_B} N_{ij} log \frac{N_{ij}N}{N_{i.}N_{.j}}}{\sum_{j=1}^{C_A} N_{i.} log \frac{N_i}{N} + \sum_{j=1}^{C_B} N_{.j} log \frac{N_{.j}}{N}}.$$ (6)

This measure is calculating by using a confusion matrix N, where rows correspond to the expected community result and the columns correspond to the obtained community structure. The value of NMI is a number between 0 and 1; if NMI is large then the result of algorithm is better.

Modularity was introduced by Newman and Girvan in 2004 as a quantity to assess whether the partition of network is good or not. The modularity Q is defined as the number of edges within communities minus expected number of such edges [26].

$$Q = \sum_i (e_{ij-a_i^2}) = Tre - ||e^2||,$$ (7)

where e is an k x k symmetric matrix whose element e_{ij} is a fraction of all edges in the network with one vertex in community i and the other in community j, $Tre = \sum_i e_{ii}$, gives the fraction of edges that connect vertices in the same community, and $a_i = \sum_i e_{ij}$ represent the fraction of edges that connect to vertices in community i The value of modularity is in between 0 and 1, and normally between 0.3 and 0.7. If the value of modularity is higher, then the used algorithm is considered better.

Table 1. Experimental results of our method on four real-world datasets. K is a number of core communities and C is the best number of communities detected.

Data set	K	C	Q	F-Measure	NMI
Zachary [35]	2	2	0.371	1	1
Dolphin [22]	3	2	0.378	0.975	0.78
Football [12]	19	10	0.574	0	0.863
Polbooks[a]	5	2	0.439	0.847	0.585

[a]See footnote 1

The algorithm is done on four real datasets combined with ground-truth in order to calculate F-Measure and NMI. Because parameter k has different values on each set of data, we can observe and select the best results the algorithm may detect, then synthetize them into Table 1. In each set of data, modularity Q, F-Measure and NMI are used to evaluate the results.

Table 2. Experimental results with different algorithms on the Zachary's karate club and Dolphin network dataset. Community C is the best number of communities detected.

Algorithm	Zachary				Dolphin			
	C	Q	F	NMI	C	Q	F	NMI
GN [26]	5	0.401	0.042	0.617	8	0.595	0.204	0.879
CNM [6]	3	0.380	0.000	0.706	6	0.549	0.005	0.708
LPA [30]	1	0.000	0.692	-	10	0.589	**0.312**	0.879
Louvain Method [2]	4	**0.418**	0.369	0.618	10	**0.604**	0.013	**0.890**
This paper	2	0.371	**1.000**	**1.000**	10	0.574	-	0.863

Table 3. Experimental results with different algorithms on the Football and PolBooks dataset. Community C is the best number of communities detected.

Algorithm	Football				PolBooks			
	C	Q	F	NMI	C	Q	F	NMI
GN [26]	8	0.595	0.204	0.879	5	**0.595**	0.074	0.561
CNM [6]	6	0.549	0.005	0.708	4	0.501	0.100	0.531
LPA [30]	10	0.589	**0.312**	0.879	4	0.503	0.087	0.544
Louvain Method [2]	10	**0.604**	0.013	**0.890**	4	0.520	0.222	0.516
This paper	10	0.574	-	0.863	2	0.439	**0.847**	**0.585**

Through the results found in the Table 1, the best one can be found in the set of data Zachary. In three set of data including Dolphin, Football, and Polbooks, the results similar to the best results can also be found. Now we compare our algorithm with some community detection methods with different results. Tables 2 and 3 show the results of Girvan and Newman (GN), Clauset (CNM), Label Propagation Algorithms (LPA) [30], Louvain Method [2] and our method on Zachary, Dolphin, Football and PolBooks. The results show that modularity of communities detected by our method is not higher than the others, but F-Measure and NMI are highest on Zachary and PolBooks. On Dolphin and Football datasets, the results are nearly highest.

In order to verify the performance of our method, we have also applied our algorithm on some datasets in the large networks. In Table 4, we show the experimental results of our method on three datasets such as Jazz musician, PowerGrid and Internet. For Jazz musicians dataset, we compared the result of Newman [25] and our method, Newmans method presented better results in terms of modularity Q was 0.442 while our work generated in terms of Q was 0.337. For PowerGrid dataset, the best result of our method in terms of Q was 0.732, which is slightly better than Q generated by FastQ [6], which generated in terms of Q was 0.452. However, we have obtained the value of modularity Q, which is not good on the Internet dataset as show in Table 4.

Table 4. Experimental results on large datasets.

Data	Node	Edge	K	C	Q
Jazz [13]	198	2,742	7	3	0.337
PowerGrid[a]	4,941	6,594	8	6	0.732
Internet[a]	22,963	48,436	5	5	0.391

[a]See footnote 1

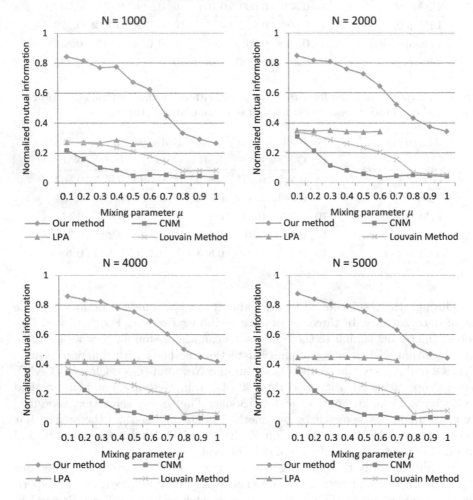

Fig. 3. The experimental results on a datasets with $N = 1000, 2000, 4000$ and 5000 nodes with four algorithms CNM, LPA, Louvain and our algorithm.

4.3   Computer-Generated Networks

In order to demonstrate the performance achievable by our method, we applied our algorithm on computer-generated networks, which we use the class of benchmark computer-generated networks proposed by Lancichinetti *et al.* [19]. Each set of data share similar properties of the real communities aiming to verify our algorithm. Each node shares a fraction $1 - \mu$ of its link with the other nodes of its community and a fraction μ with the other nodes of the network, calling μ the mixing parameter of the network. In order to check the capability of the algorithm, measurement of F and NMI is used. In the comparative tables, parameter μ has the value in between 0.1 and 1.0, and our set of value has N peaks including 1000, 2000, 4000 and 5000; other parameters have the values like $k = 20$, $maxk = 30$, $on = 0$ and $om = 0$. Experimental results show that our algorithm can detect the approximate actual community. In our algorithms results, the NMI always bigger than 0.8 when the value of parameter μ is 0.1 and 0.2 respectively. The NMI decreased to 0.3 when μ increases to 1.0 as show in Fig. 3.

5 Conclusion

In this paper, the method of detecting community structure based on node similarity is introduced. We extend the method to consider similarity of communities to create new ones. The stages of the applied algorithm include creating communities, detecting core communities, classifying communities, merging communities and selecting the best community. In the experiments, the algorithm is applied on different datasets such as real-world datasets and computer-generated networks. We compared our algorithm to many others algorithms, such as GN, CNM, LPA, and Louvains method. The results show that our algorithm is more effective than those based on F-Measure and NMI. In the future, our algorithm for analyzing complex networks will be introduced. Moreover, this algorithm can be developed to maximize the properties and attributes of each object in complex networks to detect community structures.

References

1. Albert, R., Barabási, A.L.: Statistical mechanics of complex networks. Rev. Mod. Phys. **74**(1), 47 (2002)
2. Blondel, V.D., Guillaume, J.L., Lambiotte, R., Lefebvre, E.: Fast unfolding of communities in large networks. J. Stat. Mech: Theory Exp. **2008**(10), P10008 (2008)
3. Boccaletti, S., Latora, V., Moreno, Y., Chavez, M., Hwang, D.U.: Complex networks: structure and dynamics. Phys. Rep. **424**(4), 175–308 (2006)
4. Chen, J., Zaïane, O.R., Goebel, R.: Detecting communities in social networks using max-min modularity. In: SDM, vol. 3, pp. 20–24. SIAM (2009)
5. Cilibrasi, R.L., Vitanyi, P.: The google similarity distance. IEEE Trans. Knowl. Data Eng. **19**(3), 370–383 (2007)
6. Clauset, A., Newman, M.E., Moore, C.: Finding community structure in very large networks. Phys. Rev. E **70**(6), 066111 (2004)
7. Danon, L., Diaz-Guilera, A., Duch, J., Arenas, A.: Comparing community structure identification. J. Stat. Mech: Theory Exp. **2005**(09), P09008 (2005)

8. Dinh, T.N., Thai, M.T.: Community detection in scale-free networks: approximation algorithms for maximizing modularity. IEEE J. Sel. Areas Commun. **31**(6), 997–1006 (2013)
9. Fiedler, M.: Algebraic connectivity of graphs. Czechoslovak Math. J. **23**(2), 298–305 (1973)
10. Fortunato, S.: Community detection in graphs. Phys. Rep. **486**(3), 75–174 (2010)
11. Fortunato, S., Barthelemy, M.: Resolution limit in community detection. Proc. Nat. Acad. Sci. **104**(1), 36–41 (2007)
12. Girvan, M., Newman, M.E.: Community structure in social and biological networks. Proc. Nat. Acad. Sci. **99**(12), 7821–7826 (2002)
13. Gleiser, P.M., Danon, L.: Community structure in jazz. Adv. Complex Syst. **6**(04), 565–573 (2003)
14. Good, B.H., de Montjoye, Y.A., Clauset, A.: Performance of modularity maximization in practical contexts. Phys. Rev. E **81**(4), 046106 (2010)
15. Hric, D., Darst, R.K., Fortunato, S.: Community detection in networks: structural communities versus ground truth. Phys. Rev. E **90**(6), 062805 (2014)
16. Jaccard, P.: Etude comparative de la distribution florale dans une portion des Alpes et du Jura. Impr, Corbaz (1901)
17. Kernighan, B.W., Lin, S.: An efficient heuristic procedure for partitioning graphs. Bell Syst. Tech. J. **49**(2), 291–307 (1970)
18. Lancichinetti, A., Fortunato, S., Kertész, J.: Detecting the overlapping and hierarchical community structure in complex networks. New J. Phys. **11**(3), 033015 (2009)
19. Lancichinetti, A., Fortunato, S., Radicchi, F.: Benchmark graphs for testing community detection algorithms. Phys. Rev. E **78**(4), 046110 (2008)
20. Lancichinetti, A., Kivelä, M., Saramäki, J., Fortunato, S.: Characterizing the community structure of complex networks. PloS ONE **5**(8), e11976 (2010)
21. Li, X., Ng, M.K., Ye, Y.: Multicomm: finding community structure in multidimensional networks. IEEE Trans. Knowl. Data Eng. **26**(4), 929–941 (2014)
22. Lusseau, D.: The emergent properties of a dolphin social network. Proc. Roy. Soc. London B: Biol. Sci. **270**(Suppl 2), S186–S188 (2003)
23. Newman, M.E.: Detecting community structure in networks. Eur. Phys. J. B **38**(2), 321–330 (2004)
24. Newman, M.E.: Fast algorithm for detecting community structure in networks. Phys. Rev. E **69**(6), 066133 (2004)
25. Newman, M.E.: Modularity and community structure in networks. Proc. Nat. Acad. Sci. **103**(23), 8577–8582 (2006)
26. Newman, M.E., Girvan, M.: Finding and evaluating community structure in networks. Phys. Rev. E **69**(2), 026113 (2004)
27. Pan, Y., Li, D.H., Liu, J.G., Liang, J.Z.: Detecting community structure in complex networks via node similarity. Physica A **389**(14), 2849–2857 (2010)
28. Pons, P., Latapy, M.: Computing communities in large networks using random walks. In: Yolum, I., Güngör, T., Gürgen, F., Özturan, C. (eds.) ISCIS 2005. LNCS, vol. 3733, pp. 284–293. Springer, Heidelberg (2005)
29. Radicchi, F., Castellano, C., Cecconi, F., Loreto, V., Parisi, D.: Defining and identifying communities in networks. Proc. Nat. Acad. Sci. U.S.A. **101**(9), 2658–2663 (2004)
30. Raghavan, U.N., Albert, R., Kumara, S.: Near linear time algorithm to detect community structures in large-scale networks. Phys. Rev. E **76**(3), 036106 (2007)
31. Shi, J., Malik, J.: Normalized cuts and image segmentation. IEEE Trans. Pattern Anal. Mach. Intell. **22**(8), 888–905 (2000)

32. Sorenson, T.: A method of establishing groups of equal amplitude in plant sociology based on similarity of species content. Kongelige Danske Videnskabernes Selskab 5(1–34), 4–7 (1948)
33. Yang, J., McAuley, J., Leskovec, J.: Community detection in networks with node attributes. In: 2013 IEEE 13th International Conference on Data mining (ICDM), pp. 1151–1156. IEEE (2013)
34. Yin, C., Zhu, S., Chen, H., Zhang, B., David, B.: A method for community detection of complex networks based on hierarchical clustering. Int. J. Distrib. Sens. Netw. 2015, 137 (2015)
35. Zachary, W.W.: An information flow model for conflict and fission in small groups. J. Anthropol. Res. 33(4), 452–473 (1977)
36. Zhou, T., Lü, L., Zhang, Y.C.: Predicting missing links via local information. Eur. Phys. J. B 71(4), 623–630 (2009)

Rumor Propagation Detection System in Social Network Services

Hoonji Yang[1], Jiaofei Zhong[2], Dongsoo Ha[1], and Heekuck Oh[1(✉)]

[1] Department of Computer Science and Engineering, Hanyang University,
Ansan, South Korea
bellayhj@gmail.com , hkoh@hanyang.ac.kr
[2] Department of Mathematics and Computer Science, Callifornia State University,
East Bay, Hayward, CA 94542, USA
jiaofei.zhong@csueastbay.edu

Abstract. The growing use of the smart device such as smartphones and tablets has resulted in increasing number of social network service (SNS) users recently. SNS allows a fast propagation and it is used as a tool to send information. But its negative sides need to be considered. In this paper, we analyzed actual data of malicious accounts and extracted features. Based on this results, we detect the suspected accounts that spread rumors. Firstly, we crawled actual data and analyzed feature. And we selected feature as three approaches and added a new feature as propagation approach by existing work. That is user can re-tweet influencer's tweet and edit it. We discussed it by ratio for RT. After that, we selected classification standard using average of data based on selected feature and trained it. Bayesian network is used for training. And the system may provide a new classification through re-analysis of the data. Proposed system is that the accuracy is 91.94 % and F-measure is 93.76 %.

Keywords: Social Network Services · Rumor propagation · Rumor detection · Machine learning · Bayesian Network

1 Introduction

Social Network Service (SNS) is a service that helps people to make a wide range of human network by strengthening relationships with acquaintances or forming new connections [6]. With the advent of social network service, users are able to communicate with each other regardless of location and time and two-way communication became possible by sharing own information of users.

As smart phones emerged, social network service is available at any time and the production and processing of information has became faster than ever [2]. The interactivity of social network allowed a rapid dissemination of information in real-time, however the reliability of the information has been doubted due to indiscriminate flood of information.

© Springer International Publishing Switzerland 2016
H.T. Nguyen and V. Snasel (Eds.): CSoNet 2016, LNCS 9795, pp. 86–98, 2016.
DOI: 10.1007/978-3-319-42345-6_8

Normal users also suffered from distorted information spread by malicious users while using the service. For example, social fear is created among social network service users because of false stories spread, or social matter is caused by untested slander and libel that defame celebrities [12]. A malicious users create accounts as many as the number of email addresses they have, using the simple email authentication method on Twitter or Facebook.

The easy way of authentication via email is exploited to create mass accounts and spread false rumors which then cause a bad influence in society [14]. Accordingly, removing such malicious accounts has been recognised as an important matter and a number of studies are performed to detect rumor disseminations using the structural features.

Twitter users can arbitrarily determine and report malicious accounts by clicking 'Report to @username for spam' [10]. In this process, normal users may be maliciously reported and the inconvenience of being cut off occurs. This also may allow malicious users to hide themselves skillfully to avoid the policy of Twitter.

In this paper, certain information is verified to determine whether it is a rumor or a general information and accounts that spread the information will be detected when the information is false. Malicious accounts are gradually evolving by taking a number of actions in order to avoid a policy of Twitter. Thus, it is able to detect the account that spread rumors by using Bayesian network which is one of machine learning technique to update new characteristics of internal trolls and re-classify them.

The structure of this paper is as followed. In Sect. 2, theoretical background will be discussed, followed by related research in Sect. 3. In Sect. 4, features are extracted through data crawling, and malicious accounts are detected based on classified features.

2   Related Work

The phenomenon usually referred to as 'rumors' are aspects that are becoming more universal due to the proliferation of social networks [15]. In the past when the media such as newspapers and TV broadcasts existed, there were only a small number of people who were able to express their opinions. However, everyone can express their opinions and share reprocessed information through social network services nowadays. The fact, that everyone is free to express own opinion, means that any comments or words can go up online without any judgement of the credibility. Individuals and society can be damaged when unconfirmed information without clear source is spread.

There is no one who can express an opinion with sufficient amount of information in a sophisticated modern society, therefore, it is close to impossible to determine whether the information is false or not when it is up online. This means that the information needs to be spread first in order to be verified whether it is rumor or reliable information [11,17].

In recent years, various studies have been conducted to detect and prevent the spread of the incorrect and malicious information from the SNS, and various techniques are proposed to create a method that can be applied to any network environment.

The new method, that can obscure the authenticity of rumors drifting in SNS, has been studied recently. Kwon analysed the graph of spreading rumors and general information, and classified the characteristics of rumor propagation into three categories [7].

First, rumors unlike general information has a tendency to continuously spread. General information is rarely mentioned after the wide range of dissemination, however rumors are continuously mentioned for a long period of time. Second, the dissemination of rumors consists of a sporadic involvement of any unrelated users. The path of general information is derived from the relationship between users within the online, while rumors has a characteristic that it is composed of unconnected individuals. Finally, the rumors have different linguistic characteristics to general information. This means that the ratio of words inferring doubt and denial is much higher in rumors.

Detection of account that spread rumors is similar to detection of spammers from previous studies. Spammer is an account or an individual that spam and spam includes unwanted message, commercial letter or article that is sent to a large number of unspecified recipients. In contrast, rumor is a little different to spam as it intendedly spread malicious information about an individual, a company or even a country to manipulate public opinion and create a certain atmosphere.

Gurajala proposed a spam detection technique that analyses the time of users updating the tweets and gradually reduce a set of accounts with a high probability to be fake based on the profiles of accounts [4]. Gao proposed a spam detecting technique by extracting the features of malicious messages and URL on Twitter and Facebook based on SVM (Support Vector Machine) while Bosma proposed a system that detects spams based on HITS technique [3,9].

Moreover, Jonghyuk suggested a research that detects spams based on the features of relationships between users and Kyumin proposed a social-honeypot with message training data that is a date analysis of user profiles, relationships and message information [8,10]. Benevenuto extracted 62 features based on tweet contents and behaviors of users from Twitter and Zhu proposed a spam classification model based on Matrix Factorization [1,18].

The studies above extracted feature values by analyzing the entire data. However, there is a limitation to detect malicious accounts by comprehensively reflecting the characteristics of existing malicious accounts and there was no consideration of addition of new data. Furthermore, the new features are needed to be extracted in order to detect accounts tat spread rumors. Thus, in this paper, the features of malicious accounts are extracted through actual data analysis and a efficient rumor propagation detection system that allows addition of data lively based on the analysis is proposed.

3 Preliminaries

3.1 Definition and Features of SNSs

Social network services (SNSs) is an online platform that generates and enhances social relations through the free communication and information sharing among users. In recent years, the use of smart devices such as smart phones and tablets has increased and this affected the growth of the number of social network service users.

There are social networks such as Facebook, Twitter, Instagram, and Twitter of these services has shown the fastest growth. Twitter is similar to other social networking services in a way that users meet new people or be in touch with friends and share information in real time.

However, Twitter provides a micro blogging service called 'Tweet', where users can post a relatively small data message of a short paragraph of 140 characters or a video link to other users [6]. This feature of Twitter is easily exploited by internet trolls. In order to detect the exploitation, following features of Twitter need to be discussed [16].

- Tweet: Acts of writing and posting are called "Tweet". Since the length of tweet is limited to 140 characters, users use a service which reduces the URL to post it. However, malicious users use the service to attract other users with a short URL and a small number of words.
- Following and Follower: 'Following' is an act of subscribing others' tweets and 'Follower' is a user who follows my tweets. This means that one's tweets are not uploaded on following subjects' timeline and follower's tweets are not uploaded on one's timeline. The state of following each other's tweets is called 'Matpal (mutual following)'.
- Mention: 'Mention' is a similar concept to Tweet, however it is an act of sending tweets to a specific user. It is a tweet that is created to refer a certain users and it has the form of @username.
- Retweet: This is the retransmission of tweets on timeline to one's connected followers when he wants to share it. Usually mention is added in the form of @username and there is RT which is the same concept to 'Retweet'. RT is the same as retweet, however there is a difference that users can add their opinions to tweets that are retransmitted.
- Hashtag: This is in a form of '#word' that expresses the specific topic of article and it is used to search related topics of article easily.

3.2 Bayesian Network

In this paper, we classified based on the Bayes theorem to extract the features of the account that spread rumors. Bayes theorem was derived by modifying the conditional probability, and is the arrangement showing the relationship between a prior probability and the a posterior probability of two random variables. That is, a probability of the causes of the incident after the incident occurs, can be obtained by using the information already given before the incident. Thus, the posterior probability is proportional to the product of the prior probability and likelihoods, and the expression is shown below.

$$P(h|d) = \frac{P(d|h)P(h)}{P(d)},$$

where h is hypothesis, d is data, $p(h)$ and $p(d)$ are the probabilities of h and d regardless of each other, $p(h|d)$, a conditional probability, is the probability of observing event h given that d is true, and $p(d|h)$ is the probability of observing event h given that d is true.

Bayesian Network (BN) is a useful model to solve the problem including uncertainty based on the arrangement. Bayesian network takes the form of a directed acyclic graph and nodes of the graph refer to random variables while the trunk between nodes refers to a probabilistic dependency. In other words, Bayesian network is the image of the network that is classified with calculations of the posterior probabilities for all classes using Bayse theory [5]. It is used when the problem needs to be solved with consideration of a certain dependence. This method is considered to be appropriate for the system proposed because many features are required in order to determine which one of the accounts spread malicious rumors.

4   The Proposed Rumor Propagation Detection Scheme

4.1   Data Collection and Analysis

First, we chose 20 rumor topics from the sites such as snopes.com, urbanlegends.about.com, times.com in order to collect the cases of rumors and verified information. Then, we defined a regular expression for each subject and crawled user information and hash tags of Twitter. 'twitter4j API' was used for crawling, and the number of massages in the timeline was opted to 20 while the number of followers and following were opted to 300 people each.

Recent studies on the propagation of rumors shows that the propagation of rumors takes place with a sporadic involvement of unrelated users [2]. The flow of common information is spreading via friendship between accounts, however rumors is spreading with participation of individual accounts that are not connected. This is because the accounts that spread rumors have a high probability to be unusual and less known by people. That is, the graph of the spreading rumors has more independent clusters than generic graph. Accordingly, the graph of general information and the graph of rumors are specified with the following formula.

Then, we chose 200 regular accounts that act frequently and 200 malicious accounts that repeatedly post the same false retweets in order to detect accounts that take malicious acts to spread rumors.

The information, then, will be used to analyse whether the account is spreading rumors or not, by extracting the features of general accounts and malicious

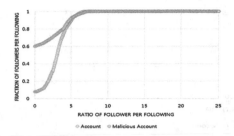

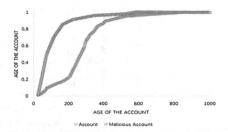

Fig. 1. Cumulative distribution graph of ratio of friend (Color figure online)

Fig. 2. Cumulative distribution graph of age of account (Color figure online)

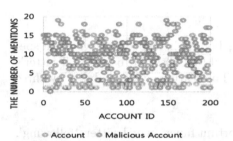

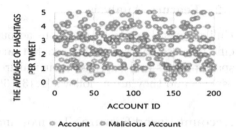

Fig. 3. Distribution graph of average of mentions (Color figure online)

Fig. 4. Distribution graph of number of hashtags (Color figure online)

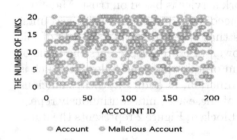

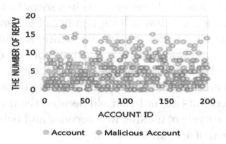

Fig. 5. Distribution graph of number of links (Color figure online)

Fig. 6. Distribution graph of number of reply (Color figure online)

accounts. Features of malicious accounts are insufficient to indicate as a few, therefore features extracted from the previous studies are used to find the accounts. Characters are as follows: the presence or absence of a profile picture, account creation date, the number of followings, the number of followers, the number of mutual followings, the number of average mention, the average number of retweets, the average number of hash tags, the number of tweets including URL.

In this paper, we classify the accounts through a Bayesian network. Bayesian networks may be efficiently used to detect.

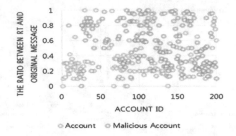

Fig. 7. Distribution graph of ratio of RT (Color figure online)

4.2   Feature Analysis

Unlike general accounts, malicious accounts behave in a specific way with a specific purpose. We selected elements that should be observed based on Twitter's spam policies and analysed their features. The features were analysed by three aspects and they are as follows [1,13,16].

Account Based Feature. The most important function of Twitter, 'following', allows users to see the other's tweets without following each other's twitter. Twitter trolls use this function to attract the other users to be interested in their accounts. It is hard for malicious accounts to be friend with normal accounts, because the relationship between users in a social network service is based on trust. Malicious accounts follow a huge number of general accounts to introduce their accounts, however it is not easy. Therefore, accounts that spread rumors have lots of followings and relatively small number of followers. Figure 1 represents the value of each account's followings divided by the sum of followings and followers which is then shown in the cumulative distribution function. In addition, these malicious accounts are not that old because the users of these accounts continuously repeat the cycle of making new account and being blocked. Figure 2 represents the duration of accounts created.

Contents Based Feature. Malicious accounts take a number of actions in order to spread rumors, and they are as follow. First to mention is the act of sending duplicate tweets to random people. Twitter trolls mention the names of celebrities or influencers on social networks to spread rumors. This is because the other person can see the tweets of malicious users by being mentioned. Second is the act of continuously posting a tweet with URL. Twitter can filter the spiteful links based on their policy, however the problem is there are malicious links that are transformed into a short URL like 'bit.ly'. And some accounts can transform external links to shorter URL to disseminate rumors and avoid the filtering. The last act is hash tagging tweets with something that has nothing to do with the content.

Malicious accounts can hash tag the most often mentioned topics or something about that people would be interested regardless of the content. This is to make people to read the tweets by making it easy to search these tweets. Therefore, in order to analyse those three actions, we looked at the number of mentions with '@' of latest 20 tweets, the number of hash tags in each tweet and the number of tweets that include 'http://'.

Figure 3 represents the number of mentions. According to this, it is clear that malicious accounts tend to have more mentions than normal accounts. Figure 4 shows an average for each account of a number of hash tags per tweet. This shows that malicious accounts have a higher average number of hash tags, but it also shows that the distribution of general accounts is wide. Figure 5 is a distribution of the number of tweets that include URL. This result shows that there is a huge number of malicious accounts and also that accounts of company for promoting are found to be in the top position among general accounts. Figure 6 is a distribution of the number of replies which is re-send to another account. And, this result shows that malicious accounts have a lower number of replies.

Propagation Based Feature. Twitter has a function of re-tweeting others' messages. Re-tweeted messages are displayed as RT @username, and Twitter trolls often re-tweet the tweets and fix them rather than tweeting directly. This method is related to the psychology of people who tends to show more trust towards tweet post brought from somewhere else than posts posted directly by others. This means that the mass psychology can be easily exploited. Thus, we looked at 20 latest tweets and the ratio of re-tweeted tweets and tweets posted by people. Figure 7 represents the number of re-tweeted tweets that are fixed. This shows that it is distributed widely regardless of which account they are, however there are more malicious accounts located on the top.

4.3   System Modeling

First, observed variable and the results value defined like that for modeling of the system. F of Eq. (1) means a set of observed variable and n means the number of variable. Using this variables, Eq. (2) represents the capable probability model of some k.

$$F = \{f_1, f_2, \ldots, f_n\} \tag{1}$$

$$p(M_k|f_1, f_2, \ldots, f_n) \tag{2}$$

We represent the equation divided n of observed feature into i of dependent feature and j of independent feature.

$$p(M_k)p(f_1, f_2, \ldots, f_i|M_k) \tag{3}$$

$$= p(M_k)p(f_1|M_k)p(f_2|M_k, f_1) \cdots p(f_n|M_k, f_1, f_2, \ldots, f_i) \tag{4}$$

Above equation is in the case of dependent feature, and independent feature is showed as follows.

$$p(M_k)p(f_1, f_2, \ldots, f_i|M_k) \tag{5}$$

$$= p(M_k)p(f_1|M_k)p(f_2|M_k) \cdots p(f_i|M_k) \tag{6}$$

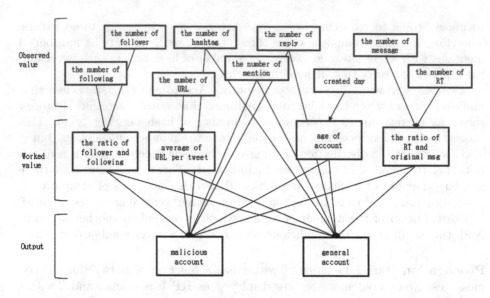

Fig. 8. Bayesian network modeling

Implemented system used two mentioned cases is modelled as follows.

$$p(M_k)p(f_1, f_2, \ldots, f_i|M_k) \tag{7}$$

$$\propto p(M_k)p(f_1|M_k)p(f_2|M_k, f_1)p(f_i|M_k, f_1, f_2,$$

$$\ldots, f_{i-1}) \cdots p(f_{i-1}|M_k)p(f_i|M - k) \tag{8}$$

$$\propto p(M_k) \prod_{n=1}^{i} p(f_n|M_k, f_1, \ldots, f_{n-1}) \prod_{n=i+1}^{j} p(f_n|M_k) \tag{9}$$

Based on this equation, Bayesian Network is constructed as Fig. 8, and a classifier from the probability model is as follows. Two equation selects the most probable hypothesis. In other words, it finds class k that has maximum probability through rule.

$$\hat{O} = argmax(M_k) \prod_{n=1}^{i} p(f_n|m_k, f_1, \ldots, f_{n-1}) \prod_{n=i+1}^{j} p(f_n|M_k)$$

4.4   System Implementation

We implemented a system that determines whether the account is spreading rumors or not when the account is inputted by putting the analysed data into the system. The purpose of the system is to allow the users to determine which account is an account that spreads rumors in a real Twitter environment (Tables 1 and 2).

The proposed system consists of three modules which are a tweeter crawling module, the feature extraction module and the module that determines whether

Table 1. Example of confusion matrix

Category		Predicted	
		Malicious account	Account
True	Malicious account	a	b
	Account	c	d

Table 2. Performance metric

Factor	Formula
Accuracy (A)	$(a+d)/(a+b+c+d)$
Precision (P)	$a/(a+c)$
Recall (R)	$a/(a+b)$
F-measure	$2PR/(P+R)$

the account carries a malicious act or not based on learned data. Three modules are implemented using java 1.8.50 and javascript. The order of process is as follows.

- Prototype software includes a user interface that can receive the account name based on the implementation of the data crawler.
- The user inputs the account name in order to find if this account is malicious or not.
- The account is crawled and classified according to the analysis based on the characteristics of contents and behaviors learned from the data.
- Analyze the trends of hash tags about rumors of recent tweets, and output the probability of being malicious account by calculating them.
- When the account is classified as a malicious account, it will be input into database and it re-learned by the system.

5 Experiment and Evalution

5.1 Evaluation Metric

The evaluation uses a confusing matrix of Table 3. Confusing matrix is known as error matrix, it means specific table layout visualized performance of algorithms 'a' represents the number of accurately classified malicious accounts and 'b' represents the number of malicious accounts that are misclassified as general accounts. 'c' is the number of general accounts that are misclassified as malicious accounts and 'd' is the number of accounts that are correctly classified as general accounts. We will evaluate them by using the confusing matrix. Rating scales are Precision, Recall, and F-measure. The performance is measured by the classification of the performance index above.

Table 3. Bayesian classification results

Category		Predicted	
		Malicious account	Account
True	Malicious account	94.16 %	5.84 %
	Account	13.34 %	86.66 %

Table 4. Machine learning algorithms scenario

No.	Algorithm
1	Support Vector Machine (SVM)
2	Decision Tree (DT)
3	Bayesian Network (BN)
4	Neural Network (NN)

Table 5. Performance results

Algorithm	Accuracy	Precision	Recall	F-measure
SVM	91.38 %	92.65 %	94.58 %	93.60 %
DT	89.72 %	92.11 %	92.50 %	92.30 %
BN	91.94 %	93.38 %	94.16 %	93.76 %
NN	90.83 %	92.94 %	93.33 %	93.13 %

5.2 Evaluation

This experiment uses WEKA which is an open source framework that contains a collection of visualization tools and algorithms for data analysis and predictive modeling. For experiment, the selected topic is a rumor called 'CallNotRegi' which is that if you do not want to receive calls coming from telemarketers, you call a specific number and register your number. The number is false and the topic is turned out to be rumor at 'snopes.com'. We crawled through hash tags, and 360 account information with 120 malicious accounts were selected. Also, the experiment processed by algorithms scenario of Table 4. And the result is as follows.

As the performance testing result is shown in Table 5, generally our proposed method improves the accuracy of the classification. However, in the case of Recall, SVM has a little better performance than our method. Moreover, DT has the lowest performance among the other algorithms.

6 Conclusion

In this paper, we proposed malicious accounts detection system based on Bayesian network to detect the action of spreading rumors. This system analyses and relearns the date based on learned features of malicious accounts. In addition, we measured the performance and classified the malicious accounts that spread rumors in a real social networks and the accuracy was 91.94 %.

References

1. Benevenuto, F., Magno, G., Rodrigues, T., Almeida, V.: Detecting spammers on Twitter. In: Collaboration, Electronic Messaging, Anti-abuse and Spam Conference (CEAS), vol. 6, pp. 12 (2010)
2. Ellison, N.B., Steinfield, C., Lampe, C.: The benefits of Facebook friends: social capital and college students use of online social network sites. J. Comput. Mediated Commun. **12**(4), 1143–1168 (2007)
3. Gao, H., Chen, Y., Lee, K., Palsetia, D., Choudhary, A.N.: Towards online spam filtering in social networks. In: Proceedings of 19th Network Distributed System Security Symposium, vol. 29, No. 23, pp. 1–10 (2010)
4. Gurajala, S., White, J.S., Hudson, B., Matthews, J.N.: Fake Twitter accounts: profile characteristics obtained using an activity-based pattern detection approach. In: Proceedings of the 2015 International Conference on Social Media and Society, p. 9. ACM (2015)
5. Jenson, F.V.: An introduction to Bayesian networks, vol. 210. UCL Press, London (1996)
6. Kwak, H., Lee, C., Park, H., Moon, S.: What is Twitter, a social network or a news media. In: Proceedings of the 19th International Conference on World Wide Web, pp. 591–600 (2010)
7. Kwon, S., Cha, M., Jung, K., Chen, W., Wang, Y.: Prominent features of rumor propagation in online social media. In: IEEE 13th International Conference on Data Mining (ICDM), pp. 1103–1108 (2013)
8. Kyumin, L., Caverlee, J., Webb, S.: Uncovering social spammers: social honeypots + machine learning. In: Proceedings of the 33rd International ACM SIGIR Conference on Research and Development in Information Retrieval, pp. 1139–1140. ACM (2010)
9. Bosma, M., Meij, E., Weerkamp, W.: A framework for unsupervised spam detection in social networking sites. In: Baeza-Yates, R., de Vries, A.P., Zaragoza, H., Cambazoglu, B.B., Murdock, V., Lempel, R., Silvestri, F. (eds.) ECIR 2012. LNCS, vol. 7224, pp. 364–375. Springer, Heidelberg (2012)
10. McCord, M., Chuah, M.: Spam detection on Twitter using traditional classifiers. In: Calero, J.M.A., Yang, L.T., Mármol, F.G., García Villalba, L.J., Li, A.X., Wang, Y. (eds.) ATC 2011. LNCS, vol. 6906, pp. 175–186. Springer, Heidelberg (2011)
11. Qazvinian, V., Radev, E., Mei, Q.: Rumor has it: identifying misinformation in microblogs. In: Proceeding of the Conference on Empirical Methods in Natural Languate Processing (EMNLP), pp. 1589–1599 (2011)
12. Starbird, K., Maddock, J., Orand, M., Achterman, P., Mason, R.M.: Rumors, false flags, and digital vigilantes: misinformation on Twitter after the 2013 Boston Marathon Bombing. In: iConference on 2014 Proceedings (2014)
13. Stringhini, G., Kruegel, C., Vigna, G.: Detecting spammers on social networks. In: Proceedings of the 26th Annual Computer Security Applications Conference, pp. 1–9 (2010)
14. Viswanath, B., Post, A., Gummadi, K.P.: An analysis of social network-based Sybil defenses. ACM SIGCOMM Comput. Commun. Rev. **41**(4), 363–374 (2011)

15. Vosoughi, S.: Automatic detection and verification of rumors on Twitter. Diss. Massachusetts Institute of Technology (2015)
16. Wang, A.H.: Don't follow me: spam detection in Twitter. In: Proceedings of the 2010 International Conference on Security and Cryptography (SECRYPT), vol. 29, no. 23, pp. 1–10 (2010)
17. Wu, K., Yang, S., Zhu, H.Q.: False rumors detection on Sina Weibo by propagation structures. In: IEEE International Conference on Data Engineering, ICDE, pp. 651–662 (2015)
18. Zhu, Y., Wang, X., Zhong, E., Liu, N.N., Li, H., Yang, Q.: Discovering spammers in social networks. In: 26th AAAI Conference on Artificial Intelligence (2012)

Detecting Overlapping Community in Social Networks Based on Fuzzy Membership Degree

Jiajia Rao$^{(\boxtimes)}$, Hongwei Du, Xiaoting Yan, and Chuang Liu

Department of Computer Science and Technology,
Harbin Institute of Technology Shenzhen Graduate School,
Shenzhen Key Laboratory of Internet Information Collaboration, Shenzhen, China
`jiajiaraohit@gmail.com`, `hwdu@hitsz.edu.cn`, `xiaotingyanhit@gmail.com`,
`chuangliuhit@gmail.com`

Abstract. Overlapping community detection in social networks is a challenging task for revealing the community structure, as one user may belong to several communities. Most previous methods of overlapping community detection ignore the belonging levels when one node belongs to several communities. The membership-degree is used to embody the belonging level. In this paper, an novel method calling *Fuzzy Membership-Degree Algorithm (FMA)* is put forward. Firstly, we propagate the membership-degree with consideration of the *nodes-attraction*, which is a new proposed definition based on topological characteristics. Then we further mine communities under the guidance of Extended Modularity (EQ). In this paper, the proposed algorithm FMA makes full use of the topological information, and membership-degree suggests the belonging level of overlapping community. Experiments on synthetic and real-world networks demonstrate that our algorithm performs significantly.

Keywords: Overlapping community · Membership-degree · Social networks

1 Introduction

With increasing number of Internet technique, such as Twitter, Facebook, Skype, people pay more attention on social networks instead of communication in real life.

Generally, social network analysis attempts to conduct research in terms of the nodes, the relationship and the network structures [1,2]. Community detection is to find latent community structures in social networks. Community, which is also called cluster, is considered as a group of nodes, in which intra-group is more similar and inter-group is more dissimilar. Nowadays, community detection has been one of the significant topics in the field of social network analysis.

Most works on community detection attempt to discover non-overlapping communities in which one node is limited to only one cluster. However, in social networks, there exist a large number of highly overlapping cohesive communities in which one node belongs to several communities. For example, in the interest

© Springer International Publishing Switzerland 2016
H.T. Nguyen and V. Snasel (Eds.): CSoNet 2016, LNCS 9795, pp. 99–110, 2016.
DOI: 10.1007/978-3-319-42345-6_9

communities, one person likes sports, food and music, which drives this person to be in several communities. As a result, it is more available to study the overlapping communities than non-overlapping communities.

Traditional methods of overlapping community detection are not supposed to definitely point out the belonging levels [3,4]. However, there exist different belonging levels to the overlapping nodes in most cases. For instance, in the interest communities, one person shows preference for music when this person is simultaneously interested in sports and food. To address the problem, the membership-degree is introduced to express the level of one node belonging to a cluster. The membership-degree is defined as an value which ranges from 0 to 1, and the membership-degree is a notion originated in fuzzy set.

In this paper, we propose a novel algorithm named FMA to detect overlapping communities. Inspired by the ideas that the membership-degree of a node is similar to its neighboring nodes, we iteratively propagate the membership-degree. And accounting that two closely linked nodes are more similar, we introduce the *nodes-attraction* to guide the propogation. Meanwhile, we further mine communities with the guidance of EQ [5]. In this step, we set a threshold for the membership-degree when the communities is with the maximum EQ. This step is to find communities with good modularity. To sum up, the main contributions in this paper are as follows:

1. In this paper, we come up with a notion of *nodes-attraction* and use it to guide the propagation. *Nodes-attraction* is proposed based on topological characteristics, which makes full use of topological information and greatly improves the clustering accuracy.
2. We further mine communities with *EQ*, which contributes to better performance. This step ensures the good modularity of the obtaining communities. And it is accepted by most researchers that communities with good modularity mean great community detection.
3. Comparing to traditional label propagation, the membership-degree propagation of proposed method is stable as there are no random choice and unstable termination.
4. The time cost of proposed algorithm is nearly linear, and the experimental results of synthetic networks and real-world networks indicate that the proposed algorithm outperforms other algorithms.

The rest of this paper is organized as follows. Section 2 discuss the related work. Section 3 explain the specific problem. Framework of FMA is described in Sect. 4. Analysis is in Sects. 5 and 6. Finally, Sect. 7 concludes this paper.

2   Related Work

Overlapping community detection in social networks has been studied over the past decades. Traditionally, the way of overlapping community detection in social networks is discrete assignment [6]. CPM [7] first finds all k-cliques in which all nodes are fully connected, then combines two cliques if they share k-1 members.

This method is suitable for dense connected networks and time cost is expensive. Gregory came up with the idea of EAGLE. It proposed the notion of extended modularity which is used for evaluating the goodness of overlapping community detection. Steve Gregory put forward the method of COPRA [8], which is based on the label propagation. Different from COPRA, the method FMA in this paper mainly relies on membership-degree propagation.

Fuzzy clustering or soft clustering is another way to detect overlapping communities, which each node assigns to a community with the membership degree. FCM [9] is the prominent method for fuzzy clustering. It iteratively updates the membership and center node based on an objection function [10]. Most previous methods for fuzzy community detection were based on an objection functions, which ignored the topological characteristic of the social networks. In this paper, we propose an novel method which combines the achievement of traditional discrete way and fuzzy membership.

3   Problem Statement

In social networks, a community is defined as a group of users who are more similar in some aspects within the group than outside [11]. For example, Interest community means that a cluster which holds the similar interests, and there exist big differences among different communities. In this section, we will state some basic notions and formulate the problem.

The social network is represented as a weighted undirected graph. The weights of edges represent the feature dissimilarity of nodes. We use $G(N, K, W)$ to denote the network, in which, N is the total number of nodes in the network, K is the total clusters number, W denotes the weight matrix. And W_{ij} denotes the weight of edge e_{ij}. And the membership u_{ij} denotes the membership-degree of node j belonging to cluster i. Obviously we can get that the membership-degree vector $\vec{u}_i$ denotes the membership degrees of nodes which belong to community i. So there exist the membership-degree matrix $U(\vec{u}_1, \vec{u}_2, \ldots \vec{u}_K)$ and the membership-degree vector $\vec{u}_i(u_{i1}, u_{i2}, \ldots u_{iN})$. $\sum u_{ij} = 1$. And the communities denote as $C(c_1, c_2, \ldots c_K)$, in which, c_i represents the community i. After above notions, we can formulate our problem as follows.

Problem Definition: Given a undirected weighted graph $G(N, K, W)$, our objection is to find communities C in which the intra-clusters are more similar and inter-clusters are more dissimilar, and we point out the membership-degree matrix U.

4   Fuzzy Membership-Degree Algorithm

In this section, we discuss the method FMA in detail. Fuzzy clustering for overlapping community detection showed up in recent years. The membership-degree is a significant notion in fuzzy clustering. Most existing fuzzy clustering for overlapping community detection are based on objective functions which ignore the topological characteristic of the network and the achievements of traditional overlapping detection algorithm. FMA is a method which combines the traditional overlapping communities detection algorithm and fuzzy clustering. The proposed algorithm incorporates the membership-degree into traditional propagating process and make full use of the EQ. Inspired by the ideas that the membership-degree of a node is similar to its neighboring nodes and the more closely the two adjacent nodes are, the more similar they are, FMA iteratively propagates the membership-degree in consideration of the *nodes-attraction* among nodes. The *nodes-attraction* will be discussed in detail in next subsections. Then, we further mine communities through EQ. This step is to keep a good modularity of communities. FMA is a learning algorithm as one node continuously learn the membership-degree from neighbor nodes.

Based on the above discussions, we conclude FMA as three main steps:

1. How to initial the membership degree?
2. How to iteratively propagate the membership-degree and terminate it?
3. How to further mine the overlapping communities under the guidance of EQ?

4.1   Initial Membership Degree

In order to initialize the membership, we firstly need to find seeds. Many studies prove that seeds can be found in communities [12,13]. Seeds are nodes which usually own high degree and are centrality of a network. In this paper, accounting that the seeds are representative for different communities at the beginning of algorithm, we choose the nodes with high degree. Moreover, seeds need to keep certain distance because different communities are dissimilar in a certain extent. Distance indicates the dissimilarity. Let $S(s_1, s_2, \ldots s_K)$ denote the seeds set, the conditions for seeds are formulate as

$$Degree(s_i > s_{i+1} \cap s_i > s_{i-1}) \tag{1}$$

and

$$Distance(|s_i - s_{i-1}| > \gamma) \tag{2}$$

in which, γ is the threshold value of the distance or dissimilarity. Then we calculate the initial membership according to selected seeds. Accounting that the membership degree is inversely proportional to the distance, we initialize the membership-degree as

$$u_{ij} = \frac{1}{1 + dist(i,j)} \tag{3}$$

where $dist(i,j)$ is the shortest path length from i to j and is computed through Dijkstra algorithm. And i is the seed s_i which represents for cluster i. And we choose the shortest path to denote the distance.

4.2   Propagation Process

Propagation process is an an learning process. Nodes continuously learn the membership-degree from their neighbor nodes, as the membership-degree of a node is mainly dependent on its neighbor nodes. For example, in social networks, the interests of one person is usually influenced by his friends instead of strangers, where friends denote as the linking nodes. Meanwhile, the closer the two people are, the more similar the interest is. That is to say, the similarity of two nodes decides the degree of learning. The *nodes-attraction* is used to express the degree of learning. And the computation of *nodes-attraction* is based on the topological characteristics. Before giving the definition of *nodes-attraction*, we need to give another proposed definition *int-similarity*.

Definition 1 (int-similarity). *The int-similarity among two nodes is defined as the ratio of common linking nodes to total linking nodes. The int-similarity is formulated as*

$$int(a,b) = \frac{N(a) \cap N(b)}{N(a) \cup N(b)} \tag{4}$$

in which, $N(i)$ denotes the number of neighboring nodes of node i, int(i, j) expresses the int-similarity between node a and node b.

Definition 2 (nodes-attraction). *The nodes-attraction is a comprehensive parameter which depends on int-similarity and weight of edge. And The nodes-attraction is proportional to the int-similarity and inversely proportional to the weight of edge.*

$$NA(i,j) = \alpha \cdot \frac{\frac{1}{W_{ij}}}{\sum\limits_{m \in N(n)} \frac{1}{W_{mn}}} + (1-\alpha) \cdot \frac{int(i,j)}{\sum\limits_{m \in N(n)} int(m,n)} \tag{5}$$

In Definition 2, $\alpha \epsilon [0,1]$, is a tuning parameter. When $\alpha = 1$, $NA(i, j)$ represents feature similarity of Node i and j. When $\alpha = 0$, $NA(i, j)$ means the *int-similarity* of node i and node j. In this paper, we set $\alpha = 0.5$.

Iteratively Propagation. The propagation is an iterative process in which we continuously updates membership-degree. After propagation, the similar nodes come together. According to the above discussion, we know that the membership-degree updates with the guidance of nodes-attraction, so we get

$$u_{ij} = \sum_{k \in N(j)} NA(j,k) \cdot u_{ik} \tag{6}$$

It is necessary to discuss about seeds propagation. In small scale network, we can usually specific the membership for seeds, often set membership-degree 1 for seeds to its representing community. In this case, we keep the membership-degree of seeds unchange. While in big scale of network, we updates membership as normal nodes.

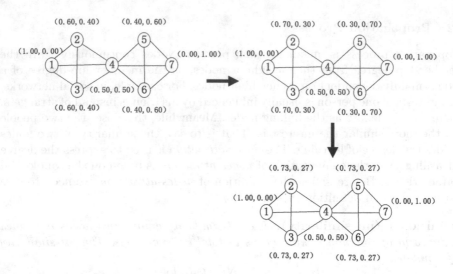

Fig. 1. The propagation process of example 1

Termination Condition. The Termination of propagation is decided by cosine similarity. Recording twice membership matrix U and U_{before} in two successive iterations. The cosine similarity between U and $Ubefore$ is defined as

$$CS(U, U_{before}) = \sum_{i,j \in [1,K]} \frac{\vec{u_i} \cdot \vec{b_i}}{|\vec{u_i}| \star |\vec{b_i}|} \qquad (7)$$

when $CS > \lambda$, we terminate the iteration. That is to say, the membership of all nodes is stable after propagation.

Example 1 of Fig. 1 is given to express the specific propagating process. Firstly according to Eqs. 1 and 2, we select seeds of Node 1 and Node 7, and compute the initialized membership with selected seeds by Eq. 3. Then we update the membership-degree through Eq. 6. When algorithm arrives to the third iteration, according to the termination condition Eq. 7, we terminate the propagation. And finally the membership tends to be instant.

4.3   Further Mine Overlapping Communities

After propagation, we further mine communities under the guide of EQ. With respect to the criteria of overlapping community, EQ shows the goodness of overlapping community composition. Aimed at weighted undirected graph, EQ gets

$$EQ = \frac{1}{2\sum W_{ij}} \sum_{i,j \in C_k} \sum (W_{ij} - \frac{d_i * d_j}{2\sum W_{ij}}) \frac{1}{O_i O_j} \qquad (8)$$

in which, O_i denotes the number of communities that node i belongs to, d_i denote the degree of node i. Selecting a threshold of membership-degree to get great overlapping communities with maximum EQ. The threshold ranges from 0.1 to the value which is the minimum of the all nodes maximum membership-degree. This process drives the algorithm to improve performance as more as possible and gets final overlapping communities. We use $u_{max}(i)$ to denote the maximum membership degree of node i. The complete procedure of FMA is shown in Algorithm 1.

Algorithm 1. Fuzzy Membership-Degree Algorithm

Input: $G = (N, K, W), \lambda$
Output: K overlapping communities, U
1: Select seeds S according to Eqs. 1 and 2
2: Initialize membership degree using Eq. 3
3: Compute $TF(i, j)$ for all neighboring nodes $pair(i, j)$
4: **while** $CS(U, U_{before}) > \lambda$ **do**
5:    Update U with Eq. 6
6:    $U_{before} \leftarrow U$
7: **end while**
8: **for** $tempVar \leftarrow 0.1$ to $minimum(u_{max}(j)), j\epsilon[1, N]$ **do**
9:    **for** All nodes in G **do**
10:       **if** $u_{ij} < tempVar$ **then**
11:          $u_{ij} \leftarrow 0$
12:       **end if**
13:    **end for**
14:    Normalize U of each nodes
15:    Compute EQ
16:    **if** $EQ > MAX(EQ)$ **then**
17:       $MAX(EQ) \leftarrow EQ$
18:    **end if**
19: **end for**
20: Normalized U with $MAX(EQ)$
21: Output final communities and U

---

5   Theoretical Analysis

Stability Analysis. In this paper, we discard the unstable termination condition and random selection process of label propagation, which makes the proposed algorithm stable. As for COPRA, there exist two latent drawbacks to result of unstability. One is the random choosing label process. However, there is no need for the proposed algorithm, as every node is with fixed number of membership. And the number of membership is fixed at K and keeps unchanged throughout the whole membership-degree propagation process. The other is the termination of COPRA. Comparing to COPRA, the termination condition Eq. 7 of FMA is stably achieved and ensure a good result.

Complexity Analysis. According to the above discussion, we know that the main time of the proposed algorithm is spent on the propagation process. We are supposed to ignore the little time on further mining with EQ when comparing to the propagation process. Suppose that the once updating operation of a node costs R, R is a constant. Then the cost of each iteration is $N * R$, and we assume the number of iterations is Q. Then the time complexity is $O(N * R * Q)$, R is a constant, so it equals to $O(N * Q)$, which is nearly linear time.

6   Experiments Analysis

To discuss the performance of FMA, we experiment the algorithm on both synthetic network and real social networks. Relatively to real network, synthetic networks provide an ground truth to evaluate the communities. In next subsections, we separately simulate on two types of network and analyze its performance.

6.1   Experiment on LFR Benchmark and Analysis

As a widely recognized measuring benchmark, LFR [14] is fully convinced to show the performance of algorithm in overlapping community detection. Make

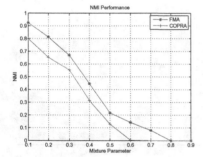

Fig. 2. Big network, $N = 2000$

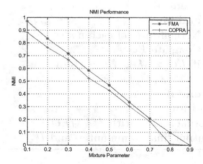

Fig. 3. Small network, $N = 50$

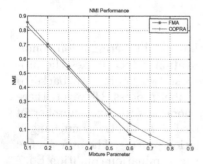

Fig. 4. Sparse network, $avek = 4$

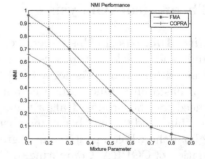

Fig. 5. Dense network, $avek = 40$

use of the ground-truth characteristic, researchers compute Normalized Mutual Information (NMI) to indicate the similarity between the our obtaining communities and the standard communities of LFR.

Experiments and Analysis on Big and Small Networks. In this experiment, we first get network through LFR benchmark with 2000 and 50 nodes. And changing the mixing parameter mu from 0.1 to 0.9, in which mixing parameter suggests the distribution of the ratio of external degree/total degree. As Figs. 2 and 3 show, NMI [15] of both networks gradually descend as the mixing parameter decaying, the reason is that a high mixing parameter means ambiguous community structure, which often drives obscure community characteristic. Meanwhile, both figures shows that NMI of FMA is always better than COPRA, whatever the mixing parameter or the network scale is. Comparing Figs. 2 and 3, we can achieve that FMA is more suitable to large scale datas than COPRA. And the subtraction of NMI among two algorithms is enlarge with the scale of network being larger.

Experiments and Analysis on Sparse and Dense Networks. As Fig. 2 shows, we compare the dense network with the sparse network by computing NMI. we change the density of network through setting the value of average degree. To highlight the distinction of density and sparsity, we separately set 40 and 4 to average-degree parameter k of 2000 nodes. In order to avoid confusion with K of above discussion, we change to use $avek$ to denote average degree of network. And comparing the NMI under different mu. On accounts of the *nodes-attraction* guiding propagation, FMA takes full consideration in the structural similarity, which greatly increases the precise of clustering. And Figs. 4 and 5 proves that NMI of dense network is better than sparse. Especially the Fig. 5 indicates that FMA has pretty great result when compares to COPRA with density networks.

Experiments to Show the Efficiency of EQ Guiding. The most obvious drawback of traditional label propagation for overlapping community detection is unstability, while FMA avoids random selection and unstable termination which eliminates the unstability of COPRA. Figures 6 and 7 show the membership

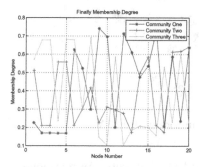

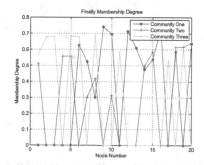

Fig. 6. U before EQ guides   **Fig. 7.** U after EQ guides

before and after setting threshold, it is a process of changing the fuzzy community detection to overlapping community detection and gets communities with better modularity. One node belongs to some communities when their membership-degree are bigger than 0, which clearly detects the overlapping communities. Taking an instance of Node 5 in Fig. 7, it attributes to community two and community three.

Table 1. Basic information of real social networks

Network	Node	Edge
Karate	34	78
Football	115	613
Email	1133	5451

6.2   Experiments on Real World Social Network

In this sections, we evaluate the ability of FMA in real-world social networks. And mainly experiment on Karate club data [16], American college football data and Emails of human interactions. Some basic information of datasets are revealed by Table 1. Karate club data is undirected weighted graph which includes 34 nodes and 78 edges. It was collected by Zachary in 1997. Football data is an undirected graph with 115 vertices. It recorded 115 college football team in regular match. Both karate and football are common datasets in social networks. Email is a dataset expressing human behaviors with 1133 nodes and average degree 3.932.

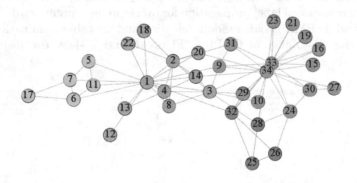

Fig. 8. Visualization of Karate communities with FMA

Our experiments are measured by computing EQ of all sets. As mentioned above, EQ is a criteria to evaluate the goodness of communities. Good communities often ranges in 0.4 to 0.7. Making comparison among datasets and show

Table 2. EQ of three data sets

EQ		
Network	FMA	COPRA
Karate	0.4236	0.3783
Football	0.4648	0.3261
Email	0.5372	0.4097

results as Table 2. Figure 8 is visualization of karate, k = 4, and Node 1, 3, 11, 28 are overlapping nodes. Due to the unstable feature of COPRA, all experiments with COPRA are running algorithm for 50 times.

7 Conclusion

FMA is proposed in this paper, it introduces the *nodes-attraction* to iteratively propagation. Then we further mine the communities with the guidance of EQ. Experimental results on synthetic and real-world networks demonstrate that our algorithm performs significantly, especially for dense and large networks. And our algorithm is effective, stable and nearly linear time cost. Meanwhile, there exist some problems, such as the way of fuzzy membership initialization and seeds selection. And these will be discussed in the future work.

Acknowledgments. This work was supported by the National Natural Science Foundation of China under grant 61370216.

References

1. Xie, J., Szymanski, B.K.: Community detection using a neighborhood strength driven label propagation algorithm. In: 2nd IEEE Network Science Workshop, pp. 188–195. IEEE Computer Society, West Point (2011)
2. Deng, D., Du, H., Jia, X., Ye, Q.: Minimum-cost information dissemination in social networks. In: Xu, K., Zhu, H. (eds.) WASA 2015. LNCS, vol. 9204, pp. 83–93. Springer, Heidelberg (2015)
3. Ding, F., Luo, Z., Shi, J., Fang, X.: Overlapping community detection by kernel-based fuzzy affinity propagation. In: 2nd International Workshop on Intelligent Systems and Applications, Wuhan, pp. 1–4 (2010)
4. Lancichinetti, A., Fortunato, S., Radicchi, F.: Benchmark graphs for testing community detection algorithms. Phys. Rev. E **78**, 561–570 (2008)
5. Shen, H., Cheng, X., Cai, K., Hu, M.B.: Detect overlapping and hierarchical community structure in networks. Phys. A Stat. Mech. Appl. **388**, 1706–1712 (2009)
6. Wang, X., Tang, L., Gao, H., Liu, H.: Discovering overlapping groups in social media. In: IEEE International Conference on Data Mining, pp. 569–578. IEEE Computer Society, Sydney (2010)
7. Palla, G., Dernyi, I., Farkas, I., Vicsek, T.: Uncovering the overlapping community structure of complex networks in nature and society. Nature **435**, 814–818 (2005)

8. Gregory, S.: Finding overlapping communities in networks by label propagation. New J. Phy. **12**, 2011–2024 (2009)
9. Bezdek, J.C.: Selected applications in classifier design. In: Bezdek, J.C. (ed.) Pattern Recognition with Fuzzy Objective Function Algorithms, vol. 22, pp. 203–239. Springer, New York (1981)
10. Nepusz, T., Petrczi, A., Ngyessy, L., Bazs, F.: Fuzzy communities and the concept of bridgeness in complex networks. Phys. Rev. E Stat. Nonlinear Soft Matter Phys. **77**, 119–136 (2008)
11. Newman, M.E.J.: The structure and function of complex networks. SIAM Rev. **45**, 167–256 (2003)
12. Nepusz, T., Petrczi, A., Ngyessy, L., et al.: Fuzzy communities and the concept of bridgeness in complex networks. Phys. Rev. E Stat. Nonlinear Soft Matter Phys. **77**, 119–136 (2008)
13. Gu, Y., Zhang, B., Zou, G., Huang, M.: Overlapping community detection in social network based on microblog user model. In: International Conference on Data Science and Advanced Analytics, pp. 333–339. IEEE, Shanghai (2014)
14. Lancichinetti, A., Fortunato, S., Kertsz, J.: Detecting the overlapping and hierarchical community structure of complex networks. New J. Phys. **11**, 19–44 (2008)
15. Lusseau, D.: The emergent properties of a dolphin social network. Proc. Roy. Soc. B Biol. Sci. **270**, 186–188 (2003)
16. Zachary, W.W.: An information flow model for conflict and fission in small groups1. J. Anthropol. Res. **33**, 452–473 (1977)

Time-Critical Viral Marketing Strategy with the Competition on Online Social Networks

Canh V. Pham[1,3(✉)], My T. Thai[2], Dung Ha[3],
Dung Q. Ngo[3], and Huan X. Hoang[1]

[1] University of Technology and Engineering, Vietnam National University,
Hanoi, Vietnam
{14025118,huanhx}@vnu.edu.vn
[2] Department of Computer and Information Science and Engineering,
University of Florida, Gainesville, USA
mythai@cise.ufl.edu
[3] Faculty of Technology and Information Security,
People's Security Academy, Hanoi, Vietnam
cvpham.vnu@gmail.com, dungha.hvan@gmail.com, quocdung.ngo@gmail.com

Abstract. According to the development of the Internet, using social networks has become an efficient way to marketing these days. The problem of Influence Maximization (IM) appeared in marketing diffusion is one of hot subjects. Nevertheless, there are no researches on propagating information whereas limits unwanted users. Moreover, recent researches shows that information spreading seems to dim after some steps. Hence, how to maximize the influence while limits opposite users after a number of steps? The problem has real applications because business companies always mutually compete and extremely potential desire to broad cart their product without the leakage to opponents.

To be motivated by the phenomenon, we proposed a problem called Influence Maximization while unwanted users limited (d-IML) during known propagation hops d. The problem would be proved to be NP-Complete and could not be approximated with the rate $1 - 1/e$ and its objective function was sub modular. Furthermore, we recommended an efficient algorithms to solve the problem. The experiments were handled via the real social networks datasets and the results showed that our algorithm generated better outcome than several other heuristic methods.

Keywords: Influence maximization · Viral marketing · Leakage information

1 Introduction

With the fast development and steady of the Online Social Networks (OSNs), such as Facebook, Twitter, Google+, etc., OSNs have become the most common vehicle for information propagation. OSNs provided a nice platform for information diffusion and fast information exchange among their users.

© Springer International Publishing Switzerland 2016
H.T. Nguyen and V. Snasel (Eds.): CSoNet 2016, LNCS 9795, pp. 111–122, 2016.
DOI: 10.1007/978-3-319-42345-6_10

The topic of Influence Maximization (IM) has received a lot of research interests in recent years. This problem is firstly proposed by Kempe et al. [5] in two diffusion models which are *Independent Cascade* (IC) model and *Linear Threshold* (LT) model, then rapidly becoming a hot topic in social network field. They also proved that influence maximization problem is NP-hard, and a natural greedy algorithm can obtain $1 - 1/e$. Although extensive related works have been conducted on the IM problem [1–3,12,13], most of them are based on such an assumption that without existence of the unwanted target users whom we do not want information come to. In reality, on OSNs exits the group of users who have opposite viewpoint and benefits with us and they create a negative impact to oppose for information received.

Considering the following example that highlights a basic need for every organization that uses OSNs. There are two mutual competitive companies A and B. The A has been deploying a large advertisement, even via the Internet. They drew a marketing blueprint on several social networks but the A tried to hide everything against every one of the B as long as its possible. Constantly, the advertising information of A can reach to the B after a time. Thus, the A needs a solution help them fast imply the marketing strategy to much many users except unwanted users (from B) to gain the best consumption more quickly than B within t hop.

Motivated by the above phenomenon, this paper proposes a new optimization problem called Maximizing Influence while unwanted target users limited (IML) that finds a set of seeding S to Maximize Influence such that influence to unwanted ones is under a certain threshold after the largest d hop propagations. The total influence is the total activated people. The unwanted ones are those whom we do not want the information come to.

Our contributions in this paper are summarized as follows:

- We first attempt to study the maximizing influence while unwanted target users limited after d hops (d-IML) under LT model and show that the objective function was *submodular*.
- We proved d-IML was NP-Complete and show it can not be approximated in polynomial time with a ratio $1 - 1/e$ unless $NP \subseteq DTIME(n^{O(\log n \log n)})$. We also designed an efficient algorithm for the problem of d-MIL.
- We conducted our experiments on real-world datasets, the results indicated that our algorithm gives better results than several other heuristic methods.

Related Work. The target is to spread the desired information for as many people as possible on OSNs. There are different related works on this topic [1–3,12,13]. Zhang et al. [1] proposed a problem to maximize the positive news in propaganda rather than maximizing the users affected. They said that to maximize positive things in many cases had more beneficial than maximizing the number of people affected. They used the Cascade Opinion (OC) model to solve the problem. On the other hands, Guo et al. [2] recommended to maximize the influence of information to a specific user by finding out the k most influential users and proved that it was NP-hard problem and the function is *submodular*. They also launched an effective approximation algorithm.

Zhuang et al. [3] have studied the *IM* problem in the dynamic social network model over time. In addition, there were several other studies: Chen et al. [12] investigated *IM* problem on a limited time; Gomez-Rodriguez et al. [13] studied *IM* problem for continuous time. Researches on IM with various contexts and various models received many attentions, but the diffusion of information problem, in addition to spreading the positive information still faced with the misinformation. How to spread the positive information while the misinformation limited? To solve it, Budak et al. [4] launched the problem selecting k users to convince them aware of good information so that after the campaign, amount of use influenced by the misinformation was the least. By using Model-Oblivious Independent Campaign Cascade, they proved the problem be NP-hard and the objective function was *submodular*. Nguyen et al. [6] gave the decontamination problem of misinformation by selecting a set of users with sources of misinformation I assumed to have existed on the social network at the rate of $\beta \in [0, 1]$ after T time. They launched the different circumstances of the I and the T, but they only solved the case I was unknown. On preventing infiltration to steal information on OSNs, Pham et al. [11] have built a Safe Community for the purpose of protecting all users in an Organization. In problems of detecting misinformation source on social networks, Nguyen et al. [8] assumed that the exist a set of misinformation sources I, they purposed of finding the largest number of users in I who started to propagate that information. Nevertheless, the predictions were likely confused because they did not know the order of real time start to spread misinformation. Zhang et al. [9] studied the problem of limited resources that often was incorrect information while maximized the positive source of misinformation on OSNs under Competitive Activation model. In this study, they were considered a model of misinformation and good information and presence on the social network, they also proved to be NP-complete problem and could not be approximated with rate $(1 - \frac{1}{e})$ unless $NP \subseteq DTIME(n^{O(\log \log n)})$.

In these researches, no one focused on the spread of information with the limiting of information to the set of ones who we did not want the information reach to (called *unwanted users*). While positive information is desired to propagate to more and more users, we also face with the existence of unlike users on OSNs. Because every time they receive the positive information, they can be able to conduct the activities, propagation strategies that opposes to our benefits.

2 Model and Problem Definition

2.1 Network and Information Diffusion Model

We are given a social network modeled as an undirected graph $G = (V, E, w)$ where the vertices in V represent users in the network, the edges in E represent social links between users and the weight represent frequency of interaction between two users. We use n and m to denote the number of vertices and edges. The set of neighbors of a vertex $v \in V$ is denote by $N(v)$. Existing diffusion models can be categorized into two main groups [5]: Threshold model and Independent Cascade model. In this work, we use the Linear Threshold (LT) model which is the one that has been extensively used in studying diffusion models among the generalizations of threshold models.

Linear Threshold (LT) Model. In this model, each node v has a threshold θ_v and for every $u \in N(v)$, edge (u, v) has a nonnegative weight $w(u, v)$ such that $\sum_{v \in N(u)} w(u, v) \leq 1$. Given the thresholds and an initial set of active nodes, the process unfolds deterministically in discrete steps. At hop t, an inactive node v becomes active if

$$\sum_{u \in N^a(v)} w(u, v) \geq \theta_v$$

where $N^a(v)$ denotes the set of active neighbors of v. Every activated node remains active, and the process terminates if no more activations are possible. Kempe et al. [5] prove that influence function $\delta(.)$ is *submodular* function.

2.2 Problem Definition

The paper we are interested in the value of the influence function after d hops. Considering that influence can be propagate at most d hops, we define the influence function $\delta_d(S)$ as total number of nodes have been active by S after d hops. We study maximizing influence while unwanted target users limited after d hops (d-IML) under LT model as follow:

Definition 1 (d-IML problem). *Given an social network represented by a directed graph $G = (V, E, w)$ and an under LT model. Let $T = \{t_1, t_2, .., t_p\}$ be the set of $|T| = p$ unwanted users and d is number of hops limited. Our goal chose the set seed user $S \subseteq V$ at most k-size that maximize $\delta_d(S)$ such that total influence user come to t_i less than threshold for prevent information leakage τ_i i.e.: $\sum_{u_i \in N^a(t_i)} t_i < \tau_i$.*

Lemma 1. *The influence function $\delta_d(.)$ is monotone and submodular for an arbitrary instance of the LT model, given any time $d \geq 1$.*

Proof. LT model is a special case of LT-M model [12] with all parameters $m(u, v) = 1$ and deadline $\tau = d$ and influence function $\delta_d(.)$ is influence function $\delta_\tau(.)$. Due to $\delta_\tau(.)$ is monotone and submodular in LT-M model, thus $\delta_d(.)$ is monotone and submodular in LT model. $\qquad\square$

3 Complexity

In this section, we first show the NP-Completeness of IML problem on LT model by reducing it from Maximum Coverage problem. By this result, we further prove the inapproximability of d-MIL which is NP-hard to be approximated within a ratio of $1 - 1/e$ unless $NP \subseteq DTIME(n^{O(\log \log n)})$.

Theorem 1. *d-IML is NP-Complete in LT model.*

Proof. We consider of the decision version of d-MIL problem that asks whether the graph $G = (V, E, w)$ contains a set $k - size$ of seed user $S \subset V$ that number active node at least K, such that $\sum_{u \in N^a(t_i)} w(u, t_i) < \tau_i$ within at most d rounds.

Given $S \subset V$, we can calculate the influence spread of S in polynomial time under LT model. This implies d-MIL is NP. Now we prove a restricted class of d-MIL instance is NP-hard, $d = 1$.

To prove that 1-MIL is NP-hard, we reduce it from the decision version of Maximum Coverage problem defined as follows.

Maximum Coverage. Given a positive integer k, a set of m element $\mathcal{U} = \{e_1, e_2, \ldots, e_m\}$ and a collection of set $\mathcal{S} = \{S_1, S_2, \ldots, S_n\}$. The sets may have common elements. The *Maximum Coverage* problem asks to find a subset $S' \subset \mathcal{S}$, such that $| \cup_{S_i \in S'} S_i |$ is maximized with $|S'| \leq k$. The decision of this problem asks whatever the input instance contains a subset S' of size k which can cover at least t elements where t is a positive integer.

Reduction. Given an instance $I = \{\mathcal{U}, \mathcal{S}, k, t\}$ of the maximum coverage, we construct an instance $G = (V, E, w, \theta)$ of 1-IML problem as follows:

- *The set of vertices:* add one vertex u_i for each subset $S_i \in \mathcal{S}$, once vertex v_j for each $e_j \in \mathcal{U}$, and a vertex x is a unwanted users.
- *The set of edges:* add an edge (v_i, u_j) for each $e_j \in S_i$ and connect x to each vertex v_j.
- *Thresholds and weights:* assign all vertices the same threshold $\theta = \dfrac{1}{m}$ and each edges (u_i, v_j) has weight $w_{u_i v_j} = \dfrac{1}{m}$. In addition, for all edges (v_j, x), we assign their weight $\dfrac{1}{m}$.
- *Threshold for prevent leakage information:* we assign threshold for prevent leakage information for vertex x is $\tau_x = \dfrac{1}{m}$.

The reduction is illustrated in Fig. 1. Finally, set $d = 1, K = t$.

Suppose that S^* is a solution to the maximum coverage instance, thus $|S^*| \leq k$ and it can cover at least t elements in $\mathcal{U}$. By our construction, we select all nodes u_i corresponding to subset $S_i \in S^*$ as seeding set S. Thus, $|S| = k$. Since S^* cover at least t elements e_j in $\mathcal{U}$ so S influence at least t vertices v_j corresponding to those e_j. Additionally, for each v_j total influence incoming based on LT model at least $w_{u_i v_j} = \theta_{u_i v_j} = w_{u_i v_j}$. Hence, there are at least $t = K$ nodes in the 1-IML has been active.

Conversely, suppose there is seeding $S, |S| = k$ such that the number of active node at least K. We see that $v_j \notin S, j = 1, 2, .., m$ because total influence incoming x at least $w_{v_j x} = \tau_x = \dfrac{1}{m}$. Thus $S \subseteq \{u_1, u_2, \ldots, u_n\}$. Then S^* can be collection of subset S_i corresponding to those $u_i \in S$. Hence the number of elements which it can cover is at least $K = t$.   □

Based on above reduction, we further show that inapproximation of IML in the following theorem.

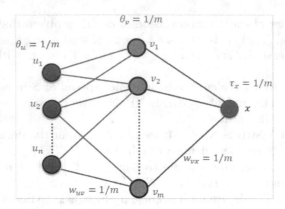

Fig. 1. Reduction from MC to 1-IML

Theorem 2. *The d-IML problem can not be approximated in polynomial time within a ratio of $\dfrac{e}{1-e}$ unless $NP \subseteq DTIME(n^{O(\log \log n)})$.*

Proof. Supposed that there exits a $\dfrac{e}{1-e}$-approximation algorithm $\mathcal{A}$ for d-MIL problem. We use the above reduction in proof of Theorem 1 then $\mathcal{A}$ can return the number of active nodes in G with seeding size equal to k. By our constructed instance in Theorem 1, we obtain the Maximum Coverage with size t if the number of active nodes in optimal solution given by $\mathcal{A}$ is K. Thus algorithm $\mathcal{A}$ can be applied to solve the Maximum Coverage problem in polynomial time. This contradict to the NP-hardness of Maximum Coverage problem in [10].

4   Methodology

4.1   ILP Formulation

One advantage of our discrete diffusion model over probabilities is that the exact solution can found be using mathematical programming. Thus, we formulate d-IML problem as an $0 - 1$ Integer Linear Programming (ILP) problem below.

The objective function (1) of the ILP is to find the number of node is active. The constraint (2) is number of set seed is bounded by k; the constraints (3) capture the propagation model; the constraint (4) limit leakage information income to unwanted user by threshold τ_i; and the constraint is simply to keep vertices active once they are activated. The number of variables and constraints of ILP are nd.

$$\text{maximize} \sum_{v \in V \setminus T} x_v^d \tag{1}$$

$$\text{st:} \sum_{v \in V \setminus T} x_v^0 \leq k \tag{2}$$

$$\sum_{v \in N(u)} x_u^{i-1}.w(u,v) + \theta_v.x_v^{i-1} \geq \theta_v.x_v^i,$$

$$\forall v \in V, i = 1..d \tag{3}$$

$$\sum_{v \in N(t_i)} x_v^d.w(v,t_i) < \tau_i \tag{4}$$

$$x_v^i \geq x_w^{i-1}, \forall v \in V, i = 1..d \tag{5}$$

$$\text{where} \quad x_v^i = \begin{cases} 1 & \text{if } x \text{ is active at round (hop) } i \\ 0 & \text{otherwise} \end{cases}$$

4.2 Meta Heuristic Algorithm

A commonly heuristic used for *IM* problem in the simple case is greedy algorithm, the idea of which choses the node is maximize *influence marginal gain* in each step:

$$\delta_d(S,v) = \delta_d(S + \{v\}) - \delta_d(S) \tag{6}$$

Although the objective function is *submodular*, propagation of influence is constraint by the leak. Hence we can not give an algorithm for approximately with the ratio $1 - 1/e$ as Kemp et al. [5]. To avoid this issue, we designed the algorithm combine the influence marginal gain and evaluation information leakage based on idea of IGC [7]. Accordingly, we used a heuristic function $f(v)$ to evaluate the fitness of user v which defined as follows:

$$f(v) = \frac{\delta_d(S,v)}{1 + \frac{1}{p}\sum_{t_i \in T} l_{t_i}(v)} \tag{7}$$

Where $L_{t_i}(S)$ is the total influence to t_i respect to seeding sets S after d hop *i.e.*, $L_{t_i}(S) = \sum_{v \in N^a(t_i)} w(v,t_i), l_{t_i}(v) = \frac{L_{t_i}(S+\{v\})}{\tau_i}$ is the normalized leakage level at t_i after adding v to seed set S. The numerator of $f(v)$ is selected to be influence marginal gain $\delta_d(S,v)$ so that the algorithm will favor users have maximizing influence, denominator of $f(v)$ will favor users with lower information leakage.

The Meta-heuristic (MH) algorithm as shown in Algorithm 1. In each iteration, firstly, we update the set of candidate users C, those whose addition to seeding set S still guarantees that the information leakage to each unwanted t_i does not exceed the threshold τ_i. The algorithms also adds one user v of candidate set C into S which has $f(v)$ is maximize until size of S no exceed k.

5 Experiment

In this section, we do a lot of experiment on three real-world datasets, and compare our algorithm with algorithms: Random method, Max degree, Greedy and ILP method.

Algorithm 1. Meta-heuristic Algorithm

Data: $G = (V, E, w, \theta, \tau)$, set users $U = \{t_1, t_2, \ldots, t_p\}$, p, k;
Result: Seeding S;
1 $S \leftarrow \emptyset$;
2 $C \leftarrow V$;
3 **for** $i = 1$ *to* k **do**
4 $\quad$ **foreach** $v \in C$ **do**
5 $\quad\quad$ **if** $\exists j : L_{t_j}(S + \{v\}) \geq \tau_j$ **then**
6 $\quad\quad\quad | \quad C \leftarrow C - \{v\}$;
7 $\quad\quad$ **end**
8 $\quad$ **end**
9 $\quad$ **if** $C = \emptyset$ **then**
10 $\quad\quad |$ Return S;
11 $\quad$ **end**
12 $\quad$ Find $v \in C$ that maximizes $f(v)$;
13 $\quad$ $S \leftarrow S + \{v\}$;
14 $\quad$ $C \leftarrow C - \{v\}$;
15 **end**
16 Return S;

5.1 Datasets

BlogCatalog. BlogCatalog is a social blog directory website. This contains the friendship network crawled and group memberships. Nodes represent users and edges are the friendship network among the users. Since the network is symmetric, each edge is represented only once [14].

ArXiv-Collaboration. The data covers papers in the period from January 1993 to April 2003 (124 months). It begins within a few months of the inception of the arXiv, and thus represents essentially the complete history of its GR-QC section. If an author i co-authored a paper with author j, the graph contains a undirected edge from i to j. If the paper is co-authored by k authors this generates a completely connected (sub)graph on k nodes [15].

Gnutella. A sequence of snapshots of the Gnutella peer-to-peer file sharing network from August 2002. There are total of 9 snapshots of Gnutella network collected in August 2002. Nodes represent hosts in the Gnutella network topology and edges represent connections between the Gnutella hosts [15].

Table 1. Basic information of Network Datasets

Network	Nodes	Edges	Type	Avg. Degree
BlogCatalog	10,312	333,983	Undirect	32.39
ArXiv-Collaboration	5,242	28,980	Direct	5.53
Gnutella	6,301	20,777	Direct	3.30

In each graph, we used the method in [5] to assign the diffusion weight to each edge and then normalize the weights of all incoming edges of a node v to let it satisfy that $\sum_{u \in N^{in}(v)} w(u, v) \leq 1$ (Table 1).

5.2   Comparision Algorithm

In this part, we describe tow comparison algorithms: random and Max degree.

1. *Random:* This is a general method used for most problem. In our problem, we chose the seeding node randomly when the information leaked to unwanted users less than the threshold leakage.
2. *Max Degree method:* The greedy algorithm chose the vertex v that had maximum degree when the information leaked to unwanted users less than the threshold leakage.
3. *Greedy algorithm (GA) method:* The method based on the idea that chose the node maximize information diffusion gain when information leaked to unwanted users less than the threshold leakage.
4. *Meta heuristic (MH) algorithm:* Here the algorithm in Sect. 4 collectively called our algorithms.
5. *ILP method:* Solve the ILP problem to compare with optimal seeding.

We solved the ILP problem on Gnutella network [15], with $d = 4$, The ILP was solve with CPLEX version 12.6 on Intel Xeon 3.6 Ghz, 16G memories and setting time limit for the solver to be 48 h. For $k = 5, 10, 15$ and 20 the solver return the optimal solution. However, for $k = 25, 30, 35, 40, 45$ and 50, the solver can not find the optimal solution within time limit and return sub-optimal solutions.

5.3   Experiment Results

Solution Quality. The number of active users changed when the number of steps d changed and fixed $k = 50$ shown in Fig. 3. The algorithm really resulted better than Max Degree and GA. The bigger k was, the better the result was than Max Degree. For example with the social network BlogCatalog, the MH generated more active users with 1.71 times ($k = 50$, $d = 4$). According to GA, when k was small, MH and GA issued the same outcome. When k was larger, the gap between MH and GA became clearer. MH produced better than GA 7.3 % with $k = 50$, using Guntella network.

Number of Activated Users. We compared the performance of MH with the others when k changed and $d = 4$. The number of active users was detailed via Figs. 1, 2 and 3. Definitely, MH generated better than Max. Even MH worked better than Maxdegree 1.7 times when $k = 10$ via the network BlogCatalog. GA generated approximately to MH when k was small ($k = 5$, 10). When k was larger, MH worked better than GA. In case of maximum of k ($k = 50$) Gnutella activated users using MH more 56 people than GA whereas in BlogCatalog, activated users of MH and GA were 56 and 44, respectively.

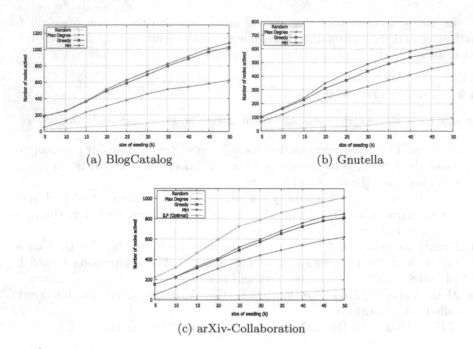

(a) BlogCatalog   (b) Gnutella

(c) arXiv-Collaboration

Fig. 2. The actived nodes when the size of seeding set varies $(d = 4)$.

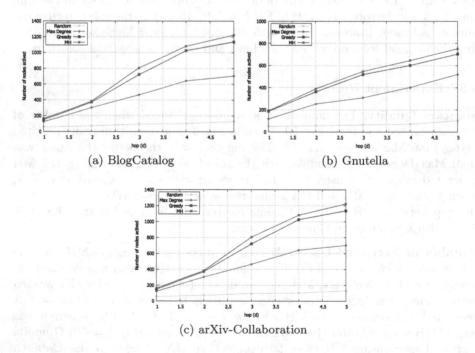

(a) BlogCatalog   (b) Gnutella

(c) arXiv-Collaboration

Fig. 3. The actived nodes when number of propagation varies $(k = 50)$.

On the whole, the estimation of the function $f(.)$ resulted better than the one of the maximization of Influence gain. Nevertheless, when k was small these ways are the same. The comparison with ILP in Arxiv Collaboration, the proportion of the solution of MH at least marked at 68 % when $k = 20$ with activate users was less than 185. The rating was smallest when $k = 50$, pointed at 80 % of sub-optimal. Note that ILP did not generate optimized solution in this case.

Number of Hops. When d was small, MH and GA issued the same outcome. When d was large, MH resulted moderately better than MD and quite better than GA. It can be seen via BlogCatalog, when $d = 5$, the largest distance between GA and MH was 86 nodes. It proved that the larger d was, the better estimation the function $f()$ had to choose the optimization values.

6 Conclusions

In order to propose a viral marketing solution while exists the competition between organizations that have benefit collisions, we built the problem of maximization influence to users whereas limits the information reach to unwanted ones in constrained time. We proved it be an NP-complete and not be approximated with $1 - 1/e$ rating number. We also recommended an efficient solution MH to solve the problem. The experiment via social networks data showed that our algorithm got the better result of object function than some ones and got desired rate in optimized solution.

References

1. Zhang, H., Dinh, T.N., Thai, M.T.: Maximizing the spread of positive influence in online social networks. In: Proceedings of the IEEE International Conference on Distributed Computing Systems (ICDCS) (2013)
2. Guo, J., Zhang, P., Zhou, C., Cao, Y., Guo, L.: Personalized influence maximization on social networks. In: Proceedings of the 22nd ACM International Conference on Conference on Information and Knowledge Management (2011)
3. Zhuang, H., Sun, Y., Tang, J., Zhang, J., Sun, X.: Influence maximization in dynamic social networks. In: Proceedings of IEEE International Conference on Data Mining (ICDM) (2013)
4. Budak, C., Agrawal, D., El Abbadi, A.: Limiting the spread of misinformation in social networks. In: Proceedings of the 20th International Conference on World Wide Web (WWW 2011), pp. 665–674. ACM, New York, NY, USA (2011)
5. Kempe, D., Kleinberg, J., Tardos, E.: Maximizing the spread of influence through a social network. In: Ninth ACM SIGKDD International Conference on Knowledge Discovery and Data Mining, KDD 2003, New York, NY, USA, pp. 137–146 (2003)
6. Nguyen, N.P., Yan, G., Thai, M.T., Eidenbenz, S.: Containment of misinformation spread in online social networks. In: Proceedings of ACM Web Science (WebSci) (2012)
7. Dinh, T.N., Shen, Y., Thai, M.T.: The walls have ears: optimize sharing for visibility and privacy in online social networks. In: Proceedings of ACM International Conference on Information and Knowledge Management (CIKM) (2012)

8. Nguyen, D.T., Nguyen, N.P., Thai, M.T.: Sources of misinformation in online social networks: who to suspect? In: Proceedings of the IEEE Military Communications Conference (MILCOM) (2012)
9. Zhang, H., Li, X., Thai, M.: Limiting the spread of misinformation while effectively raising awareness in social networks. In: Proceedings of the 4th International Conference on Computational Social Networks (CSoNet) (2015)
10. Feige, U.: A threshold of ln n for approximating set cover. J. ACM (JACM) **45**(4), 634–652 (1998)
11. Pham, C.V., Hoang, H.X., Vu, M.M.: Preventing and detecting infiltration on online social networks. In: Thai, M.T., Nguyen, N.T., Shen, H. (eds.) CSoNet 2015. LNCS, vol. 9197, pp. 60–73. Springer, Heidelberg (2015)
12. Chen, W., Wei, L., Zhang, N.: Time-critical influence maximization in social networks with time-delayed diffusion process. http://arxiv.org/abs/1204.3074
13. Gomez-Rodriguez, M., Song, L., Nan, D., Zha, H., Scholkopf, B.: Influence estimation and maximization in continuous-time diffusion networks. ACM Trans. Inf. Syst. **34**, 2 (2016). doi:10.1145/2824253
14. Tang, L., Liu, H.: Relational learning via latent social dimensions. In: Proceedings of the 15th ACM SIGKDD Conference on Knowledge Discovery and Data Mining (KDD 2009), pp. 817–826 (2009)
15. Leskovec, J., Kleinberg, J., Faloutsos, C.: Graph evolution: densification and shrinking diameters. ACM Trans. Knowl. Disc. Data (ACM TKDD) **1**(1), 2 (2007)

Analysis of Viral Advertisement Re-Posting Activity in Social Media

Alexander Semenov[1], Alexander Nikolaev[2], Alexander Veremyev[3],
Vladimir Boginski[3,4(✉)], and Eduardo L Pasiliao[5]

[1] University of Jyvaskyla, P.O. Box 35 40014 Jyvaskyla, Finland
`alexander.v.semenov@jyu.fi`
[2] University at Buffalo, 312 Bell Hall, Buffalo, NY 14260, USA
`anikolae@buffalo.edu`
[3] University of Florida, 1350 N Poquito Road, Shalimar, FL 32579, USA
`{averemyev,vb}@ufl.edu`
[4] University of Central Florida, 12800 Pegasus Dr., Orlando, FL 32816, USA
`vladimir.boginski@ucf.edu`
[5] Munitions Directorate, Air Force Research Laboratory, Eglin AFB, Valparaiso, FL 32542, USA
`pasiliao@eglin.af.mil`

Abstract. More and more businesses use social media to advertise their services. Such businesses typically maintain online social network accounts and regularly update their pages with advertisement messages describing new products and promotions. One recent trend in such businesses' activity is to offer incentives to individual users for re-posting the advertisement messages to their own profiles, thus making it visible to more and more users. A common type of an incentive puts all the re-posting users into a random draw for a valuable gift. Understanding the dynamics of user engagement into the re-posting activity can shed light on social influence mechanisms and help determine the optimal incentive value to achieve a large viral cascade of advertisement. We have collected approximately 1800 advertisement messages from social media site VK.com and all the subsequent reposts of those messages, together with all the immediate friends of the reposting users. In addition to that, approximately 150000 non-advertisement messages with their reposts were collected, amounting to approximately 6.5 M of reposts in total. This paper presents the results of the analysis based on these data. We then discuss the problem of maximizing a repost cascade size under a given budget.

Keywords: Social media · Reposts · Information cascades · Viral advertisements · Influence maximization

1 Introduction

Social media sites host large amounts of data about users' social connections, preferences and decisions. The largest one, Facebook.com, has more than 1.3 B registered users, Twitter has 316 M users active in any given month, and VK.com has over 300 M registered users, with over 70 M visiting the site on a daily basis (https://vk.com/about). There

© Springer International Publishing Switzerland 2016
H.T. Nguyen and V. Snasel (Eds.): CSoNet 2016, LNCS 9795, pp. 123–134, 2016.
DOI: 10.1007/978-3-319-42345-6_11

are hundreds of other social media sites, created for different purposes. In [1] social networking sites are defined as "web-based services that allow individuals to (1) construct a public or semi-public profile within a bounded system, (2) articulate a list of other users with whom they share a connection, and (3) view and traverse their list of connections and those made by others within the system". The latter connections may be called "friends", "followers", etc. A majority of social media sites offers to its users the ability to share posts (or links to 3rd party site) with other users. Such exchanges are typically done through one's public front page, often referred to as "wall".

Importantly, the shared content can be shared further, reaching larger and larger audience in a cascading manner, and this phenomenon can be strategically exploited. The set of social media site users includes businesses and public figures; usually, the sites provide special interfaces for such entities, where they create and manage "public pages". However, at some sites (e.g. Twitter), the businesses are provided with exactly the same type of the page as regular users. Nowadays, many businesses maintain their presence in the social media, aiming at reaching more customers through it. Typically, businesses regularly update their social media pages with descriptions of new products and services. Social media sites offer different possibilities of promotion of these pages; one of the most widely offered types is purchasing of impressions of the post to other users in a special part of social media site page intended for ads. Part of the impressions would attract users to go to the public page of the business being advertised. Then, some users may share the posts they saw on business public page to their own profile, so that their friends would see them, too. In some cases that may lead to viral message spread, and long reposting cascades.

One of the most recent trends in social media advertising is to offer incentives to the users, who would repost the message to their own profile, thus making it visible to their peers. Often, the incentive is the possibility to participate in a random draw for some valuable gift. Sometimes, a guaranteed incentive can be offered instead – such payments typically target special users with a large follower base, – however, this paper focuses on the analysis of the messages that explicitly mention the gift that is up for the winning to anyone.

The reported analyses are conducted with the data of the social media site VK.com. It is one of the most popular social networking resources in Russia and post-USSR countries. Its former name is "Vkontakte" (translated as "in touch"), and it is known as a "Russian Facebook". VK.com is the 2nd in Alexa ranking for Russia, and 21st in the global Alexa ranking (http://www.alexa.com/topsites/countries/RU).

Each user of VK.com has a profile, searchable by a numeric identifier. The profile front page can include the user's personal information: the first and last names, profile picture, gender, education details, etc. Each profile page has a "wall" – a part of the page that houses the user's posts. Users can place posts to each other's walls, making these posts visible to any visitor of the respective wall. In addition to that, VK.com allows its users to "repost" the posts they see; when a user reposts a post, it appears on their wall with the reference to both the author of the original post and the reposting user.

Users of VK.com may add other VK.com users to their friend lists. Moreover, users may create community pages, which can include information about the community and the community wall. There are two types of communities: groups, and public pages.

VK.com users may subscribe to community pages. Each logged-in user of VK.com may see their own news feed, containing combination of recent updates on the walls of all her friends, with all the updates on the walls of communities the user is subscribed to. A user may comment on and/or repost the posts seen in the feed. When the user reposts a post, their friends see this repost in their feeds; importantly, in the case when an already reposted message is reposted again, VK.com allows its users to see the intermediaries, i.e., enabling the tracking of the chain of reposts all the way up to the originator. This feature makes VK.com reposts valuable for different analyses. For example, Twitter.com does not store or reveal the intermediaries, and hence, if someone retweets an already retweeted message, the information of all the re-tweeters except for the last one is lost: Twitter would show that this re-tweeter re-tweeted directly the original tweet.

Summarizing the terms: wall is a part of the profile page, which stores the posts of users or community public; post is the public message sent by VK.com user or community to her own wall (it is equivalent to tweet on Twitter.com), repost is a post reposted by some user from the wall of another user/community to her own wall (it is equivalent to retweet on Twitter.com); xth level repost is the repost is the repost of the message, which is a repost of $(x - 1)$th level itself. For example, a 1st level repost is a repost of the original message, while a 2nd level repost, is a repost of the 1st level repost.

The first goal of the present research is to perform descriptive analyses of the VK.com repost data, perform the analysis of the user base responses to the offered incentives for reposting their advertisement posts (analyze number of reposts, depth of cascades, and cost of the incentives), and formulate and solve the problem of maximizing a repost cascade size under a given budget. Next, we look to design a model capturing the cascade evolution and find a strategy to exploit cascades.

2 Related Work

In recent years, the increased availability of online social interaction data provided opportunities to analyze information cascades in various domains. Although there is an extensive body of work investigating such information diffusion processes (see the review of news sharing literature [2] and one recent survey [3]), we briefly mention some of the relevant notable studies analyzing the structural properties of information cascades in social networks such as Facebook [4–8], Twitter [9, 10], LinkedIn [11], viral marketing [12] and others [13, 14]. Paper [8] provides a large-scale and extensive study (anatomy) of the photo resharing cascades in Facebook. In [12], the person-to-person recommendation network and the dynamics of viral marketing cascades have been investigated. Paper [14] describes the diffusion patterns (mainly, tree depth and tree size) arising from seven online domains, ranging from communications platforms to networked games to microblogging services, each involving distinct types of content and modes of sharing. Differences in the mechanics of information diffusion across topics: idioms, political hashtags, and complex contagion on Twitter are described in paper [9]. Although, the most widely reported statistics of diffusion cascades are size and depth (tree size and tree depth), other interesting characteristics of cascades have also been analyzed. For example, in [10], the Weiner index (average distance) is proposed as a structural characteristic of cascades and analyzed on Twitter data including propagation of news stories, videos, images, and petitions. In [7], the authors further investigate Weiner index and other cascade characteristics to address cascade prediction problems on Facebook photo sharing data. The invisible or exposed audience has also been considered, for example, in [5, 6]. For more details on the related work we refer the reader to the aforementioned studies and the references therein.

3   Data Collection

VK.com provides application programming interface (API), through which external clients can query details about users, their friends, public messages, community details and messages, and so on. Queries to VK.com API can be made through HTTP protocol, and VK.com returns the responses in JavaScript Object Notation (JSON) format. Objective of data collection was to gather messages from community pages, and collect the reposts of these messages, along with friends of users who reposted the messages. We picked 8 popular VK.com communities: (news, music, sports, popular press, etc.), and added 3 communities, which concentrate on posting advertisement messages. These communities formed our seed set. Then, we collected all the messages, posted by these communities; after that, we collected all the reposts of these messages up to the 14th level, and collected the identifiers of community members, and the friends of those who reposted messages. After the collection, about 25 GB (in text) of reposts were collected, about 4 GB (in text) of friends and followers; there were 194,630 wall posts (from 11 groups) with about 7.2 M of reposts. Out of these posts, approximately 1800 contained offers of some gifts for the reposts. Overall, 169,535 unique posts produced 6,585,496 of first-level reposts; 64,4850 posts produced 639,897s-level reposts; 13,862 posts produced 46,998 third-level reposts; 2,693 posts produced 6,053 fourth-level reposts; 614 posts produced 1231 fifth-level reposts. In total, these reposts were made by approximately 1,7 M different users. Friends and followers for those 1,7 M of people were collected, that led to 296,049,591 friends (not unique). We used distributed data collection software [15], using 11 servers: one server in Finland (having 16 cores, 10 TB Disk space, and 192 GB RAM), and ten servers spread over Europe (mainly, located in Frankfurt and Amsterdam). Collection of the data took about 10 days.

4   Findings

Table 1 shows the main topic discussed in each group for all 11 groups, as well as the number of messages gathered from each groups, the number of reposts of these messages (taken from VK.com counters), and the average number of reposts per message for the group. The three groups labeled "free gifts" turn out to have the highest "reposts/posts" ratio.

Figure 1 shows a log-log plot of the probability mass function (PMF) of the repost counts (per post): it reveals a power law, with very few post reposted more than 10000 times, and many boasting between zero and ten reposts. Table 2 shows the types of top ten most reposted messages. The first eight of these posts were originated by the advertisement-oriented communities and had gifts offered for reposting them.

Table 1. "Reposts per Post" relation

Group topic	Posts	Reposts	Avg. # reposts
IT related news	62,053	1,892,074	30
News (RT)	34,392	1,679,802	48
News	37,176	1,705,571	45
Music radio	32,029	2,018,262	62
Sports (Olympic games)	7,772	1,022,205	131
Black and white photos	6,832	171,880	29
Popular magazine	4,607	376,966	81
Musicians	2,722	22,837	11
Free gifts	846	231,942	274
Free gifts	755	178,677	236
Free gifts	182	21,526	118

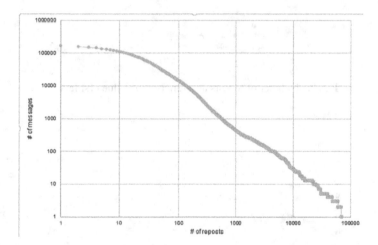

Fig. 1. Log-Log plot of the PMF of the repost counts

Table 2. Top ten most reposted posts

# of reposts	text
59,329	Apple iPhone 6 as a gift
56,823	Apple iPhone 6 as a gift
40,768	MacBook Air as a gift
40,376	iPad mini as a gift
39,106	iPad mini as a gift
29,249	Apple iPhone 6 as a gift
28,667	Apple iPhone 6 as a gift
28,222	10 kg of chocolate
27,347	PS 4 as a gift
24,688	Russian skiers got Gold medal at Olympics

We have read all the advertising posts and labeled them with the estimates of the worth/value of each gift in US dollars (USD). The most expensive gift was a MacBook Air. The average value over all the gifts was 33 USD. In total, there were about 1800 of such advertisement messages. Figure 2 shows the number of reposts at level one, as a function of gift value. Figure 3 shows the impact of the gift value on the 2nd level repost volume.

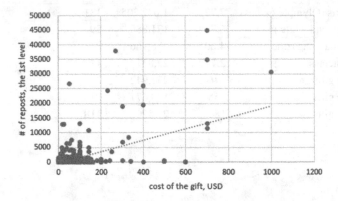

Fig. 2. Number of reposts as a function of value (1st level)

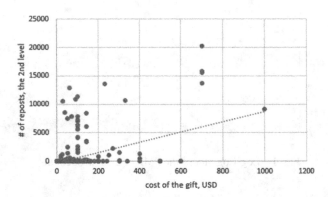

Fig. 3. Number of reposts as a function of value (2nd level)

Observe that for the 1st level reposts, the number of reposts grows with the value of an offered gift; however, on the 2nd level, many messages have a rather large number of the reposts.

Figure 4 shows the dependence of the number of 2nd level reposts on the number of the 1st level. It can be observed that many messages having a large number of the reposts had only few reposts on the 2nd level, and surprisingly, vice versa.

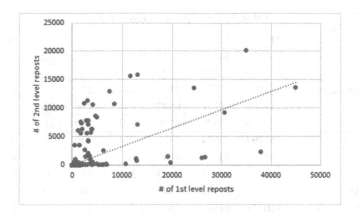

Fig. 4. Dependency of the # of 2nd level reposts on the 1st

Figure 5 reveals the number of reposts that can be gained on average per 1 USD as a function of the gift value. The messages, offering the most expensive gifts, can "buy" about 50 reposts for 1 USD. The highest number of reposts per USD in our data had the message with 700 reposts per 1 USD, that was the message offering 10 kg of chocolate. Table 3 shows number of the reposts on different levels of the cascade. Columns represent level of the cascade (depth of repost tree, from 1 to 8), rows represent types of the groups.

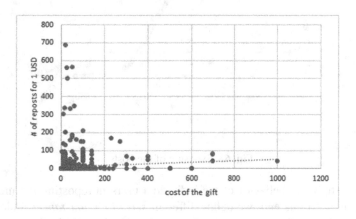

Fig. 5. Number of reposts for 1 USD as a function of value

It can be observed that the group concentrating on posting news (labeled "News") has 22 reposts on the 8th level, the second largest number of reposts on this level belongs to the popular magazine (8); meanwhile, the number of reposts for groups with free gifts is smaller.

Table 3. # of reposts on levels 1–8

	#1	#2	#3	#4	#5	#6	#7	#8
Free gifts	203436	54158	2775	237	30	10	5	2
Free gifts	460869	221745	8533	504	53	15	7	5
Sports (Olympic games)	861976	39704	3144	400	74	13	10	0
Radio	1536657	88319	6942	736	136	25	9	3
News (RT)	738606	48447	6146	962	150	31	13	0
IT related news	551794	23024	2439	424	109	27	8	2
Musicians	19119	3313	464	50	6	5	3	2
Free gifts	309978	42331	2393	225	41	16	7	2
Popular magazine	205950	11847	1641	343	102	50	18	8
Black and white photos	160441	9198	741	113	25	9	1	1
News	1536670	97809	11781	2055	500	179	68	22

Figure 6 shows exposure of the posts: total number of users who could have seen the post in their feed, not including those who are members of the communities which submitted the post. Minimum value of exposure equals to 1, maximal exposure is 27393223.

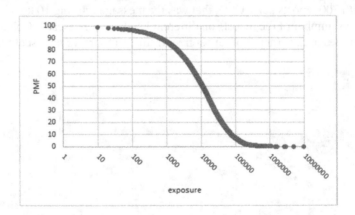

Fig. 6. Exposure of the posts

Next, we study the behavior of the individual users in reposting the messages of different nature. To this end, we define different *sets of circumstances* under which a user may get an opportunity to repost a message from their wall (the front page of their account): each set of circumstances is defined by how exactly (through whom) a message appears on the user's wall. Set1 includes the situations where a user – referred to as *ego* – receives a message (as a repost) from another user, typically a friend or a person the ego follows – referred to as *parent*. Set 2 includes the situations where an ego user receives a message directly from a group that creates it; such is the case when the ego is subscribed to the message-originating group. Set 3 includes the situations where an ego receives a message (as a repost) through a non-message-originating group. Note that the groups involved in the situations in Set 2 can be viewed as authorities, while the groups involved in the situations in Set 3 can be viewed as hubs: the former groups create original content in their main topic of choice, while the latter ones serve to expose their users to the original content from multiple authorities, effectively filtering the diverse content to better appeal to the subscriber base.

Our objective is to compare the behavior of the users under the circumstances falling in Sets 1, 2 and 3. More specifically, we look to distill the drivers of the *repost probability* for two kinds of messages – paid ad messages, originated by "Free gift" groups, and non-incentivized topical messages that news, sports, or other topical groups would originate. To this end, we collect and organize the information about all the reposting opportunities given to VK.com users for three topical groups and three "Free gift" groups from the beginning of Year 2015 until 01 September 2015. Using these reposting opportunities as data points, we partition them into the three sets of circumstances described above. Then, within each set, we fit a logistic regression model with the repost indicators as dependent variable and the ego user and circumstance characteristics as the predictors; here's a list of the predictors used:

- EGO_SEX – sex of the ego (i.e., the user who has an opportunity to repost a message from their wall), with category 1 for females and 2 for males,
- PARENT_FF – the decimal logarithm of the sum of the number of friends and the number of followers of the parent (i.e., the user via whom a message reaches the ego's wall), rounded down,
- PARENT_SEX – sex of the parent, with category 1 for females and 2 for males,
- VALUE – the value of a gift offered to be won in a draw that the ego would enter if they reposted the message in question (only for messages originated by "free gift" groups),
- EGO_REP – the decimal logarithm of the total number of the reposts the ego did over the period of the study, rounded down,
- REP_SEEN – the number of the reposts collected by the message in question at the parent level at the time the message appeared on the ego's wall.

In Table 4 we provide a representative summary of one such model run, the one for a "Free gift" group, circumstance:

Table 4. Representative summary of a model run

Predictor	Estimate	Std.	Pr(> |z|)
(Intercept)	−2.559	0.0807	< 2e − 16 ***
EGO_SEX2	−0.375	0.0323	< 2e − 16 ***
PARENT_SEX	0.294	0.0366	1.05e − 15 ***
PARENT_FF	−1.212	0.0215	< 2e − 16 ***
VALUE	0.001	0.0001	< 2e − 16 ***
REP_SEEN	1.663	0.0323	< 2e − 16 ***
EGO_REP	0.445	0.018	< 2e − 16 ***

The above results can be interpreted as follows. The baseline log-odds ratio is −2.559 for reposting on an opportunity presented to a female ego by a message that came to her wall through a "free gift" group a female peer. If the ego was a male, that would reduce the log-odds of reposting by 0.375. Also, if the message came through a male, that would increase the log-odds of reposting by 0.294. If the parent had many friends/followers, then per any log-unit in their number, the log-odds of the ego reposting the message would decrease by 1.212, signaling that the ego would be less attentive to any single post from this parent. Every 1 USD increase in the value of the gift, offered for reposting the message, increases the log-odds of reposting by 0.001; note that this can be a substantial increase for the gifts worth several hundred dollars. Now, if the parent's repost is already reposted by one of its friends/followers (besides the ego) then the log-odds of the ego following suit increases by 1.663. Finally, per any log-unit in the number of reposts the ego contributed during the study period (reflecting the ego's overall reposting activity), the log-odds of the ego reposting this particular message increases by 0.445.

We proceed with a detailed summary of the observations made by comparing the regression outputs across "Free gift" and "non-Free gift" groups and across the circumstance Sets 1, 2, and 3.

First, we report the observation of user behavior in reposting the messages that reached them through other users. We find that the users in "Free gift" groups are less likely to repost a message with minimal expected reward, compared to the (always un-incentivized) messages from topical groups, per the models' baselines. Males are less likely to repost messages than females, in general, and even more so in "Free gift" groups. Yet, if an "Free gift" message reaches an ego's wall through a male peer, then it is more likely to be reposted, which is not the case for un-incentivized messages: in the latter case, the sex of a parent matters much less, with the reposts from females slightly more likely to be further reposted. Messages arriving to the ego through from bulk reposters (i.e., those with many friends/followers) are less likely to be reposted, in general, – and this effect is about the same with "Free gift" and "non-Free gift" groups. The effect of reposts at a parent level attracting more reposts is equally strong for the messages originated by groups of all types. Bulk reposters (i.e., those who repost a lot) appear to value "Free gift" messages more than the non-bulk reposters.

Second, we discuss the insights of user behavior in reposting the messages that come to them through groups. We begin by looking at the members of the authority groups (original message creators). The baseline log-odds of reposting is about the same for such groups' members, no matter if the group is a "Free gift" or "non-Free gift" one. Sex difference is less of an effect for group members, but still, males are less likely to repost messages than females. The volume of reposts the ego makes has about the same effect on the repost log-odds for "Free gift" and "non-Free gift" group members, which is different from the behavior of the users who do not subscribe to authority groups. Finally, we turn attention to the users whom the message in question reaches through hub groups. Under such circumstances, the baseline log-odds of reposting a message from a "Free gift" group is higher than that of reposting an un-incentivized message. Sex difference has a much less pronounced effect on reposting activity among the hub subscribers. Quite interestingly, volume reposters are less likely to repost an incentivized post coming through their hub group than an un-incentivized post – this is in contrast with the user behavior in reposting the messages that reach them through peers (i.e., not through groups).

Third, we compare the user behavior across the three sets of repost opportunity circumstances. We find that the repost baseline for the hub groups is overall higher than for authority groups (both for incentivized and un-incentivized message reposting). Moreover, in turn, the baseline for reposting from friends (i.e., for the messages received through other users as opposed to directly from/through groups) is even higher.

5   Conclusions

In this paper we analyze reposting cascades of the posts sent by 11 groups in social media site VK.com. Groups post messages on different topics; three groups out of 11 concentrate on posting advertisement messages, where incentive is offered to the users who would repost it in their own wall. We have gathered all messages sent by 11 groups, and data on reposts of those messages. Average number of reposts per post for the groups offering incentives for the repost is higher than of those which do not offer any incentives. Out of ten mostly reposted messages in our dataset, nine belong to groups offering incentives, however mostly the reposts are done by users who are members of these groups, and depth of the cascades are less than in the other groups (such as those which post news).

Acknowledgements. This research was supported in part by the 2015 U.S. Air Force Summer Faculty Fellowship Program, sponsored by the AFOSR. Research of Alexander Semenov was supported in part by the Academy of Finland, grant nr. 268078 (MineSocMed project).

References

1. Boyd, D., Ellison, N.: Social network sites: definition, history, and scholarship. J. Comput. Mediat. Commun. **13**, 210–230 (2007)
2. Kümpel, A.S., Karnowski, V., Keyling, T.: News sharing in social media: a review of current research on news sharing users, content, and networks. Soc. Media Soc. **1**, 2056305115610141 (2015)
3. Guille, A., Hacid, H., Favre, C., Zighed, D.A.: Information diffusion in online social networks: a survey. ACM SIGMOD Rec. **42**, 17–28 (2013)
4. Sun, E., Rosenn, I., Marlow, C., Lento, T.M.: Gesundheit! modeling contagion through Facebook news feed. In: ICWSM, San Jose, California (2009)
5. Bakshy, E., Rosenn, I., Marlow, C., Adamic, L.: The role of social networks in information diffusion. In: Proceedings of the 21st International Conference on World Wide Web, pp. 519–528. ACM (2012)
6. Bernstein, M.S., Bakshy, E., Burke, M., Karrer, B.: Quantifying the invisible audience in social networks. In: Proceedings of the SIGCHI Conference on Human Factors in Computing Systems, pp. 21–30. ACM (2013)

7. Cheng, J., Adamic, L., Dow, P.A., Kleinberg, J.M., Leskovec, J.: Can cascades be predicted? In: Proceedings of the 23rd International Conference on World Wide Web, pp. 925–936. ACM (2014)
8. Dow, P.A., Adamic, L.A., Friggeri, A.: The anatomy of large Facebook cascades. In: ICWSM (2013)
9. Romero, D.M., Meeder, B., Kleinberg, J.: Differences in the mechanics of information diffusion across topics: idioms, political hashtags, and complex contagion on Twitter. In: Proceedings of the 20th International Conference on World Wide Web, pp. 695–704. ACM (2011)
10. Goel, S., Anderson, A., Hofman, J., Watts, D.: The structural virality of online diffusion. Manag. Sci. **62**, 180–196 (2015)
11. Anderson, A., Huttenlocher, D., Kleinberg, J., Leskovec, J., Tiwari, M.: Global diffusion via cascading invitations: structure, growth, and homophily. In: Proceedings of the 24th International Conference on World Wide Web, pp. 66–76. International World Wide Web Conferences Steering Committee (2015)
12. Leskovec, J., Adamic, L.A., Huberman, B.A.: The dynamics of viral marketing. ACM Trans. Web TWEB **1**, 5 (2007)
13. Li, J., Xiong, J., Wang, X.: The structure and evolution of large cascades in online social networks. In: Thai, M.T., Nguyen, N.T., Shen, H. (eds.) CSoNet 2015. LNCS, vol. 9197, pp. 273–284. Springer, Heidelberg (2015)
14. Goel, S., Watts, D.J., Goldstein, D.G.: The structure of online diffusion networks. In: Proceedings of the 13th ACM Conference on Electronic Commerce, pp. 623–638. ACM (2012)
15. Semenov, A., Veijalainen, J., Boukhanovsky, A.: A generic architecture for a social network monitoring and analysis system. In: Barolli, L., Xhafa, F., Takizawa, M. (eds.) The 14th International Conference on Network-Based Information Systems, pp. 178–185. IEEE Computer Society, Los Alamitos (2011)

Structure and Sequence of Decision Making in Financial Online Social Networks

Valeria Sadovykh and David Sundaram[✉]

Department of Information Systems and Operations Management, University of Auckland,
Auckland, New Zealand
{v.sadovykh,d.sundaram}@auckland.ac.nz

Abstract. Online Social Communities and Networks (OSNs) have become a popular source of information for those seeking advice for everyday decision making. A key benefit of OSNs is known to be the provision of free and easy access to a wide range of information, largely unconstrained by geographical barriers and free of charge. This paper specifically addresses the potential use of OSNs as a support tool for financial decision making. The key objectives of this paper are to explore and identify the structure and sequence of decision-making phases in financial OSNs (FOSNs). This research uses Netnography - a qualitative research method to achieve these objectives. Key results suggest that most of the decision-making phases identified by Simon and Mintzberg are present in FOSNs and that the sequence of these phases tends to be anarchical.

Keywords: Decision making · Decision-making structure and sequence · Netnography · Online social networks · Financial online social networks

1 Introduction

Online social networks in the recent past have started to gain the attention of those in the financial sectors [2]. There has been a substantial increase in the amount of financial information, advice, services and tools that can be accessed online [21, 22]. Finance (encompassing money, financial wellbeing) is considered to be one of the most important elements in everyday life [16]. Financial issues or finance-related questions have been identified as one of the top ten most commonly researched topics on the internet [3].

Despite being a relatively new phenomenon, FOSNs play a significant role in the day-to-day dissemination of financial information and decision making for individuals and professionals within the financial industry. FOSNs provide many sources of online information that can include official listed companies, financial wealth management advisers and experts, financial institutions, stock traders, and others that can distribute investment information, including real-time market data, research, and trading recommendations. Therefore, it is understandable that FOSNs have become popular virtual space for individuals seeking information on personal finance, budgeting, investment strategies, stock market trading, or simply a place for self - education on financial matters [22]. Therefore, FOSNs have become a decision-making tool that is used to support

© Springer International Publishing Switzerland 2016
H.T. Nguyen and V. Snasel (Eds.): CSoNet 2016, LNCS 9795, pp. 135–146, 2016.
DOI: 10.1007/978-3-319-42345-6_12

different types of decision making, ranging from the instrumental to the emotional and informational.

Regardless of advances in related technology, when faced with a DM situation, it is reasonable to assume that the core processes are similar. However, given the social and behavioral nature of the DM process, it is necessary to find evidence in reality to confirm the existence of a common core. Nowadays, the concept of reality is also related to the online world of human interaction. With the explosion of FOSNs and the potential wealth of information contained therein, we are interested in considering FOSNs as a support tool for DM. The primary research objectives are to explore, identify and understand (1) the structure of DM phases that is supported in FOSNs and (2) the sequence of DM phases assisted by the use of FOSNs.

2   Research Motivation and Objectives

Finance and financial-orientated research has caught the attention of the public and academics for centuries. By nature, it is a sensitive, personal and globally important topic. Even though the research into FOSNs has not gained the same popularity as, say, online branding and online shopping yet, but the significance of this topic cannot be underestimated. It is important to state that there is no lack of research with regard to OSNs, decision making, and finance as stand-alone subjects of interest. Despite these research topics being discussed within both academia and industry, the synergy of these themes provides an innovative and unique perspective. There is a research gap in how FOSNs support the decision making of key stakeholders: individuals, professional investors, listed firms and financial institutions as a decision-making information source, whether to seek for financial advice and/or to analyse market news, trends and fluctuations. Therefore this research study uses this unique opportunity and tries to discover a niche that has not been overly researched yet.

To overcome the problem stated above, we propose a set of objectives and requirements that should be addressed and further employed. This study not only wants to observe stakeholders' behaviour within FOSNs and conduct an analysis of their participation, but also aims to concurrently investigate the two following objectives: (1) Determine the structure of DM phases in FOSNs, specifically what decision-making phases are supported and influenced by FOSNs and identify any construct(s) that may not have been identified before in this context. (2) Explore the sequence of DM phases in FOSNs. This includes the proposal and validation of concepts relating to the sequence of the FOSN-based DM processes. It may be possible, through zooming in, to understand the relationship between the decision-makers and the process they undertake by using FOSNs.

The requirements for these objectives are to understand and define the DM processes, phases and sequence that are supported by FOSNs. This will be accomplished by using the chosen qualitative research methodology, Netnography, and conducting a Netnographic study across various categories of FOSNs. In the following section we look at DM processes, theories and concepts (Sect. 3). Thereafter, we proceed with the definition of Netnography as a chosen research method and the Netnography research process to follow (Sect. 4). This will result in a detailed description of how the Netnography research process is undertaken in this study (Sect. 5). We continue with the discussion of findings from Netnography and how the research objectives and requirements have been met in Sect. 6. Section 7 concludes this paper by discussing the overall findings and the potential contributions of this research to the theory and practice of DM, OSNs and FOSNs, as well as potential future studies.

3   Decision Making

The history of decision-making (DM) research is long, rich and diverse. In terms of quantity, there is no shortage of frameworks, taxonomies, approaches and theories. Decision making is a complex field; it can involve the adoption of various technologies, in addition to having to accommodate the different psychological perspectives of individuals. One of the foundational and impactful theories in the field of behavioral studies in human decision making has been developed by Herbert Simon [4]. Simon [20] suggested that the decision-making process can be structured and ordered in three phases: intelligence, design, choice. Later, Huber and McDaniel [8] extended this model by adding two further phases: implementation and monitoring. Figure 1 presents the view on the decision-making process by Simon [20] with additions by [8].

Fig. 1.   Decision making process (adapted from Langley et al. [11])

Intelligence is where the decision-maker is collecting information about the problem, identifying the problem and its cause. The design phase is where recognition and understanding of possible alternatives and consequences of the future decision occur. Choice is where identified alternatives and options are narrowed down to the best utility option that leads to a decision-maker's choice. The implementation phase is about actual execution of the chosen option, while monitoring relates to the consequences of the implemented option.

Other researchers, for example, Cooke and Slack [6] developed the sequential model based on Simon's model to explain decision-making as a cyclical process that focuses around the problem. Problem solving in their theory is not merely the three phases of the Simon model, but a continuous process of identifying the best alternatives and course of actions. Mintzberg et al.'s [15] model follows the linear structure from Simon's rational decision-making process and reflects the repetitive elements and incoherent phases of decision making. In this model, the decision-maker comes with recognition of a problem or tangible request that requires an action, with the solution coming in the form of different repetitive DM stages that do not necessarily follow the sequence.

Unlike rational and sequential models, decision-making theories emerged into an anarchical problem-solving process that is driven by events. There is no sequence for decision phases and there is no established process to follow. There are chaotic and incoherent phases of decision making that build on need. In other words, this model is a free decision-making process that is more intuitive than the rational one developed by Langley et al. [11]. The decision-making process driven by events is similar to Cohen et al.'s [5] garbage can model of decision choice. The four streams that interplay in Cohen's model are problems, solutions, participants, and choice opportunities. Sinclair and Ashkanasy [19] developed a model of integrated analytical and intuitive decision making that supports two mechanisms of decision making: first, the decision-making process follows an intuitive behavior that is driven by events [5, 11]; and second, decision making is rational and structured in a logical order toward problem solving.

4 Research Methodology

This research primarily follows Kozinets' guidance in how to conduct Netnography [9, 10]. Netnography is a new approach in conducting exploratory research through the use of ethnographic principles that combines archival and online communications work, participation and observation, with new forms of digital and network data collection, analysis and research representation [10]. This method helps us to gain an understanding of human experience from online social interaction and/or content. The undertaken research steps and their description is shown in Table 1.

Table 1. The netnography research process

#	Steps	Description
1	Planning and entrée	Definition, identification, selection of research questions; communities; conversations of interest and categorisation of networks and participants
2	Data selection and collection	Observation, participation and engagement; filtering process, review and revisiting of conversation selection; collection challenges; obtaining and selection of a large sets of data for reading, analysis, and coding
3	Data analysis	Data interpretation process with the use: discourse, content and textual analysis, coding and noting
4	Discussion and findings	Representation and reporting of research findings and theoretical implications for the research objectives

5 Adapted Netnography Research Process

5.1 Planning and Entrée

The planning step requires the research questions and objectives to be defined. This has been done and stated in the earlier sections of this paper. The entrée involves the actual choice of networks of interest where the observation takes place first and then the researcher can proceed with data collection. There are networks and communities that are specifically designed to support the general public in finding answers to questions on financial matters (e.g. everyday budgeting, saving tips, retirement plan options, passive investment strategies and financial news on economy and market updates). There are also more sophisticated financial services offered online, such as online platforms for trading currencies and shares, wealth management firms for providing financial planning advice and a variety of online money management tools offered to the public with diverse needs and requirements (e.g. mobile applications to track daily expenditure, mirror trading (following a financial expert's trading strategy or investing in their portfolio via online routes) etc.).

For the purposes of this paper, we have adopted and modified the categorisation of FOSNs provided by Mainelli [12]. These three categories of FOSN are retail, support services and professional. Within these three categories we have identified communities of interest and their main topics; the mapping between FOSN categories and communities of interest (with the web links as an example) outlined in Table 2.

Table 2. Planning and entre: Mapping of FOSN categories and communities of interest

FOSN category	FOSN community and topics	Example
Retail	Investment options and strategies, Online wealth management service, robo-advising, insurance and retirement plans	www.nerdwallet.com
		www.boards.fool.com
		www.barrons.com
		www.wealthmanagement.com
Support services	Saving and budgeting tips, Retirement advice, non-professional investment advice	www.savingadvice.com
		www.reddit.com
Professional	Professional Investment - Forex and share trading (i.e. mirror trading)	www.oanda.com/
		www.fxstreet.com

One of the other important steps in the planning and entrée phase of netnography is an understanding of the participants on selected networks. Because the Internet has already been in existence for a substantial period, researchers have categorised online participants into various groups and come up with specific nomenclature [7]. The categorisation of OSN participants used in this study have been developed by Kozinets [9], who defined OSN users as being either Advisers, Seekers, Experts, or Observers.

Observers are less associated with community life, and are searching for the right information to support their decision or simply to find some clues to, or interest in, questions or answers. Observers are silent members of the group and the percentage of observers of a particular network/community cannot be easily identified. Rodan et al. [17] indicated that there is an approximately four to one ratio between people who have accessed the site and those who have posted in the communities. *Seekers* do not always have strong ties with an associated group. They are confident and brave enough to ask questions, start a thread on the topic of their interest, and look for support. Seekers are interested in immediate results – advice provided by advisers. Once seekers get the information or find answers, their relationships with the community might dissolve. *Advisers* have strong ties with a group, a high rate of participation, and take a strong interest in the group. Advisers are those who provide support to seekers in order to solve a problem. There are always two parts to the story: advisers can support decision-makers and at the same time mislead them. *Experts* have strong ties within the community and their respect mainly depends on their profile, on which can be displayed their expertise, education, volume and past history of participation. Some advisers can fall into the category of experts, especially the ones with a high presence in the network, or, in some cases, 'experts' can be acting as administrators of networks. In this study, experts did not exist in every community that has been followed; but experts are important, especially in FOSNs.

5.2 Data Selection and Collection

The second step from the netnographic research framework involves data selection and collection. This is considered to be a delicate and important procedure that serves many purposes in the research approach [1]. Kozinets [9] recommends obtaining three different types of data during the data collection process, namely: archival, elicited, and field notes. This study took two types of data for the collection process. First, the written communications between different stakeholders that occurred in the online communities (archival) have been directly downloaded and saved for analysis. Second are the researcher's self-authored field notes, in which the observation ideas and comments were recorded and synthesised in the analysis section.

Field notes can provide the first fresh research perspective on the data collected. Those that were taken from the observational process were mainly about the participants' behaviour, FOSN design (i.e. background style, font style, graphical presentation, and features of the webpage), conversations styles for each community, financial tools that are offered to users (i.e. mortgage calculator, currency converter, financial adviser matching questionnaires, etc.). All of these can significantly amplify or attenuate the DM process of FOSN users.

The screening process proved to be a lengthy one: prior to screening, a list was drawn up of factors that are essential to this research and could help in identification and filtering the relevant communities and conversations from the pool of hundreds of FOSN websites. The search for suitable posts was conducted using search engines like Google and Yahoo Finance using keywords such as: top financial forums, top financial online investing platforms, best financial virtual communities, and etc.

After the identification of the networks of interest, the researcher tried to become familiar with the network by reading terms and conditions, searching for popular topics, topics by last dates modified, observing member behaviour, understanding login requirements and any other website design components that could influence the immediate attitude towards the network and subsequently affect the decision-making process. As a result, 10 websites were chosen (at least three per FOSN category) that had conversations related to the research questions and conversations showing the presence of the DM process and phases. A total of more than 30 posts were selected from the chosen websites. Statistically, one website could provide more than three conversations for analysis purposes. This analysis ensured that the conversations were not on a single subject and not conducted by the same participants; therefore the decision-making process is not repetitive, and different participants represent the voice of different demographics. Conversations were collected and separated according to the subject of interest (indicated in Table 2), phases of decision making and conversation headings. Each post was assigned a specific code that indicates post subject correlated with the subject of interest.

One of the challenges of using online networks as a source of data collection is the abundance of data available. After the observation period, it was evident that the themes of conversations in FOSNs are repetitive and the major difference in participants' behaviour and the way conversations are structured and their sequence is dependent on the FOSN category, whether professional, retail or support services. Therefore, the number of conversations is not that essential: what was important was the variety of conversations and questions for decision making.

5.3 Data Analysis

Once conversations, posts and websites that are directly related to the research objectives are identified and the data collection process has taken place, the researcher starts sorting the data. That is where the data analysis process really starts. Data analysis process in a grounded theory approach has a generic sequence of common qualitative steps. We followed the procedure described by Milles and Huberman [14] that considers four steps that are applied to this study: (1) Generalising and Theorising: high level analysis of collected data; (2) Coding and Noting: identification of main categories and research concepts; (3) Abstracting and Comparing: supporting data, evidence, facts, identification of new categories; and (4) Checking and Refinement: revision of findings and developing of new concepts and insights.

More detailed analysis started when the collected data started to fall into particular categories and the pattern of findings could be identified. These categories came from the research theory and research objectives. The initial categories were taken from Simon's DM model [20] in which he concluded that a rational human mind goes through the phases of a DM process, those of Intelligence, Design, Choice, Implementation and Monitoring. In the repetitive process of analysis, with additional exploration of data, literature and pre-formed findings: the additional phases of the DM process were established, as well as new categories for further analysis and coding. The new categories were based on the existing phases of Simon's DM model [20], but with the addition of the use of FOSNs. This was achieved by observation and comparison of multiple conversations and networks.

The difference between real and online worlds is how people present and describe their future decisions or the experience of DM processes they already have. After revising the existing categories of phases of the DM process, the new categories within the adviser model emerged. Most of the time, the adviser provides a model of decision (design phase) based on their own previous experience of decisions made in the past, knowledge and insights. The adviser model can include options or alternatives provided. Therefore, the previous experience of the adviser expands the categories of the DM process.

Through revising the conversations it was also identified that seekers, in contrast to advisers, follow a real life conversation situation. By entering the community, they familiarise themselves with it and introduce themselves (in some cases) and their question, and provide a background on their decision-making situation; for example, seekers introduce a situation with relevant background information that might assist advisers to advise on a solution, so they can make a good choice or leave without taking any responsibility for the decision to be made. In finance-related conversations, all the information that is provided by participants is usually relevant or closely correlated with the future/past decision making.

The background information provided by a seeker can be identified as an entry step into the DM process. In most of the observed conversations, the information provided followed the logical explanation, if relevant. There is not much human introduction in FOSNs as there would be in a real conversation: rarely will participants tell you where they are from or what they do for living. In the case of advisers, they mainly provide options and models at first, and only then is this followed with the background introduction. In most cases, their advice is based on their previous experience or existing knowledge. What is common and interesting about advisers and seekers is that both types of participants provide enough background information for DM, whether that be a seeker posting an inquiry or an adviser proposing a model, options, or alternatives.

Figure 2 illustrates the data analysis on the DM conversation phases, structure and sequence from the three categories of FOSN that this study observed. It shows how the

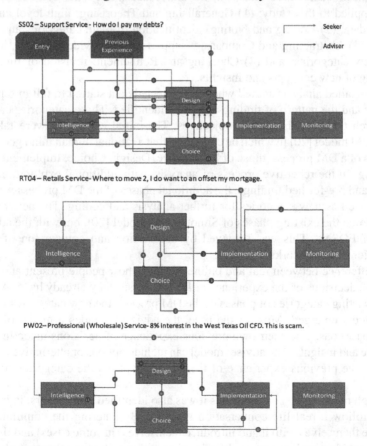

Fig. 2. Analysis on DM phases, structure and sequence in FOSN (Conversations Analysis)

phases of DM process are interconnected in the online environment that makes the DM process follow the anarchical structure. The phases of the DM process are visible, but the sequence in which the conversations move between them is unstructured and appears random. In the Fig. 2 - SS02 conversation, it is interesting to note that most advisers started their conversation by stating the choice – the "decision" to be made first - and only then proceeded with the explanation of the advice provided.

Another aspect that has emerged during the analysis process is advertising. In FOSNs, online advertising undoubtedly affects the decision-making process [18]. Advertisement posts in FOSNs can be easily recognised and be identified by readers; most of the time people are openly advertising their services (i.e. financial brokers, asset management firms) with relevant credentials and experience. However, advertising or self-advertisement as a service does not fit into any of the phases of the DM process. It can be an influential factor, and therefore has been indicated as an additional step present in the DM process, especially in an online environment.

6 Structure and Sequence of Decision Making in FOSN

To understand the structure and sequence of FOSN conversations, this study coded the collected conversations to Simon's DM-process phases.

Intelligence Phase (I): With the use of an FOSN, the decision-maker is capable of retrieving information in real time in a matter of seconds. FOSNs are not standard search engines, but can provide information according to the search query or problem. Furthermore, it was evident that, through the use of an FOSN, the decision-maker can find similar problems and already-developed solutions that have been tested and evaluated by other members of networks. Therefore, FOSNs can enhance the intelligence phase of the decision-making process by providing access to a variety of data sources and different formats of data (visual, textual, mathematical, and graphical) [13].

Design Phase (D): The design phase is all about alternatives and models of outcomes and consequences and additional questions that might lead to a better design option for DM. An FOSN provides an opportunity for decision-makers to explore alternatives by simply asking for advice or browsing through the different FOSNs of interest. An FOSN also can attenuate this phase by simply presenting already-developed models of solutions that were provided by other members of the FOSN. Decision makers are not required to accept the provided models, but they can evaluate them and find them useful or irrelevant; the selection process leading to the choice of the right alternative is one of the sub-processes of the design phase, before making a choice.

Choice Phase (C): The choice phase in FOSN was found to be present, specifically in professional networks where investors could replicate the adviser's strategy and show their financial gains or losses; and also it could be seen in the posts where a seeker returns to the thread to post the choice made or acknowledge that the thread had been reviewed and used in a real-life environment.

Implementation and Monitoring (IM and M): The implementation phase was found to be partially present in FOSNs, even though originally it was anticipated that it would be difficult to observe. Monitoring could be detected in professional or retail networks, mainly when seekers were coming back to share the results of the decision made and the consequences, or some part of the adviser's options or models.

The FOSN can also help the decision-maker in identifying and providing tools and resources that can assist in the DM process (i.e. the use of a budgeting spreadsheet, mobile application apps for everyday monitoring of spending, online investment portfolio accounts where performance is monitored online). Moreover, an FOSN can assist users in conducting a post - analysis evaluation of the financial decision made (e.g. review of a report, or analysis of an asset wealth management service provider).

If the decision is viewed from the perspective of the 'initial issue' such as conversations between seekers and advisers in FOSN, then the phases of the DM process do not have a sequence and do not follow any logical process. What is interesting is the difference between how an advisers post their choices made in the past as part of their previous experience and how seekers provide background information based on their experiences of past decisions. Advisers, when suggesting a choice to make, usually start the conversation with a clear statement - the choice to be made - and then proceed with a description of their advice and reasoning (Design-Model (D) – Intelligence (I)), while seekers usually follow the opposite sequence when explaining their DM. It usually started with Intelligence (I) – background information on the decision to be (already) made; C (Choice) and/or need; and D (Design-Model) – options and alternatives available to them.

7   Conclusion

From the discussion above, it is apparent that an FOSN is used as a support tool which helps to (1) find relevant information, understand alternatives, options, choices and consequences; (2) observing and sharing the DM process experience; and (3) identifying the necessary resources for implementation and evaluation of outcomes from decisions taken. Based on our analysis using netnography as a research methodology, it is evident that online conversations support most of the phases of the decision-making process identified by Simon and Mintzberg; however, our results indicate that the phases in online conversation do not follow Simon's [20] sequence of a rational decision-making process and that the sequence of these phases tends to be anarchical.

However, Simon's study mainly concentrated on analysing the behaviour of rational decision makers in the DM process. The main thoughts, discussions and considerations that took place in this study were focused around the subject of the DM process and how it can be supported by an FOSN. Rational models of decision-making emphasize structure and sequence while anarchical models of decision-making imply that there is no structure and sequence in many real-world decision making contexts. However our results challenge both these models by suggesting that decision making on FOSNs exhibits structure but not sequence.

The findings of this research study suggest that certain DM processes observed in FOSNs, from an overall perspective, are in some way related to the well-known model of an anarchical DM process driven by events developed by [11]. The impulsive phases of the DM process are recognised and evolved as needs of the decision maker arise. Also, the observation validates the pattern of the DM process as being the interplay of four streams (problems, participants, solutions and choice opportunities) [5]. A decision is generated by various opportunities, alternatives, associated problems and people. Discussions in FOSN environments involve individuals (as advisers or seekers) and models of choice, with alternatives and possible options that can be recommended by people or provided from their experience. The observed FOSN DM process was found to have no structure and displayed anarchic behavior; it also exhibited characteristics similar to the Mintzberg et al. [15] model of the DM process as an iterative sequence.

Using discourse and conversational analysis for the data interpretation from netnography, it was observed that not every phase of DM was present in every conversation. It was not surprising to realise that FOSN are more structured and result-orientated networks and are always about figures, statistical analysis and predictions. On the other hand, this phenomenon could not be found to exist in the Support Service FOSN category, nor, more specifically, in everyday budgeting and retirement conversations. It was also observed that whether in support-orientated or retail FOSNs, advertisements were found to be present; in some cases, the advertisers were targeting seekers specifically by using FOSN as a tool. Therefore, the key results suggest that most of the decision-making phases identified by Simon and Mintzberg are present in an FOSN and that the sequence of these phases tends to be anarchical.

One of the unique findings of this study that will be further explored in detail in future studies is the use of online financial tools provided by FOSNs, either for free or for cost. That is where the rise of technological advances could be predominantly seen. This factor has been recognised across most of the categories of FOSNs: even regulated websites and networks have provided links or easily downloadable tools for managing a specific matter of interest. Other considerations for future research include: first, an expansion of the scope of the research categories of FOSNs - the number of posts and websites analysed. This might contribute to the identification of new phases of the DM process; second, a future study should carefully consider the import of FOSN location because each country has different investment schemes, retirement plans and financial regulations, legislation and obligations pertaining to professional and personal use. This also affects the decision-making style of the decision makers; and a third area of investigation for future study is global market manipulation by FOSNs. After conducting this study, it will be beneficial to understand how and if the conversations posted online on financial matters might lead to overall market fluctuations or, in some cases, manipulation.

Virtual financial communities are real, significant and growing. Organisations have only started to scratch the surface of how technology can help to build these communities. It is not the technological capability that is important; it is the ability of new technology ideas to secure communities' trust, i.e. managing risk and reward. The technology is here. What is needed are novel ideas for using that technology. Ideas for building virtual financial communities will succeed if they attract, engage and retain people, build trust and spread to new people.

References

1. Armstrong, J.S., Overton, T.S.: Estimating nonresponse bias in mail surveys. J. Mark. Res. **14**(3), 396–402 (1977)
2. DasGupta, S.: Encyclopedia of Virtual Communities and Technologies. IGI Publishing, Hershey (2006)
3. Campbell, J.: Investor empowerment or market manipulation in financial virtual communities. In: Dasgupta, S. (ed.) Encyclopedia of Virtual Communities and Technologies, pp. 296–301. IGI Publishing, Hershey (2006)
4. Campitelli, G., Gobet, F.: Herbert Simon's decision-making approach: investigation of cognitive processes in experts. Rev. Gen. Psychol. **14**(4), 354 (2010)
5. Cohen, M.D., March, J.G., Olsen, J.P.: A garbage can model of organizational choice. Adm. Sci. Q. **17**(1), 1–25 (1972)
6. Cooke, S., Slack, N.: Making Management Decisions. Prentice Hall, Englewood Cliffs (1991)
7. Harridge-March, S., Quinton, S.: Virtual snakes and ladders: social networks and the relationship marketing loyalty ladder. Mark. Rev. **9**(2), 171–181 (2009)
8. Huber, G.P., McDaniel, R.R.: The decision-making paradigm of organizational design. Manag. Sci. **32**(5), 572–589 (1986)
9. Kozinets, R.V.: Netnography: Doing Etnographic Research Online. Sage Publications, Thousand Oaks (2010)
10. Kozinets, R.V.: Netnography: Redefined. Sage, Thousand Oaks (2015)
11. Langley, A., Mintzberg, H., Pitcher, P., Posada, E., Saint-Macary, J.: Opening up decision making: the view from the black stool. Organ. Sci. **6**(3), 260–279 (1995)
12. Mainelli, M.: Risk/reward in virtual financial communities. Inf. Serv. Use **23**(1), 9–17 (2003)
13. Mayer, A.: Online social networks in economics. Decis. Support Syst. **47**(3), 169–184 (2009)
14. Miles, M.B., Huberman, M.A.: Qualitative Data Analysis: An Expanded Sourcebook, 2nd edn. Sage Publications, Thousand Oaks (1994)
15. Mintzberg, H., Raisinghani, D., Théorêt, A.: The structure of "unstructured" decision processes. Adm. Sci. Q. **21**(2), 246–275 (1976)
16. Rath, T., Harter, J., Harter, J.K.: Wellbeing: The Five Essential Elements. Simon and Schuster, New York (2010)
17. Rodan, D., Uridge, L., Green, L.: Using nicknames, pseudonyms and avatars on HeartNET: a snapshot of an online health support community. In: ANZCA Conference, Canberra, Australia (2010)
18. Senecal, S., Kalczynski, P.J., Nantel, J.: Consumers' decision-making process and their online shopping behavior: a clickstream analysis. J. Bus. Res. **58**(11), 1599–1608 (2005)
19. Sinclair, M., Ashkanasy, N.M.: Intuition. Manag. Learn. **36**(3), 353–370 (2005)
20. Simon, H.A.: Rational decision making in business organizations. Am. Econ. Rev. **69**(4), 493–513 (1979)
21. Tumarkin, R., Whitelaw, R.F.: News or noise? Internet postings and stock prices. Financ. Anal. J. **57**(3), 41–51 (2001)
22. Wysocki, P.D.: Message boards speak volumes-and-volatility. Futures: News Anal. Strat. Futures Options Deriv. Traders **29**(14), 42 (2000)

Kirchhoff Centrality Measure
for Collaboration Network

Vladimir V. Mazalov[1(✉)] and Bulat T. Tsynguev[2]

[1] Institute of Applied Mathematical Research,
Karelian Research Center, Russian Academy of Sciences,
11, Pushkinskaya St., Petrozavodsk, Russia 185910
vmazalov@krc.karelia.ru
[2] Transbaikal State University,
30, Aleksandro-Zavodskaya St., Chita, Russia 672039
btsynguev@gmail.com

Abstract. This paper extends the concept of betweenness centrality based on Kirchhoff's law for electric circuits from centrality of nodes to centrality of edges. It is shown that this new measure admits analytical definition for some classes of networks such as bipartite graphs, with computation for larger networks. This measure is applied for detecting community structure within networks. The results of numerical experiments for some examples of networks, in particular, *Math-Net.ru* (a Web portal of mathematical publications) are presented, and a comparison with *PageRank* is given.

Keywords: Kirchhoff centrality measure · Betweenness centrality · Weighted graph · Community structure

1 Introduction

Betweenness centrality is an efficient concept for analysis of social and collaboration networks. A pioneering definition of the betweenness centrality for node i was given in [9] as the mean ratio of the total number of geodesics (shortest paths) between nodes s and t and the number of geodesics between s and t that i lies on, considered for all i, j.

A shortcoming of betweenness centrality is analyzing only the shortest paths, thereby ignoring the paths being one or two steps longer; at the same time, the edges on such paths can be important for communication processes in a network. In order to take such paths into account, Page and Brin developed the well-known PageRank algorithm for ranking of all pages in the Web's graph based on the Markov chain limit theorem. Jackson and Wolinsky [11,12] proposed the model of communication game in which the utility depends of the structure of the network. They apply the Myerson value [1,13] to analyse the betweenness of the nodes in the network. Other allocation rules were proposed in [5,6,16,17]. Brandes and Fleischer [8] and Newman [14] introduced the concept of current

© Springer International Publishing Switzerland 2016
H.T. Nguyen and V. Snasel (Eds.): CSoNet 2016, LNCS 9795, pp. 147–157, 2016.
DOI: 10.1007/978-3-319-42345-6_13

flow betweenness centrality (CF-centrality). In [2,8], a graph was treated as an electrical network with edges being unit resistances. The CF-centrality of an edge is the amount of current flowing through it with averaging over all source-destination pairs, when one unit of current is induced at the source and the destination (sink) is connected to the ground.

In [3], the authors developed this concept further, suggesting to ground all nodes equally; as a result, averaging runs only over the source nodes, reducing computational cost. This new approach will be referred to as beta current flow centrality (βCF-centrality). Moreover, in contrast to the works [2,8,14,15], the weighted networks were considered in [3], mostly in the context of the βCF-centrality of nodes. The present paper focuses on the βCF-centrality of edges. Using this measure, a community detection method is proposed for the networks. The method is tested on the collaboration network presented by *Math-Net.ru*, a Web portal of mathematical publications.

2   Beta Current Flow Centrality Based on Kirchhoff's Law

Recall in brief the concept of βCF-centrality proposed in [3]. Consider a weighted graph $G = (V, E, W)$, with V denoting the set of nodes, E the set of edges and W the matrix of weights, i.e.,

$$W(G) = \begin{pmatrix} 0 & w_{1,2} & \cdots & w_{1,n} \\ w_{2,1} & 0 & \cdots & w_{2,n} \\ \vdots & \vdots & \ddots & \vdots \\ w_{n,1} & w_{n,2} & \cdots & 0 \end{pmatrix}.$$

Here $w_{i,j} \geqslant 0$ is the weight of an edge connecting nodes i and j, $n = |V|$ gives the number of nodes. Note that $w_{i,j} = 0$ if nodes i and j are not adjacent. By assumption, G represents an undirected graph, viz., $w_{i,j} = w_{j,i}$.

For the weighted graph G, the Laplacian matrix $L(G)$ is defined by

$$L(G) = D(G) - W(G) = \begin{pmatrix} d_1 & -w_{1,2} & \cdots & -w_{1,n} \\ -w_{2,1} & d_2 & \cdots & -w_{2,n} \\ \vdots & \vdots & \ddots & \vdots \\ -w_{n,1} & -w_{n,2} & \cdots & d_n \end{pmatrix} \quad (1)$$

where $d_i = \sum_{j=1}^{n} w_{i,j}$ yields the sum of weights of the edges adjacent to node i in the graph G.

Let a graph G' be obtained from the graph G by adding single node $n + 1$ connected to all nodes of the graph G via the links of a constant conductance β.

Thus, the Laplacian matrix for the modified graph G' has the form

$$L(G') = D(G') - W(G') = \begin{pmatrix} d_1 + \beta & -w_{1,2} & \cdots & -w_{1,n} & -\beta \\ -w_{2,1} & d_2 + \beta & \cdots & -w_{2,n} & -\beta \\ \vdots & \vdots & \ddots & \vdots & \vdots \\ -w_{n,1} & -w_{n,2} & \cdots & d_n + \beta & -\beta \\ -\beta & -\beta & \cdots & -\beta & \beta n \end{pmatrix}.$$

Suppose that a unit of current flows into node $s \in V$ and node $n+1$ is grounded. Let φ_i^s be the electric potential at node i when an electric charge is concentrated at node s. By Kirchhoff's current law, the vector of all potentials $\varphi^s(G') = [\varphi_1^s, \ldots, \varphi_n^s, \varphi_{n+1}^s]^T$ at the nodes of the graph G' satisfies the following system of equations:

$$L(G')\varphi^s(G') = b'_s, \tag{2}$$

where b'_s is the $(n+1)$-dimensional vector

$$b'_s(i) = \begin{cases} 1 & i = s, \\ 0 & \text{otherwise.} \end{cases} \tag{3}$$

The Laplacian matrix (1) is singular. The potentials can be determined up to a constant. Hence, without loss of generality, assume that the potential at node $n+1$ is 0 (a grounded node). Then it follows from (2) that

$$\tilde{\varphi}^s(G') = \tilde{L}(G')^{-1}b_s, \tag{4}$$

where $\tilde{\varphi}^s(G')$, $\tilde{L}(G')$ and b_s are obtained from (2) by deleting the last row and column that correspond to node $n+1$.

Thus, the vector $\tilde{\varphi}^s(G')$ can be considered as the vector of potentials at the nodes of graph G, i.e.,

$$\tilde{\varphi}^s(G) = [L(G) + \beta I]^{-1}b_s.$$

According to Ohm's law, for the link $e = (i, j)$ the let-through current is $x_e^s = |\varphi_i^s - \varphi_j^s| \cdot w_{i,j}$. Define the βCF-centrality of edge e as

$$CF_\beta(e) = \frac{1}{n} \sum_{s \in V} x_e^s. \tag{5}$$

Given that the electric charge is concentrated at node s, the mean value of the current flowing through node i is

$$x^s(i) = \frac{1}{2}(b_s(i) + \sum_{e:i \in e} x_e^s). \tag{6}$$

And finally, define the beta current flow centrality (βCF-centrality) of node i in the form

$$CF_\beta(i) = \frac{1}{n} \sum_{s \in V} x^s(i). \tag{7}$$

3   Unweighted Network of Eleven Nodes

Let us begin with a simple example of an eleven-node network, which elucidates the weak properties of flow centrality based on geodesics (see Fig. 1). A graph contains two sets of nodes connected by three nodes *1*, *2* and *11*. Nodes *2* and *11* have higher values of centrality, as all geodesics between the two sets pass through these nodes. Hence, node *1* has zero centrality based on geodesics. Actually, node *1* plays an important role in information distribution. Information can be transmitted not only directly from one node to another, but also via an additional node.

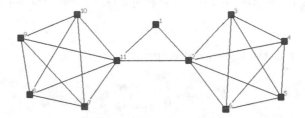

Fig. 1. Unweighted network of eleven nodes.

First, compute the βCF-centrality of the nodes in this network with $\beta = 0.5$. This method ranks nodes *2* and *11* as higher ones with centrality value 0.291 and node *1* as third with centrality value 0.147. The other nodes have centrality values 0.127.

Then, calculate the βCF-centrality of the edges in this graph. The βCF-centrality of the edge $(2, 11)$ is 0.137, and the centrality of edges $(1, 2), (1, 11)$ is 0.101. The other edges have centrality 0.0647.

In fact, the centrality of nodes *2* and *11* is twice as great as that of node *1*. At the same time, the centrality of node *1* and adjacent edges exceeds the centrality of the other nodes and edges in the network.

4   Bipartite Unweighted Graph

Consider a bipartite graph G with n nodes divided into two sets V_1 and V_2 so that a node from V_1 is connected only with a node from V_2, and vice versa (see Fig. 2). Denote this graph by $K_{|V_1|, |V_2|}$. Here all links have unit weights.

4.1   Bipartite Graph $K_{2,n-2}$

Let $V_1 = \{v_1, v_2\}$, $v' \in V_2$. Then the Laplacian matrix is

$$D(G) - W(G) + \beta I$$

$$= \begin{pmatrix} n-2+\beta & 0 & -1 & -1 & \cdots & -1 \\ 0 & n-2+\beta & -1 & -1 & \cdots & -1 \\ -1 & -1 & 2+\beta & 0 & \cdots & 0 \\ -1 & -1 & 0 & 2+\beta & \cdots & 0 \\ \vdots & \vdots & \vdots & \vdots & \ddots & \vdots \\ -1 & -1 & 0 & 0 & \cdots & 2+\beta \end{pmatrix}.$$

Its inverse has the form

$$(D(G) - W(G) + \beta I)^{-1}$$

$$= \frac{1}{\beta(n+\delta)} \begin{pmatrix} \frac{n-2+n\beta+\beta^2}{n+\beta-2} & \frac{n-2}{n+\beta-2} & 1 & 1 & \cdots & 1 \\ \frac{n-2}{n+\beta-2} & \frac{n-2+n\beta+\beta^2}{n+\beta-2} & 1 & 1 & \cdots & 1 \\ 1 & 1 & \frac{2+n\beta+\beta^2}{2+\beta} & \frac{2}{2+\beta} & \cdots & \frac{2}{2+\beta} \\ 1 & 1 & \frac{2}{2+\beta} & \frac{2+n\beta+\beta^2}{2+\beta} & \cdots & \frac{2}{2+\beta} \\ \vdots & \vdots & \vdots & \vdots & \ddots & \vdots \\ 1 & 1 & \frac{2}{2+\beta} & \frac{2}{2+\beta} & \cdots & \frac{2+n\beta+\beta^2}{2+\beta} \end{pmatrix}.$$

For $s = v_1$, the current is:

$$x^s(v_1) = \frac{1}{2}\left(1 + \frac{(\beta + n - 1)(n-2)}{(n+\beta)(n+\beta-2)}\right),$$

$$x^s(v_2) = \frac{(n-2)}{2(n+\beta-2)(\beta+n)},$$

$$x^s(v') = \frac{1}{2(n+\beta-2)}.$$

Nodes v_1 and v_2 are symmetrical; and so, for $s = v_2$ we obtain

$$x^s(v_1) = \frac{(n-2)}{2(n+\beta-2)(\beta+n)},$$

$$x^s(v_2) = \frac{1}{2}\left(1 + \frac{(\beta + n - 1)(n-2)}{(n+\beta)(n+\beta-2)}\right),$$

$$x^s(v') = \frac{1}{2(n+\beta-2)}.$$

For $s = v'$,

$$x^s(v_1) = x^s(v_2) = \frac{\beta + 2n - 4}{2(2+\beta)(\beta+n)},$$

$$x^s(s) = \frac{1}{2}\left(1 + \frac{2(\beta + n - 1)}{(2 + \beta)(\beta + n)}\right),$$

$$x^s(v') = \frac{1}{(2 + \beta)(\beta + n)},$$

which yields

$$CF_\beta(v_1) = CF_\beta(v_2) = \frac{1}{2n}\left(1 + \frac{n - 2}{n + \beta - 2} + \frac{(n - 2)(\beta + 2n - 4)}{(2 + \beta)(\beta + n)}\right),$$

$$CF_\beta(v') = \frac{1}{2n}\left(1 + \frac{2}{n + \beta - 2} + \frac{2(\beta + 2n - 4)}{(2 + \beta)(\beta + n)}\right).$$

4.2 Bipartite Graph $K_{3,n-3}$

Let $V_1 = \{v_1, v_2, v_3\}$, $v' \in V_2$.

In this case, the Laplacian matrix is

$$D(G) - W(G) + \beta I$$

$$= \begin{pmatrix}
n - 3 + \beta & 0 & 0 & -1 & -1 & \cdots & -1 \\
0 & n - 3 + \beta & 0 & -1 & -1 & \cdots & -1 \\
0 & 0 & n - 3 + \beta & -1 & -1 & \cdots & -1 \\
-1 & -1 & -1 & 3 + \beta & 0 & \cdots & 0 \\
-1 & -1 & -1 & 0 & 3 + \beta & \cdots & 0 \\
\vdots & \vdots & \vdots & \vdots & \vdots & \ddots & \vdots \\
-1 & -1 & -1 & 0 & 0 & \cdots & 3 + \beta
\end{pmatrix}.$$

Its inverse has the form

$$(D(G) - W(G) + \beta I)^{-1} = \frac{1}{\beta(n + \beta)}$$

$$\times \begin{pmatrix}
\frac{n-3+n\beta+\delta^2}{n+\beta-3} & \frac{n-3}{n+\beta-3} & \frac{n-3}{n+\beta-3} & 1 & 1 & \cdots & 1 \\
\frac{n-3}{n+\beta-3} & \frac{n-3+n\beta+\delta^2}{n+\beta-3} & \frac{n-3}{n+\beta-3} & 1 & 1 & \cdots & 1 \\
\frac{n-3}{n+\beta-3} & \frac{n-3}{n+\beta-3} & \frac{n-3+n\beta+\delta^2}{n+\beta-3} & 1 & 1 & \cdots & 1 \\
1 & 1 & 1 & \frac{3+n\beta+\delta^2}{3+\beta} & \frac{3}{3+\beta} & \cdots & \frac{3}{3+\beta} \\
1 & 1 & 1 & \frac{3}{3+\beta} & \frac{3+n\beta+\beta^2}{3+\beta} & \cdots & \frac{3}{3+\beta} \\
\vdots & \vdots & \vdots & \vdots & \vdots & \ddots & \vdots \\
1 & 1 & 1 & \frac{3}{3+\beta} & \frac{3}{3+\beta} & \cdots & \frac{3+n\beta+\beta^2}{3+\beta}
\end{pmatrix}.$$

By analogy with $K_{2,n-2}$, for $K_{3,n-3}$ it follows that

$$CF_\beta(v_1) = CF_\beta(v_2) = CF_\beta(v_3)$$

$$= \frac{1}{2n}\left(1 + \frac{(n-3)(\beta+n+1)}{(n+\beta-3)(\beta+n)} + \frac{(n-3)(\beta+2n-5)}{(3+\beta)(\beta+n)}\right),$$

$$CF_\beta(v') = \frac{1}{2n}\left(1 + \frac{3(\beta+n+1)}{(n+\beta-3)(\beta+n)} + \frac{3(\beta+2n-5)}{(3+\beta)(\beta+n)}\right).$$

4.3   Bipartite Graph $K_{r,n-r}$

Consider the general case where $v \in V_1$, $v' \in V_2$ and $r = |V1|$, $n - r = |V2|$.

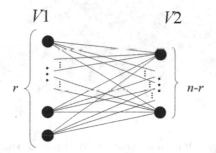

Fig. 2. Bipartite graph $K_{r,n-r}$.

By induction, for the general graph $K_{r,n-r}$,

$$CF_\beta(v) = \frac{1}{2n}\left(1 + \frac{(n-r)(\beta+n-2+r)}{(n+\beta-r)(\beta+n)} + \frac{(n-r)(\beta+2n-2-r)}{(r+\beta)(\beta+n)}\right),$$

$$CF_\beta(v') = \frac{1}{2n}\left(1 + \frac{r(\beta+n-2+r)}{(n+\beta-r)(\beta+n)} + \frac{r(\beta+2n-2-r)}{(r+\beta)(\beta+n)}\right).$$

Observe that all edges have the same βCF-centrality for the bipartite graph $K_{r,n-r}$, i.e.,

$$CF_\beta(e) = \frac{1}{n}\left(\frac{\beta+n-2+r}{(n+\beta-r)(\beta+n)} + \frac{\beta+2n-2-r}{(r+\beta)(\beta+n)}\right).$$

5   The Results of Computer Experiments with *Math-Net.ru*

Figure 3 shows the subgraph associated with *Math-Net.ru*, a Web portal of mathematical publications. The total amount of the authors on the portal is 78839.

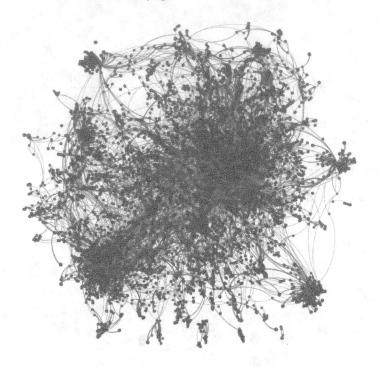

Fig. 3. Subgraph of mathematical Web portal *Math-Net.ru*.

We will consider only one connected component of this graph with 7606 mathematicians and 10747 publications coauthored by them. The nodes of the graph describe the authors and the link weights give the number of coauthored publications. Actually, the publications having more that 6 coauthors are ignored.

For simplicity, all links with the weights smaller than 7 are deleted, see the result in Fig. 4. Clearly, nodes *40, 34, 56* and *20* represent the centers of "local" stars and, consequently, must have a high centrality. Note that node *32* also must have a high centrality, as connecting two separate components.

Table 1 combines the ranking results for the first 11 nodes of the graph using βCF-centrality (formula (8) with the parameter $\beta = 1$), the PageRank algorithm with the parameter $\alpha = 0.85$ and electric centrality (CF-betweenness) developed in [7,8].

As supposed, nodes *40, 34, 56* and *20* have high centrality in all ranking methods considered. But PageRank assigns a low rank (34) to node *32*.

Now, let us detect the community structure of the network adhering to the approach developed in [10]. The whole idea of this approach lies in the following. If a network contains communities or groups that are only weakly connected via a few internal edges, then the edges connecting the communities have a high βCF centrality. By removing these edges, the groups are separated from each other, and the underlying community structure of the network is revealed.

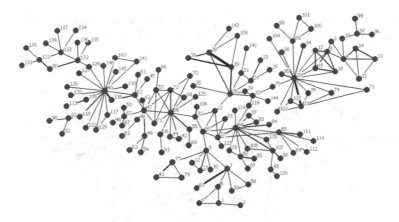

Fig. 4. Main component of subgraph associated with *Math-Net.ru*.

Table 1. Ranking results for nodes of *Math-Net.ru*

Node	Centrality (CF_β)	Node	PageRank	Node	CF-betwenness centrality
40	0.15740	40	0.04438	56	0.54237
34	0.14981	34	0.03285	32	0.53027
20	0.13690	20	0.03210	47	0.48222
47	0.12566	56	0.02774	22	0.41668
56	0.12518	47	0.02088	33	0.41361
26	0.10880	39	0.01874	34	0.39517
30	0.09098	28	0.01824	30	0.39426
9	0.08149	21	0.01695	52	0.37421
33	0.08024	65	0.01632	40	0.36946
32	0.07959	26	0.01552	26	0.35259
22	0.07903	107	0.01424	20	0.34413

First, calculate the βCF-centrality of all edges in the network. Find an edge with the higher centrality (actually, edge (32,56)) and remove it from the graph. Next, recalculate the βCF centrality of all edges in the modified network. Again, find an edge with the higher centrality and remove it from the graph, etc. The described process is continued until no edges remain.

The results of these computations are presented below.

$$(32, 56), (9, 30), (47, 52), (20, 75), (22, 26), (34, 119), (128, 132), (9, 11), (4, 5), \ldots$$

After ranking of all edges, all nodes of the network can be divided into the communities (clusters). Figure 4 shows the resulting community structure of the collaboration network on the Web portal *Math-Net.ru*. The graph splits into 7 communities corresponding to different fields of mathematics, namely, coding, discrete mathematics, mathematical physics, functional analysis, algebra and topology, optimal control, and probability theory (Fig. 5).

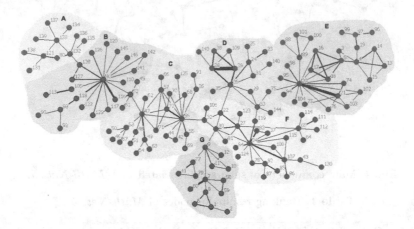

Fig. 5. Community structure of the subgraph associated with *Math-Net.ru*.

6   Conclusion

This paper has investigated the community structure of networks using a new concept of betweenness centrality measure. The βCF-centrality measure of the nodes in a network depending on the parameter β was introduced earlier in [3] based on electric circuit interpretation. In the present paper, this measure has been extended to the edges in a network. Moreover, the measure has been applied for detecting the community structure of networks. The proposed method has been have tested on the graph of mathematical publications available at the Web portal *Math-Net.ru*.

Acknowledgements. This research was supported by the Russian Foundation for Basic Research (project no. 16-51-55006), the Russian Humanitarian Science Foundation (project no. 15-02-00352) and the Division of Mathematical Sciences of the Russian Academy of Sciences.

References

1. Aumann, R., Myerson, R.: Endogenous formation of links between players and coalitions: an application of the Shapley value. In: Roth, A. (ed.) The Shapley Value, pp. 175–191. Cambridge University Press, Cambridge (1988)
2. Avrachenkov, K., Litvak, N., Medyanikov, V., Sokol, M.: Alpha current flow betweenness centrality. In: Bonato, A., Mitzenmacher, M., Prałat, P. (eds.) WAW 2013. LNCS, vol. 8305, pp. 106–117. Springer, Heidelberg (2013)

3. Avrachenkov, K.E., Mazalov, V.V., Tsynguev, B.T.: Beta current flow centrality for weighted networks. In: Thai, M.T., Nguyen, N.T., Shen, H. (eds.) CSoNet 2015. LNCS, vol. 9197, pp. 216–227. Springer, Heidelberg (2015)
4. Borgatti, S.P., Everett, M.G., Freeman, L.C.: Ucinet for Windows: Software for Social Network Analysis. Analytic Technologies, Harvard (2002)
5. Borm, P., Owen, G., Tijs, S.: On the position value for communication situations. SIAM J. Disc. Math. **5**(3), 305–320 (1992)
6. Borm, P., van den Nouweland, A., Tijs, S.: Cooperation and communication restrictions: a survey. In: Gilles, R.P., Ruys, P.H.M. (eds.) Imperfections and Behavior in Economic Organizations. Kluwer Academic Publishers, Boston (1994)
7. Brandes, U.: A faster algorithm for betweenness centrality. J. Math. Sociol. **25**, 163–177 (2001)
8. Brandes, U., Fleischer, D.: Centrality measures based on current flow. In: Diekert, V., Durand, B. (eds.) STACS 2005. LNCS, vol. 3404, pp. 533–544. Springer, Heidelberg (2005)
9. Freeman, L.C.: A set of measures of centrality based on betweenness. Sociometry **40**, 35–41 (1977)
10. Girvan, M., Newman, M.E.J.: Community structure in social and biological networks. Proc. Natl. Acad. Sci. U.S.A. **99**(12), 7821–7826 (2002)
11. Jackson, M.O.: Allocation rules for network games. Games Econ. Behav. **51**(1), 128–154 (2005)
12. Jackson, M.O.: Social and Economic Networks. Princeton University Press, Princeton (2008)
13. Mazalov, V.V., Trukhina, L.I.: Generating functions and the Myerson vector in communication networks. Disc. Math. Appl. **24**(5), 295–303 (2014)
14. Newman, M.E.J.: A measure of betweenness centrality based on random walks. Soc. Netw. **27**, 39–54 (2005)
15. Opsahl, T., Agneessens, F., Skvoretz, J.: Node centrality in weighted networks: generalizing degree and shortest paths. Soc. Netw. **32**, 245 251 (2010)
16. Slikker, M., Gilles, R.P., Norde, H., Tijs, S.: Directed networks, allocation properties and hierarchy formation. Math. Soc. Sci. **49**(1), 55–80 (2005)
17. Talman, D., Yamamoto, Y.: Average tree solutions and subcore for acyclic graph games. J. Oper. Res. Soc. Jpn. **51**(3), 187–201 (2008)

Trust Evaluation Based Friend Recommendation in Proximity Based Mobile Social Network

Fizza Abbas, Ubaidullah Rajput, Hasoo Eun, Dongsoo Ha, Taeseon Moon,
Wenhui Jin, Hyunjun Back, Honglae Jo, Sul Bang, Seung-ho Ryu,
and Heekuck Oh$^{(\boxtimes)}$

Department of Computer Science and Engineering, Hanyang University,
Ansan, South Korea
{fizza2012,hkoh}@hanyang.ac.kr

Abstract. Proximity Based Mobile Social Networks (PMSN) is a special type of Mobile Social Network (MSN) where a user interacts with
other users present in near proximity in order to make social relationships
between them. The users' mobile devices directly communicate with each
other with the help of Bluetooth/Wi-Fi interfaces. The possible presence
of a malicious user in the near proximity poses significant threats to the
privacy of legitimate users. Therefore, an efficient trust evaluation mechanism is necessary that enables a legitimate user to evaluate and decide
about the trustworthiness of other user in order to make friendship. This
paper proposes a protocol that evaluates various parameters in order
to calculate a trust value that assists a PMSN user in decision making
for building new social ties. This paper describes various components of
trust such as friend of friend, credibility and the type of social spot where
trust evaluation is being performed and then calculates these parameters in order to compute the final trust value. We utilize semi-trusted
servers to authenticate users as well as to perform revocation of malicious users. In case of an attack on servers, real identities of users remain
secure. We not only hide the real identity of the user from other users
but also from the semi-trusted servers. The inherent mechanism of our
trust evaluation protocol provides resilience against Sybil and Man-In-
The-Middle (MITM) attacks. Towards the end of the paper, we present
various attack scenarios to find the effectiveness and robustness of our
trust evaluation protocol.

1 Introduction

With the growth of Online Social Networks (OSN), there is a new emerging
paradigm that uses mobile devices for social networking. This new paradigm is
known as Mobile Social Network (MSN) and enables users to not only connect
with their social ties but also make new social relationships at any time and
place. There are many applications of MSN such as finding jobs, make friendship, missed connection services for example craigslist (just to name a few) [1].
Taking a step further, there is another class of MSN, known as proximity-based
mobile social network (PMSN) that enables a user to interact with people in

© Springer International Publishing Switzerland 2016
H.T. Nguyen and V. Snasel (Eds.): CSoNet 2016, LNCS 9795, pp. 158–169, 2016.
DOI: 10.1007/978-3-319-42345-6_14

near proximity [2,11]. For example, Jambo network and Nokia Sensor are such networks that are created on the go on locations such as conferences, exhibitions and concerts [3]. Because the users of such networks do not have any prior interaction, therefore, the inherent threat of a malicious user in the near proximity requires some sort of authentication and trust evaluation of users in near proximity. This trust evaluation helps a user to make relationships with other users while keeping the malicious users at bay [4].

Various trust models have been presented in the literature. The authors in [5] propose a model for trust in P2P MSN. The trust is evaluated based on the recommendation of existing communities and therefore a straight forward disadvantage for a user with no prior interaction with other users of some community. Goncalves et al. [6] utilize a trust model that is based on reputation of a node. They use direct and indirect factors where direct factor comprises of prior interaction of nodes while the indirect factor is based on opinion of other nodes. In [3] authors propose a mechanism that is based on direct and indirect models. Direct trust is obtained by monitoring the traffic while the indirect trust is based on trust evaluated by other nodes for the target node. Furthermore, in [7] authors design a trust model based on factors like prior trust between users, opinion of the third parties and reflexive relationship between users that is computed with interests similarity and mutual friends. However, we argue that the opinion of third parties might not reflect the actual reputation of a user unless the context of interaction is known. Furthermore, the protocol seems to require knowledge of a user's friends list and in practice the user might not want to disclose his/her friends list until unless some mutual friends are found. Recently authors propose GenTrust model which investigates the use of genetic programming on trust management using P2P system [8].

This paper proposes a trust evaluation protocol that is based on three factors. The first is based on secure and privacy preserving evaluation of mutual friends of two users. The second factor is based on credibility of a user that indicates the honesty of a user in his/her previous interactions. The third and final factor is based on the environment where the communications is taking place such as user's work place, exhibition or subway (just to name a few). We argue that this information plays an important role in trust evaluation of a person. A user evaluating trust at his/her workplace or institution where he/she is studying will feel more trusted about other users than a place like subway. The main contributions of our proposed protocol are as follows:

- We derive a novel trust equation that is based on Friend of Friend (FoF), the credibility of candidate user and the social spot where the communication is being taken place.
- We use semi-honest servers to assist the users of the system. The servers do not keep the real identities of the users and only keep the encrypted real identities. In case of a server compromise, real identities of the users remain secure.
- Our proposed protocol provides resilience against Sybil and MITM attacks.
- The trust based equation enables a user to give variable weightage to each of the trust evaluation parameters according to his/her preference.

The remainder of this paper is organized as follows. Section 2 provides preliminaries. Section 3 explains the proposed trust evaluation protocol. Section 4 gives analysis of our protocol. Section 5 provides conclusion and future work.

2   Preliminaries

This section includes the system model, threat model, design goals, assumptions, notations and the cryptographic tools.

2.1   System Model

The mobile devices of the users can be identified by a unique identifier such as International Mobile Equipment Identity. A user might have different mobile devices and therefore, it is difficult to prevent a Sybil attack. Therefore, we propose the use of social security number (SSN) of a user. However, due to the sensitive nature of this information, an efficient mechanism is required that should not only prevent any possible theft of this information but also any kind of misuse should not be allowed. In this regard, we propose the use of two separate servers for user registration and revocation and expect an honest-but-curious behavior from them. Following are the description of each of the system participants.

- **Registration Server (RegS):** The responsibilities of RegS include the user registration and data verification, managing database of users that includes encrypted data such as SSN, verifying incoming user authentication requests and providing encrypted SSN for revocation to the Revocation Server (RevS) in case of a malicious activity. The SSN is then updated as malicious in RegS database (DB).
- **Revocation Server (RevS):** When a user is registered with RegS, the most sensitive data item, the SSN, is encrypted by RegS with the public key of RevS and the plaintext SSN is deleted by the RegS. After that RevS has no access to this encrypted data. In case of detecting a malicious user, the RegS sends a revocation request along with encrypted SSN to RevS. The RevS decrypts SSN and sends the SSN to RegS.
- **Initiator (I):** The users of our system model first need to register themselves with the registration server. Upon registration, each of them receives an authentication token (referred as *Auth_Token*). The initiator is the one who first initiates the protocol and provide his/her *Auth_Token* to the other user (Responder). Both users authenticate each other with the help of this token. Initiator interacts with other users, exchange messages with them and evaluate various parameters in order to calculate final trust value. If the result of the resultant trust computation is greater than a threshold then the users exchange a friend token and become friends.
- **Responder (R):** The first job of a Responder is to verify the legitimacy of the incoming request. The responder sends the received *Auth_Token* of

the Initiator to the RegS in order to verify that the *Auth_Token* is not a result of a replay or Man-In-The-Middle (MITM) attack. The Responder also provides his *Auth_Token* to the Initiator as well. After verification, both users follow the protocol till the end. Like Initiator, Responder also evaluates trust parameters and takes decision about friendships.

2.2 Threat Model

Our threat model consists of two types of threats. The first type of threats comes from semi-honest (also referred as honest-but-curious) users. These are the users who follow the protocol but try to learn extra information such as keys and values of ciphertexts. The second types of threats come from an attacker. These attacks involve disrupting and deviating from the protocol by actively manipulating the protocol steps. These attacks include (but not limited to) Sybil attack, sabotaging the servers and using fake information such as presenting identities for finding mutual friends who are not actual friends of an attacker.

2.3 Design Goals

The design goals are as follows:

1. Modeling a trust equation that includes various user based parameters and subsequent evaluation of these parameters in order to assist a user in decision making regarding friendship.
2. The parameters that involve a user's personal data are needed to be evaluated in a privacy preserving manner. Moreover, only the mutual data should be revealed to both the parties without any significant advantage to one over the other.
3. Trust assumptions should be kept to a minimum.
4. Sybil attacks should be prevented in the mobile environment.

2.4 Assumptions

Following are the assumptions of proposed protocol.

1. The trust evaluation protocol, once started, cannot be terminated until its completion.
2. Users and servers keep their private keys safe.
3. The servers, RegS and RevS, do not collude with each other or with other users.
4. During the registration phase, the RegS performs cross checks on a user's Social Security Number (SSN) in order to verify it.
5. Both the servers communicate through a secure channel.

2.5 Notations

The notations used in our protocol are presented in Table 1.

Table 1. Notations

Notation	Explanation
I	The protocol initiator
R	Responder
RegS	Registration server
RevS	Revocation server
N_i, N_r	Initiator's and responder's random number
Cr	Credibility
id_{I_i}	Initiator's ith friend
id_{R_j}	Responder's jth friend
a, b	Random exponent number generated by initiator and responder respectively
PK, SK	ECC public and private keys
PK'	Paillier public key of RegS

2.6   Cryptographic Tools

Three separate cryptosystems have been used in our protocol. For simple public key infrastructure (PKI) operations the protocol utilizes elliptic curve cryptography (ECC). While for the probabilistic encryption, paillier homomorphic cryptosystem is used. Moreover, for matching purpose we use commutative encryption.

2.6.1   Elliptic Curve Cryptography (ECC)

The cubic equation of an elliptic curve has the form $y^2 + axy + by = x^3 + cx^2 + dx + e$, where a, b, c, d, & e are all real numbers. In an elliptic curve cryptography (ECC) system, the elliptic curve equation is defined as the form of $E_p(a, b) : y^2 = x^3 + ax + b(modp)$, over a prime finite field F_p, where a, $b \in F_p$, $p > 3$, and $4a^3 + 27b^2 \neq 0(modp)$ [9].

2.6.2   Commutative Encryption

Commutative encryption states that $E_{k_1}(E_{k_2}(m)) = E_{k_2}(E_{k_1}(m))$. It implies that if m is encrypted with a secret key and the resultant ciphertext is encrypted again by another secret key then changing the order of the decryption keys will not have any effect on the result. Therefore, $E_{k_1}(E_{k_2}(m)) = E_{k_2}(E_{k_1}(m'))$ iff $m = m'$ [10].

2.6.3   Paillier Encryption

A paillier encryption is a probabilistic cryptosystem [12]. In probabilistic encryption, the encryption results are probabilistic instead of deterministic. The same plaintext may map to two different ciphertexts at two different probabilistic encryption processes:

$$C_1 = E_k(M), C_2 = E_k(M), C_3 = E_k(M), \ldots, C_n = E_k(M)$$

The paillier cryptosystem provides significant facilities to guarantee the security of the designed protocols.

3   Proposed Protocol

Our protocol consists of five phases namely (1) System Initialization (2) User Registration (3) Trust Parameters Evaluation (4) Trust Computation and (5) Revocation. Following are the details regarding each of the phases. Figure 1 shows overall working of proposed protocol.

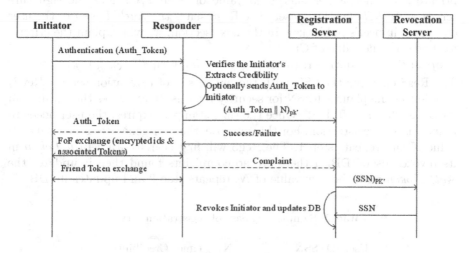

Fig. 1. Overall working of proposed protocol

3.1   System Initialization

First of all the system is initialized by the RegS by publishing its paillier public key (n, g). The Registration Server also establishes the ECC domain parameters p, a, b, G, n and h.

1. Let the field is defined by p.
2. The cyclic group is defined by its base point G.
3. n is the order of G.
4. a, b are curve constant
5. cofactor $h = 1/n \, |E(E_p)|$

All the users download these parameters. RegS randomly choose $x \in Z_{p^*}$ as its private secret key and announces it public key. Similarly other participants generate their ECC public private key pairs.

3.2   User Registration

Each user needs to be registered with the RegS. In order to do that, the user generates a 128 bit random number N, a user defined identity id, user's SSN, user's ECC public key and sends this to the RegS encrypted in RegS's ECC public key.

- **Step 1:** Initiator → RegS: $(SSN, id, N, PK_i)_{PK_{RegS}}$
  The RegS creates an authentication token for the user referred as $Auth\_Token$ in our protocol. In order to create this token, the RegS generates a time stamp T for the user, creates a parameter for the user named as Cr that is initialized with a value of zero, encrypts the value of N in its paillier public key and digitally signs this information along with user generated id with its ECC private key. The paillier encrypted value of N, id, T, Cr and the signature serves as $Auth\_Token$ for a user. The RegS sends the $Auth\_Token$ to the user encrypted in user's public key. In the next section, we will explain the criteria for updating the value of Cr.
- **Step 2:** RegS → Initiator: $(h((N)_{PK'_{RegS}} \parallel id \parallel Cr \parallel T)_{SK_{RegS}})_{PK_i}$
  The RegS encrypts the SSN with the public key of revocation server (RevS) and deletes the plaintext SSN for security reasons. RegS stores these values in its DB as shown in Table 2. Once the time stamp T expires, the user needs to renew his/her registration. For this purpose, the user generates a new random value N and repeats step 1. The RegS will first searches the $Auth\_Token$ in its revoked users' DB, if there are no complaints found then it updates the $Auth\_Token$ with the new value of N, repeats step 2 and updates its DB.

Table 2. Sample database of registration server

User ID	SSN	N	Time	Credibility
id_A	$(SSN)_{PK_{RevS}}$	N_A	T_A	Cr_A
id_B	$(SSN)_{PK_{RevS}}$	N_B	T_B	Cr_B
-	-	-	-	-

3.3 Trust Parameter Evaluation

In this phase, two users, namely Initiator and Responder interact with each other in order to evaluate various parameters that are later used to compute the final trust value. These parameters are Credibility (Cr), Friend of Friend (FoF) of two users, and the particular Social Spot Type (SST) where the communication is being taken place.

3.3.1 Credibility

Each $Auth\_Token$ of a user contains a value referred as Cr in our protocol. This value is updated in DB every time by the RegS when a user successfully authenticates himself to another user during the protocol. The RegS updates this value in $Auth\_Token$ at each renewal when timestamps T expires. Section 3.3.2 explains the updation of this value in detail.

3.3.2   Friend of Friend (FoF)

In the start of the communication, Initiator generates a random r from Z_{n^*} and encrypts the value of N in the paillier public key of RegS. Due to the property of paillier cryptosystem, the ciphertext of N will be different from the ciphertext of same N contained in the *Auth_Token*. The Initiator then appends this ciphertext with the *Auth_Token* and sends this to the Responder.

- **Step 1:** I $\rightarrow$ R: $(Auth\_Token)||(N_i)_{PK'_{RegS}}$
  To make things clear, we denote Initiator's N with N_i and Responder's N with N_r. Similarly the Responder sends his/her *Auth_Token* along with his/her paillier encrypted N_r.
- **Step 2:** R $\rightarrow$ I: $(Auth\_Token)||(N_r)_{PK'_{RegS}}$
  Initiator and Responder extract the value of Cr from the *Auth_Token* and forward the message to the RegS. It is important to note that unless the users cheat, this communication with the RegS is only required once during the interaction between both users.
  The RegS decrypts both the paillier encryptions. The one contained in the *Auth_Token* as well as the one appended with it. If the decryption of both reveals the same value of N then the RegS reply the user with *Success* message and increments the Cr value of the user in its DB. Otherwise sends the user a *Failure* message (that shows a possible Replay or MITM attack).
- **Step 3:** RegS $\rightarrow$ I : Success/Failure
  Following is the secure evaluation of FoF. We have adopted the approach proposed by [10] and further enhanced by [15] for users' interests matching.
  After getting assurance from the server, both the parties proceed. The Initiator generates an exponent a, and exponentiate each of his/her friends' ids with a as $id^a_{I_i}$ (id of the Initiator's ith friend). The Initiator then prepares a message containing the exponentiated ids of friends, appends Responder's *Auth_Token*, digitally signs entire message with his/her ECC private key and sends this message to Responder. The signature on the message provides non-repudiation. Similar procedure is followed by the Responder after generating his exponent b and the resultant set will consists of $id^b_{R_j}$ values (id of the Responder's jth friend).
- **Step 4:** I $\rightarrow$ R: $(\forall i \in (0,m) : id^a_{I_i}||Auth\_Token_r)_{SK_i}$.
- **Step 5:** R $\rightarrow$ I: $(\forall i \in (0,n) : id^b_{R_j}||Auth\_Token_i)_{SK_r}$.
  In the next step, the Initiator commutatively exponentiate values sent by Responder. After that he/she prepares a commitment by taking hash of the pair $id^b_{R_j}$, $(id^b_{R_j})^a$, *Auth_Token*, signature on entire message and sends this commitment to Responder.
- **Step 6:** I $\rightarrow$ R: $h(\forall i \in (0,n) : (id^b_{R_j}, (id^b_{R_j})^a)||Auth\_Token_r)_{SK_i}$.
  In reply, Responder prepares the message consists of the pair $id^a_{I_i}$, $(id^a_{I_i})^b$, *Auth_Token*, signature on the entire message and sends it to Initiator.
- **Step 7:** R $\rightarrow$ I: $(\forall i \in (0,m) : (id^a_{I_i}, (id^a_{I_i})^b)||Auth\_Token_i)_{SK_r}$.
  Initiator sends value of commitments.
- **Step 8:** I $\rightarrow$ R: $(\forall i \in (0,n) : (id^b_{R_j}, (id^b_{R_j})^a)||Auth\_Token_r)_{SK_i}$.

- **Step 9:** Initiator computes $(id^b_{R_j})^a \cap (id^a_{I_i})^b$.
  - If there is a match, the Initiator identifies this match with the help of corresponding $id^a_{I_i}$ in the pair sent by the Responder in step 7.
- **Step 10:** Responder computes $(id^a_{I_i})^b \cap (id^b_{R_j})^a$.
  - If there is a match, the Responder identifies the match with the help of corresponding $id^b_{R_j}$ in the pair sent by the Responder in step 8.
  - For each value of $id_{R_j} = id_{I_i}$, there must be a match. Therefore, Initiator and Responder must have same number of matches.
  - Finally, both the Initiator and Responder are required to provide the proof that they have a valid friendship with corresponding matched id_{I_i} (in case of Initiator) and id_{R_j} (in case of Responder).
  - A valid proof means to provide the *Friend Token*. Whenever two users have a successful trust relationship, each of them exchanges *Friend Token* with each other. Suppose A and B are two users who become friends after a successful trust evaluation then the *Friend Token* given by A to B will have the following structure. $(id_A || id_b || Auth\_Toekn_A)_{SK_A}$.
  - Therefore, for each of the id_{I_i}, where the corresponding id_{R_j} is a match, both the users need to provide the corresponding Friend Token for that id. Failing to do so will cause the victim to complain the RegS with the recordings of the protocol and subsequent revocation of the guilty party. At the end of this sub phase, each party has the common number of FoF.

3.3.3 Social Spot Type

Mobile social networks such as developed in [13], enable a user's device to strike up communication at any time and place: on the bus, in a bar, at a conference, in a subway, in a shopping mall or at work place. We argue that the environment where a communication is being taken place can contribute to a user's over all trust. For example, a user A communicating with a user B within the proximity of his/her work environment will feel more trusted about B than a user C who is striking a communication with A on subway. Therefore, in our protocol, we allow a user to give a rating/value to a particular social spot. These values range from 1 to 10. For example the user gives a higher value of 10 to its work environment while give a lowest value of 1 to subway while traveling. However, a user who travels more frequently might put a higher value to a travel related social spot.

3.4 Trust Computation

After evaluating the values of various components, the final trust equation can be given as:

$$Trust = \alpha * FoF + \beta * Cr + \gamma * SST$$

Where $\alpha + \beta + \gamma \leq 1$ are adjustable user preference parameters and can be adjusted according to a user's preference given to each of the evaluated trust components. If the Trust $\geq \tau$ (user defined threshold) then user requests other user to exchange *Friend Token*.

It is important to note that this trust equation considers a user's opinion in the final trust computation. A user might be more comfortable to give a higher weightage to friend of friend feature while the other might consider the credibility of a user a more important while computing overall trust.

3.5   Revocation

If any of the participants cheat, then the victim presents the protocol recordings to the RegS. Let's suppose Responder is the victim.

- The victim sends the protocol recordings to the RegS.
  Responder → RegS : Recordings
  RegS checks the signatures on the messages and the evidence of cheating. For example, the malicious party denied sending the *Friend Token*. The RegS will send a request of revocation to the RevS. This request consists of encrypted SSN of the malicious user. The *id* of a user in *Auth_Token* leads to the encrypted SSN in RegS DB.
- RegS → RevS : Revoke_Request$||(SSN)_{PK_{RevS}}$.
  The RevS will decrypt the SSN and will give it back to the RegS.
- RevS → RegS: SSN.
  The RegS will mark corresponding id/SSN as *Revoked* in its database.

4   Analysis of Proposed Protocol

This section presents the security analysis of the protocol. We analyze the protocol with respect to various design goals as well as various types of attacks.

Scenario 1: In our protocol, the users should only know about mutual friends. No other information except the matched elements should be disclosed.

Solution: Each user encrypts friend's identity with a secret exponent. After that each of them commutatively encrypts other's data and matches the resultant ciphertexts. Due to the property of commutative encryption as described in Sect. 2.6.2, the commutative ciphertexts match iff they have a common identity in their data set. Therefore, both the users only find the mutual friends between them.

Scenario 2: During FoF matching, if an eavesdropper tries to know the encrypted communication.

Solution: Agrawal et al. [10] proved that given the encryption $f_a(m)$ with the secret exponent a, one cannot compute $f_a(m)^{-1}$ without knowing a. Therefore, the communication is secure.

Scenario 3: If an honest-but-curious user wants to know the encrypted identities of friends.

Solution: In case of an Initiator, all the identities Id_{I_i} are encrypted with secret exponent a of the Initiator. According to Agrawal et al. [10], $f_a(m)$ is irreversible without the key a. Therefore, without the necessary key information, the Responder cannot learn $f_a(m)^{-1}$.

Scenario 4: During the initial authentication, an attacker plays a Man-In-The-Middle attack by sending some initiator's *Auth_Token* to a Responder.

Solution: A user's *Auth_Token* comprises of a secret value N that is encrypted in RegS's paillier public key and only known to the owner of *Auth_Token*. During initial authentication, a user is required to encrypt the value of N again with RegS's paillier public key along with the *Auth_Token*. The responder verifies this value by sending both the ciphertexts to the RegS. Unless the value of N is known, an attacker cannot launch such attack.

Scenario 5: A user is able to take part indefinitely after revocation.

Solution: The *Auth_Token* is digitally signed and issue by the RegS. It contains a timestamp T and it must be re-acquired by a user after the T expires. In case a revoked user tries to re-acquire the *Auth_Token*, he/she needs to presents the previous *Auth_Token* that contains user's identity. The RegS maintains a DB of revoked users and finds the identity of a revoked user upon arrival of a *Auth_Token* issue request. Therefore, the RegS will reject the request of a revoked user.

Scenario 6: In our protocol, if the RegS or the RevS is compromised.

Solution: During the initial registration of a user, his/her SSN is encrypted by the RegS with the public key of RevS and the plaintext data is deleted. This idea is adopted from one of our earlier work [14]. Therefore, in an unlikely case of a RegS compromise, no user identity data is revealed to the attacker. Similarly, In case RevS compromise, there is no loss of data because RevS has no user data.

5   Conclusions and Future Work

This paper proposes a secure protocol to evaluate trust in PMSN. The paper proposes various trust evaluation parameters and allows a user to securely evaluate these values in order to find the final trust value. The paper also proposes a secure authentication mechanism that prevents replay and MITM attacks. In order to provide authentication and revocation, the paper proposes two semi-trusted servers that collaborate with each other to provide efficient authentication and revocation. In case of a compromise of a server, no valuable user identity data is revealed. In the future work, we aim to extend and implement the paper by developing a mobile application, evaluate our trust mechanism more comprehensively and compare the protocol to similar approaches.

Acknowledgment. This research was supported in part by the MSIP (Ministry of Science, ICT and Future Planning), Korea, under the ITRC (Information Technology Research Centre) support program (IITP-2016-H8501-16-1007) and (IITP-2016-H8501-16-1018) supervised by the IITP (Institute for Information and communications Technology Promotion).

This work was also supported in part by the NRF (National Research Foundation of Korea) grant funded by the Korea government MEST (Ministry of Education, Science and Technology) (No. NRF-2015R1D1A1A09058200).

References

1. Najaflou, Y., Jedari, B., Xia, F., Yang, L.T., Obaidat, M.S.: Safety challenges and solutions in mobile social networks. IEEE Syst. J. **99**, 1 21 (2013)
2. Kayastha, N., Niyato, D., Wang, P., Hossain, E.: Applications, architectures, and protocol design issues for mobile social networks: a survey. In: Proceedings of the IEEE, vol. 99, pp. 2130–2158 (2011)
3. Li, J., Zhang, Z., Zhang, W.: Mobitrust: trust management system in mobile social computing. In: IEEE 10th International Conference on Computer and Information Technology (CIT), pp. 954–959 (2010)
4. Sherchan, W., Nepal, S., Paris, C.: A survey of trust in social networks. ACM Comput. Surv. **45**(4), 47:1–47:33 (2013)
5. Qureshi, B., Min, G., Kouvatsos, D.: M-Trust: a trust management scheme for mobile P2P networks. In: IEEE/IFIP International Conference on Embedded and Ubiquitous Computing, pp. 476–483 (2010)
6. Goncalves, M., Dos Santos Moreira, E., Martimiano, L.: Trust management in opportunistic networks. In: IEEE 9th International Conference on Networks (ICN), pp. 209–214 (2010)
7. Huerta-Canepa, G., Lee, D., Han, S.Y.: Trust me: a trust decision framework for mobile environments. In: IEEE 10th International Conference on Trust, Security and Privacy in Computing and Communications (TrustCom), pp. 464–471 (2011)
8. Tahta, U.E., Sen, S., Can, A.B.: GenTrust: a genetic trust management model for peer-to-peer systems. Appl. Soft Comput. **34**, 693–704 (2015)
9. Certicom Research: Sec 1: Elliptic Curve Cryptography. http://www.secg.org/sec1-v2.pdf
10. Agrawal, R., Evfimievski, A., Srikant, R.: Information sharing across private databases. In: ACM International Conference on Management of Data (SIGMOD), pp. 86 97 (2003)
11. Abbas, F., Rajput, U., Hussain, R., Eun, H., Oh, H.: A trustless broker based protocol to discover friends in proximity-based mobile social networks. In: Rhee, K.-H., Yi, J.H. (eds.) WISA 2014. LNCS, vol. 8909, pp. 216–227. Springer, Heidelberg (2015)
12. Paillier, P., Pointcheval, D.: Efficient public-key cryptosystems provably secure against active adversaries. In: Lam, K.-Y., Okamoto, E., Xing, C. (eds.) ASIACRYPT 1999. LNCS, vol. 1716, pp. 165–179. Springer, Heidelberg (1999)
13. Eagle, N., Pentland, A.: Social serendipity: mobilizing social software. IEEE Pervasive Comput. **4**(2), 28–34 (2005)
14. Rajput, U., Abbas, F., Eun, H., Hussain, R., Oh, H.: A two level privacy preserving pseudonymous authentication protocol for VANET. In: IEEE 11th International Conference on Wireless and Mobile Computing, Networking and Communications (WiMob), pp. 643–650 (2015)
15. Xie, Q., Hengartner, U.: Privacy-preserving matchmaking for mobile social networking secure against malicious users. In: IEEE 9th International Conference on Privacy, Security and Trust (PST), pp. 252–259 (2011)

Integrating with Social Network to Enhance Recommender System Based-on Dempster-Shafer Theory

Van-Doan Nguyen$^{(\boxtimes)}$ and Van-Nam Huynh

Japan Advanced Institute of Science and Technology (JAIST),
1-1 Asahidai, Nomi, Ishikawa 923-1292, Japan
nvdoan@jaist.ac.jp

Abstract. In this paper, we developed a new collaborative filtering recommender system integrating with a social network that contains all users. In this system, user preferences and community preferences extracted from the social network are modeled as mass functions, and Dempster's rule of combination is selected for fusing the preferences. Especially, with the community preferences, both the sparsity and cold-start problems are completely eliminated. So as to evaluate and demonstrate the advantage of the new system, we have conducted a range of experiments using Flixster data set.

Keywords: Recommender system · Collaborative filtering · Social network · Community preference · Dempster-Shafer theory

1 Introduction

Recently, recommender systems [1,2] have been developed to satisfy both online users (customers) and online suppliers. For online users, recommender systems help to deal with information overload by providing a list of suitable items (products or services) to a specific person [1]. On the other hand, for providers, the systems are employed as an effective tool for increasing sale growths, improving the user satisfaction and fidelity, as well as better understanding what a user wants [3]. For this reason, recommender systems have been widely applied in e-commerce applications [4,5].

According to the literature, recommendation techniques can be classified into six main categories: content-based, collaborative filtering (CF), demographic, knowledge-based, community-based, and hybrid [3]. Among these, CF is referred to as "people to people correlation" [6] and considered to be the most popular and widely implemented technique [7]. In order to generate suitable recommendations for an active user, commonly, CF systems try to find other users who have similar preferences with this user, and then use their existing ratings for calculating predictions for the user. However, CF systems are limited due to the sparsity

This research work was supported by JSPS KAKENHI Grant No. 25240049.

© Springer International Publishing Switzerland 2016
H.T. Nguyen and V. Snasel (Eds.): CSoNet 2016, LNCS 9795, pp. 170–181, 2016.
DOI: 10.1007/978-3-319-42345-6_15

and cold-start problems [1]. The first problem happens because each user only rates a very small subset of items; and this issue is considered to be significantly affects performances of recommendations [8]. The second problem is caused by new items and new users. When new items have been just added, the systems do not have information about people's preferences on these items; in such a case, it is difficult to recommend them to users. Additionally, when new users have just joined in, the systems have no knowledge about preferences of the users; so, it is also difficult to generate recommendations for them.

So far, many researchers focus on solving the sparsity as well as cold-start problems in CF systems, and a variety of methods have been proposed to overcome these problems. A popular method is Matrix Factorization [9] that exploits latent factors and applys them for predicting all unprovided ratings. In [10,11], the authors combined CF technique with other one as content-based or demographic. Besides, some authors suggested using additional information from other sources, such as implicit preferences inferred from users' actions that relate to specific item [12], or context information [13,14].

Furthermore, these days social networks are growing rapidly as well as playing a vital role on the Internet. In general, these networks are known as an effective communication and collaboration medium that can connect many people. In fact, social networks consist of huge amount of information that could be useful to improve quality of recommendations [15]. Because of this, integrating recommender systems with social networks has emerged as an active research topic [16,17]. Up to now, various CF systems based on social networks have been developed; and most of these systems employ social trust [18–20] or community context information [21] for tackling the sparsity and cold-start problems.

Naturally, in a social network, people are formed into some communities; and in each of which, members frequently interact with one another [22] as well as discussing or sharing information about a variety of topics including the ones about items. Before buying an item, commonly, people tend to ask for experience or consult for advice from their relatives or friends in the same community. Moreover, most of people believe in opinion or recommendations from other members in the community rather than from the ones outside.

Under such an observation, in this paper, we develop a new CF recommender system that exploits community preferences extracted from a social network containing all users for improving the quality of recommendations as well as dealing with both the sparsity and cold-start problems. Additionally, in the new system, user and community preferences are modeled using Dempster-Shafer (DS) theory [23,24]; and with this characteristic, the system is capable of representing the preferences with uncertain, imprecise, and incomplete information as well as fusing information from different sources easily.

The rest of this paper is arranged as follows. In the next section, background information of DS theory is presented. Then, details of the proposed system are described. Next, experiments and results are shown. Finally, conclusions are drawn in the last section.

2   Dempser-Shafer Theory

This theory [23,24] offers a particularly convenient framework for modeling uncertain, imprecise and incomplete information. Let us consider a problem domain represented by an exhaustive and finite set Γ containing L elements, called a frame of discernment (FoD) [24]. A mass function $m \colon 2^\Gamma \to [0,1]$ is defined on Γ as follows

$$m(\emptyset) = 0; \quad \sum_{A \subseteq \Gamma} m(A) = 1. \tag{1}$$

A subset $A \subseteq \Gamma$ with $m(A) > 0$ is called a focal element of mass function m. In case $m(\Gamma) = 1$ and $\forall A \neq \Gamma, m(A) = 0$, mass function m is called to be vacuous. In addition, when the information source providing mass function m has probability δ of reliability, this mass function needs to be discounted by a discount rate $1 - \delta \in [0,1]$ as below

$$m^\delta(A) = \begin{cases} \delta \times m(A), & \text{for } A \subset \Gamma; \\ \delta \times m(\Gamma) + (1 - \delta), & \text{for } A = \Gamma. \end{cases} \tag{2}$$

Belief function $Bel \colon 2^\Gamma \to [0,1]$ and plausibility function $Pl \colon 2^\Gamma \to [0,1]$ are derived from mass function m, as follows

$$Bel(A) = \sum_{\emptyset \neq B \subseteq A} m(B);$$

$$Pl(A) = \sum_{A \cap B \neq \emptyset} m(B). \tag{3}$$

A probability distribution Pr such that $Bl(A) \leq Pr(A) \leq Pl(A), \forall A \in 2^\Gamma$ is said to be compatible with mass function m, and pignistic probability distribution Bp [25] is a typical one, as below

$$Bp(\gamma) = \sum_{\{A \subseteq \Gamma | \gamma \in A\}} \frac{m(A)}{|A|}.$$

Assuming that we have two mass functions m_1 and m_2 defined on the same FoD Γ. These two mass functions can be combined by using Dempster's rule of combination [24], denoted by $\odot$, as follows

$$(m_1 \odot m_2)(\emptyset) = 0;$$

$$(m_1 \odot m_2)(A) = \frac{1}{1 - K} \sum_{\{B,C \subseteq \Gamma | B \cap C = A\}} m_1(B) \times m_2(C), \tag{4}$$

where $K = \sum_{\{B,C \subseteq \Gamma | B \cap C = \emptyset\}} m_1(B) \times m_2(C) \neq 0$, and K represents basic probability mass associated with conflict.

3 Proposed System

3.1 Data Modeling

Let $\mathfrak{U}$ be the set of M users and $\mathfrak{I}$ be the set of N items. Each rating of a user $u \in \mathfrak{U}$ on an item $i \in \mathfrak{I}$ is represented as a mass function $r_{u,i}$ spanning over a rating domain Γ consisting of L preference levels, and Dempster's rule of combination, described in Eq. (4), is selected to apply for fusing information.

Context information can be considered as concepts which may significantly influence user preferences on items [13,14,21]. For example, when evaluating a cell phone product, '*Iphone 6s*', users will focus on its characteristics such as '*color*', '*internal memory*', '*battery*', '*shape*'; thus, these characteristics can be viewed as concepts. Additionally, each concept can consist of a number of groups, e.g. for '*Iphone 6s*', concept '*color*' can consist of several groups as '*silver*', '*gold*', '*space gray*'. Formally, assuming that, in the system, context information consists of P concepts denoted by $\mathfrak{C} = \{C_1, C_2, ..., C_P\}$; and each concept $C_p \in \mathfrak{C}$ contains at most Q_p groups, denoted by $C_p = \{g_{p,1}, g_{p,2}, ..., g_{p,Q_p}\}$.

For a concept $C_p \in \mathfrak{C}$, a user $u \in \mathfrak{U}$ can be interested in several groups, and an item $i \in \mathfrak{I}$ can belongs to one or some groups of this concept. Groups in which user u is interested and groups to which item i belong are identified by mapping functions $^\mathfrak{U}f_p$ and $^\mathfrak{I}f_p$ respectively, as follows

$$\begin{aligned}
^\mathfrak{U}f_p : \mathfrak{U} &\rightarrow 2^{C_p} \\
u &\mapsto {}^\mathfrak{U}f_p(u) \subseteq C_p; \\
^\mathfrak{I}f_p : \mathfrak{I} &\rightarrow 2^{C_p} \\
i &\mapsto {}^\mathfrak{I}f_p(i) \subseteq C_p.
\end{aligned} \tag{5}$$

We assume that all users together join to a social network representing as an undirected graph $\mathfrak{G} = (\mathfrak{U}, \mathfrak{F})$, where $\mathfrak{U}$ is the set of nodes (users) and $\mathfrak{F}$ is the set of edges (friend relationships). In this network, users can form into several communities, and a user can belong to one or several communities at the same time. One of existing methods, such as removal of high-betweenness [26], mimicking human pair wise communication [27], analysing graph structures [28,29] and so on, can be adopted to detect the overlapping communities in the network. Supposing that, after detecting, we achieve T communities in total.

Generally, the recommendation process of the proposed system is illustrated in Fig. 1. As can be seen in this feature, main tasks such as extracting information, dealing with the problems, computing user-user similarities, selecting neighborhoods and generating recommendations are performed in each community independently. Note that, in the remainder of this section, we will present about details of these tasks in a community.

3.2 Performing in a Community

The rating matrix of all members in the community is denoted by $\mathfrak{R} = \{r_{u,i}\}$ with $r_{u,i}$ is the rating of user u on item i. In addition, all items rated by user u as well as all users who have rated item i are denoted by $^\mathfrak{I}\mathfrak{R}_u$ and $^\mathfrak{U}\mathfrak{R}_i$, respectively.

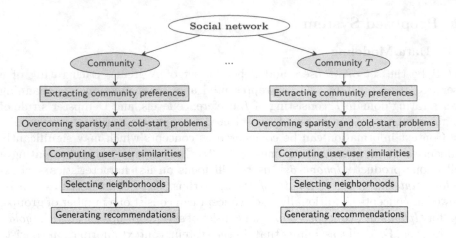

Fig. 1. Recommendation process

Extracting Community Preferences. Let us consider an item i. If this item belongs to group $g_{p,q}$ of a concept C_p then each rating of user u, who is also interested in group $g_{p,q}$, on this item can be considered as a piece of community preference on item i regarding group $g_{p,q}$, denoted by $^G m_{i,p,q}$. Consequently, $^G m_{i,p,q}$ can be obtained by combining all related pieces as below

$$^G m_{i,p,q} = \bigodot_{\{u \in {}^{u}\mathfrak{R}_i | g_{p,q} \in {}^{u}f_p(u) \cap {}^{f}f_p(i)\}} r_{u,i}. \tag{6}$$

Since $g_{p,q} \in C_p$, $^G m_{i,p,q}$ reflects a piece of community preference on item i regarding concept C_p, denoted by $^C m_{i,p}$. Thus, $^C m_{i,p}$ is computed as follows

$$^C m_{i,p} = \bigodot_{\{q | g_{p,q} \in C_p, g_{p,q} \in {}^{f}f_p(i)\}} {}^G m_{i,p,q}. \tag{7}$$

Obviously, $^C m_{i,p}$ is regarded as a piece of community preference on item i as a whole, denoted by $^C m_i$. Therefore, $^C m_i$ is calculated as below

$$^C m_i = \bigodot_{\{p | \exists g_{p,q} \in C_p, g_{p,q} \in {}^{f}f_p(i)\}} {}^C m_{i,p}. \tag{8}$$

Besides, community preference on item i regarding group $g_{p,q}$, $^G m_{i,p,q}$, is viewed as a piece of community preference regarding group $g_{p,q}$ overall all items, denoted by $^G m_{p,q}$. As a result, $^G m_{p,q}$ is obtained as follows

$$^G m_{p,q} = \bigodot_{\{i \in \mathfrak{I} | \exists g_{p,q} \in C_p, g_{p,q} \in {}^{f}f_p(i)\}} {}^G m_{i,p,q}. \tag{9}$$

Overcoming the Sparsity Problem. In the system, unprovided ratings need to be created in order to deal with the sparsity problem. Supposing that, user u has not rated item i; the process for generating unprovided rating $r_{u,i}$ contains five steps as follows:

– Firstly, as mentioned earlier, the preference of user u can be influenced by preference of the community. Thus, according to a concept $\mathcal{C}_p$, if user u is interested in a group $g_{p,q}$ to which item i belong, user u's preference on item i regarding group $g_{p,q}$, denoted by $^G m_{u,i,p,q}$, is signed by community preference on item i regarding the same group $g_{p,q}$, as below

$$^G m_{u,i,p,q} = {}^G m_{i,p,q}. \tag{10}$$

– Secondly, it can be seen that $^G m_{u,i,p,q}$ reflects a piece of user u's preference on item i regarding concept $\mathcal{C}_p$, denoted by $^C m_{u,i,p}$. Consequently, $^C m_{u,i,p}$ is computed as follows

$$^C m_{u,i,p} = \bigodot_{\{q|g_{p,q}\in\mathcal{C}_p, g_{p,q}\in {}^\mathcal{U}\!f_p(u)\cap {}^\mathcal{I}\!f_p(i)\}} {}^G m_{u,i,p,q}. \tag{11}$$

– Thirdly, $^C m_{u,i,p}$ is considered to be a piece of user u's preference on item i as a whole, denoted by $^C m_{u,i}$. As a result, $^C m_{u,i}$ is achieved as below

$$^C m_{u,i} = \bigodot_{\{p|\exists g_{p,q}\in\mathcal{C}_p, g_{p,q}\in {}^\mathcal{U}\!f_p(u)\cap {}^\mathcal{I}\!f_p(i)\}} {}^C m_{u,i,p}. \tag{12}$$

– Next, unprovided rating $r_{u,i}$ is assigned user u's preference on item i, as shown below

$$r_{u,i} = {}^C m_{u,i}. \tag{13}$$

– Finally, if the context information does not affect user u and item i, in other words $\forall p, {}^\mathcal{U}\!f_p(u) \cap {}^\mathcal{I}\!f_p(i) = \emptyset$, unprovided rating $r_{u,i}$ cannot be generated by using Eqs. (10), (11), (12) and (13). In this case, we propose that $r_{u,i}$ is assigned community preference on item i, as follows

$$r_{u,i} = {}^C m_i. \tag{14}$$

Overcoming the Cold-Start Problem: New Items. Let us consider a new item i'. According to a concept $\mathcal{C}_p$, for each $g_{p,q} \in \mathcal{F}_p(i')$, the community preference on a group $g_{p,q}$ is considered to be the community preference on item i' regarding group $g_{p,q}$, as below

$$^G m_{i',p,q} = {}^G m_{p,q}. \tag{15}$$

Then, we can apply Eqs. (10), (11), (12), (13) and (14) to generate the unprovided ratings on item i' for all users in the community.

In the special situation, the groups to which item i' belongs are very new for the community; in order words, $^G m_{p,q}$ corresponding to $\forall g_{p,q} \in \mathcal{F}_p(i')$ does not exist. If there are some users, who are not interested in any $g_{p,q} \in \mathcal{F}_p(i')$ but have rated the item, the unprovided rating of user u on item i' is assigned by combining all existing ratings on the item as follows

$$r_{u,i'} = \bigodot_{\{u' \in {}^{u}\mathfrak{R}_{i'}|g_{p,q} \notin {}^{u}\mathcal{F}_p(u'), g_{p,q} \in \mathcal{F}_p(i')\}} r_{u',i'}. \tag{16}$$

Otherwise, if nobody in the community has rated this item, ${}^{u}\mathfrak{R}_{i'} = \emptyset$, then for each user u in the community, $r_{u,i'}$ is assigned by vacuous.

Overcoming the Cold-Start Problem: New Users. Let us consider a new user u'. In case the profile of user u' contains information about the groups of each concept $\mathcal{C}_p$, in which u' is interested, the unprovided ratings of this user on each item i are generated as follows:

– Community preference on item i regarding a group $g_{p,q} \in {}^{u}\mathcal{F}_p(u')$ is considered as user u''s preference on item i regarding group $g_{p,q}$, as below

$$^{G}m_{u',i,p,q} = {}^{G}m_{i,p,q}. \tag{17}$$

– Applying the Eqs. (11), (12), (13) and (14) for user u', the unprovided rating $r_{u',i}$ will be created.

Otherwise, if the information about the groups in which user u' is interested is not available, the unprovided rating of this user on item i is assigned community preference on item i as follows

$$r_{u',i} = {}^{\mathcal{C}}m_i. \tag{18}$$

At this point, all unprovided ratings related new items as well as new users have been created. In the next tasks, these items are treated the same as any other one; and there is no difference between the new users and the other ones in terms of being recommended. As a result, the cold-start problem is eliminated in the system.

Computing Similarities. The method proposed in [13] is used for computing user-user similarities in the system. With this method, the distance between two users u and u', denoted by $D(u, u')$, is computed as below

$$D(u, u') = \sum_{i=1}^{N} \mu(x_{u,i}, x_{u',i}) \left(\ln \max_{\gamma \in \Gamma} \frac{Bp_{u',i}(\gamma)}{Bp_{u,i}(\gamma)} - \ln \min_{\gamma \in \Gamma} \frac{Bp_{u',i}(\gamma)}{Bp_{u,i}(\gamma)} \right), \tag{19}$$

where $Bp_{u,i}$ and $Bp_{u',i}$ are the pignistic probability distributions according to ratings of user u and user u' on item i respectively; and $\mu(x_{u,i}, x_{u',i}) \in [0, 1]$ is a reliable function referring to the trust of the evaluation of both user u and user u' on item i. Here, $x_{u,i} \in \{0, 1\}$ and $x_{u',i} \in \{0, 1\}$ equal to 1 if $r_{u,i}$ and $r_{u',i}$ are provided ratings respectively; otherwise, $r_{u,i}$ and $r_{u',i}$ are predicted ratings. The function $\mu(x_{u,i}, x_{u',i})$ can be computed as follows

$$\mu(x_{u,i}, x_{u',i}) = 1 - \rho_1 \times (x_{u,i} + x_{u',i}) - \rho_2 \times x_{u,i} \times x_{u',i}, \qquad (20)$$

where ρ_1 and ρ_2 are the reliable coefficients representing the state when a user has rated an item and two users together have rated an item, respectively [13].

The user-user similarity between users u and u', denoted by $s_{u,u'}$, is computed as follows

$$s_{u,u'} = e^{-\lambda \times D(u,u')}, \text{ where } \lambda \in (0, \infty). \qquad (21)$$

With the higher value of $s_{u,u'}$, the user u is closer to user u'. Finally, the user-user similarities among all users are represented in a matrix $S = \{s_{u,u'}\}$.

Selecting Neighborhoods. Let us consider item i which has not been rated by active user u. A set containing K nearest neighborhoods of this user is denoted by $\mathfrak{N}_{u,i}$ and selected by using the method in [30]. Firstly, a set of users who already rated item i and whose similarities with user u are equal or greater than a threshold τ is chosen; this set is denoted by $\mathcal{N}_{u,i}$ and obtained by the following equation

$$\mathcal{N}_{u,i} = \{u' \mid i \in {}^{\mathfrak{I}}\mathfrak{R}_{u'}, s_{u,u'} \geq \tau\}. \qquad (22)$$

Note that, the condition $i \in {}^{\mathfrak{I}}\mathfrak{R}_{u'}$ is removed if ${}^{\mathfrak{I}}\mathfrak{R}_{u'} = \emptyset$. Secondly, all of members in $\mathcal{N}_{u,i}$ is sorted in ascending by $s_{u,u'}$ and top K members are selected as the neighborhood set $\mathfrak{N}_{u,i}$.

Generating Recommendations. The members in $\mathfrak{N}_{u,i}$ are considered to have the similar taste with user u on item i. Therefore, the estimated rating of user u on item i, denoted by $\hat{r}_{u,i}$, is calculated by the following equation

$$\hat{r}_{u,i} = r_{u,i} \odot \tilde{r}_{u,i}, \qquad (23)$$

where $\tilde{r}_{u,i}$ is the overall preference of all members in $\mathfrak{N}_{u,i}$. The rating of each user $u' \in \mathfrak{N}_{u,i}$ on item i need to be discounted by a discount rate $1 - s_{u,u'}$ [14]. As a result, $\tilde{r}_{u,i}$ can be calculated as below

$$\tilde{r}_{u,i} = \bigodot_{\{u' \in \mathcal{N}_{u,i}\}} \dot{r}_{u',i}^{s_{u,u'}}, \qquad (24)$$

$$\text{with } \dot{r}_{u',i}^{s_{u,u'}} = \begin{cases} s_{u,u'} \times r_{u',i}(A), \text{ for } A \subset \Gamma; \\ s_{u,u'} \times r_{u',i}(\Gamma) + (1 - s_{u,u'}), \text{ for } A = \Gamma. \end{cases}$$

Note that, in case user u belongs to several communities simultaneously, the last estimated rating of this user on item i is achieved by fusing the estimated ratings item i in the communities to which user u belongs.

The proposed system offers both hard and soft decisions. To generate a hard decision on a singleton $\gamma \in \Gamma$, the pignistic probability is applied, and then the singleton having the highest probability is selected as the preference label. Additionally, to generate a soft decision, the maximum belief with overlapping interval strategy (maxBL [31]) is selected [14].

4 Experiment and Result

Flixster data set [21] was selected to use in experiments. It consists of 3,827 users with 49,410 friend relationships, 1210 movies, and 535,013 hard ratings with rating domain $\Gamma = \{0.5, 1, 1.5, 2, 2.5, 3, 3.5, 4, 4.5, 5\}$. Since the proposed system works with soft ratings, the mapping function suggested in [14] was adopted for transforming hard ratings into soft ratings. Note that, in this function, the trust factor and dispersion factor were selected as 0.9 and 2/9, respectively. In the data set, each movie can belong to at most 19 genres. However, the genres in which a user is interested are not available. Thus, we assume that a user is interested in a genre if this user has rated at least 11 movies belonging to that genre. In addition, so as to discover overlapping communities in the social network whose nodes are linked by undirected friendships, the algorithm developed in [27] was selected; and after executing this algorithm, 7 overlapping communities were detected.

Table 1. Experimental results

K	DS-MAE		DS-Precision		DS-Recall	
	Baseline	New system	Baseline	New system	Baseline	New system
5	0.95978	**0.85227**	0.20442	**0.22506**	0.17956	**0.18872**
10	0.95263	**0.85241**	0.20533	**0.22495**	0.18101	**0.18878**
15	0.94846	**0.85241**	0.20561	**0.22498**	0.18174	**0.18884**
20	0.94643	**0.85236**	0.20580	**0.22511**	0.18202	**0.18894**
25	0.94492	**0.85235**	0.20596	**0.22516**	0.18231	**0.18895**
30	0.94378	**0.85231**	0.20596	**0.22514**	0.18247	**0.18895**
35	0.94259	**0.85231**	0.20589	**0.22517**	0.18263	**0.18898**
40	0.94151	**0.85229**	0.20600	**0.22522**	0.18281	**0.18902**
45	0.94057	**0.85229**	0.20594	**0.22522**	0.18295	**0.18905**
50	0.94007	**0.85229**	0.20595	**0.22525**	0.18300	**0.18908**
55	0.93964	**0.85227**	0.20597	**0.22531**	0.18308	**0.18912**
60	0.93919	**0.85226**	0.20600	**0.22536**	0.18315	**0.18916**
65	0.93900	**0.85226**	0.20599	**0.22540**	0.18317	**0.18918**
70	0.93891	**0.85226**	0.20600	**0.22539**	0.18320	**0.18918**
75	0.93869	**0.85226**	0.20602	**0.22541**	0.18322	**0.18919**
80	0.93853	**0.85227**	0.20605	**0.22538**	0.18325	**0.18919**

The system [21], developed based on DS theory, was selected as a baseline for performance comparison. In this baseline, community context information extracted from the social network is used for dealing with the sparsity problem; and for new items as well as users who are not affected by the context information, unprovided ratings are signed by vacuous. Additionally, evaluation methods *DS-MAE* [14], and *DS-Precision*, *DS-Recall* [32] were chosen to measure recommendation performances.

In each community, all values in user-user similarity matrix S were sorted in ascending order, and then, the value of $s_{u,u'}$ that can retain top 50 % of highest values in S was selected for parameter τ. The other parameters were set as following: $\rho_1 = 0.3, \rho_2 = 0.1$ and $\lambda = 10^{-4}$. In addition, this data set was separated into two parts, testing data and training data; the first one contains random selections of 5 ratings of each user, and the other consists of the remaining ratings.

The experimental results are summarized and illustrated in Table 1. In this table, performances of two systems according to the selected evaluation criteria vary with neighborhood size K; and the bold values indicate the better ones. As observed, the proposed system is more effective than the baseline in all cases. These results show that integrating with community preferences is capable of improving the quality of recommendations.

5   Conclusion

It can be seen that performances in accuracy of CF systems are limited by the sparsity and cold-start problems. In this paper, we have introduced a new CF system that exploits the community preferences extracted from the social network for dealing with these problems. In the future, we will focus on improving the quality of community preferences.

References

1. Adomavicius, G., Tuzhilin, A.: Toward the next generation of recommender systems: a survey of the state-of-the-art and possible extensions. IEEE Trans. Knowl. Data Eng. **17**(6), 734–749 (2005)
2. Bobadilla, J., Ortega, F., Hernando, A., Gutiérrez, A.: Recommender systems survey. Knowl. Based Syst. **46**, 109–132 (2013)
3. Ricci, F., Rokach, L., Shapira, B.: Introduction to recommender systems handbook. In: Ricci, F., Rokach, L., Shapira, B., Kantor, P.B. (eds.) Recommender Systems Handbook, pp. 1–35. Springer, New York (2011)
4. Al-hassan, M., Lu, H., Lu, J.: A semantic enhanced hybrid recommendation approach: a case study of e-government tourism service recommendation system. Decis. Support Syst. **72**, 97–109 (2015)
5. Shambour, Q., Lu, J.: A hybrid trust-enhanced collaborative filtering recommendation approach for personalized government-to-business e-services. Int. J. Intell. Syst. **26**(9), 814–843 (2011)
6. Schafer, J.B., Konstan, J.A., Riedl, J.: E-commerce recommendation applications. Data Min. Knowl. Discov. **5**(1/2), 115–153 (2001)
7. Jannach, D., Zanker, M., Ge, M., Gröning, M.: Recommender systems in computer science and information systems – a landscape of research. In: Huemer, C., Lops, P. (eds.) EC-Web 2012. LNBIP, vol. 123, pp. 76–87. Springer, Heidelberg (2012)

8. Huang, Z., Chen, H., Zeng, D.: Applying associative retrieval techniques to alleviate the sparsity problem in collaborative filtering. ACM Trans. Inf. Syst. **22**(1), 116–142 (2004)
9. Koren, Y., Bell, R.M., Volinsky, C.: Matrix factorization techniques for recommender systems. IEEE Comput. **42**(8), 30–37 (2009)
10. Lucas, J.P., Luz, N., García, M.N.M., Anacleto, R., de Almeida Figueiredo, A.M., Martins, C.: A hybrid recommendation approach for a tourism system. Expert Syst. Appl. **40**(9), 3532–3550 (2013)
11. Shambour, Q., Lu, J.: An effective recommender system by unifying user and item trust information for B2B applications. J. Comput. Syst. Sci. **81**(7), 1110–1126 (2015)
12. Grčar, M., Mladenič, D., Fortuna, B., Grobelnik, M.: Data sparsity issues in the collaborative filtering framework. In: Nasraoui, O., Zaïane, O.R., Spiliopoulou, M., Mobasher, B., Masand, B., Yu, P.S. (eds.) WebKDD 2005. LNCS (LNAI), vol. 4198, pp. 58–76. Springer, Heidelberg (2006)
13. Nguyen, V.D., Huynh, V.N.: A reliably weighted collaborative filtering system. In: Destercke, S., Denoeux, T. (eds.) Symbolic and Quantitative Approaches to Reasoning with Uncertainty. LNCS, vol. 9161, pp. 429–439. Springer, Switzerland (2015)
14. Wickramarathne, T.L., Premaratne, K., Kubat, M., Jayaweera, D.T.: Cofids: a belief-theoretic approach for automated collaborative filtering. IEEE Trans. Knowl. Data Eng. **23**(2), 175–189 (2011)
15. He, J., Chu, W.W.: A social network-based recommender system (SNRS). In: Memon, N., Xu, J.J., Hicks, D.L., Chen, H. (eds.) Data Mining for Social Network Data. Annals of Information Systems, vol. 12, pp. 47–74. Springer, US (2010)
16. Konstas, I., Stathopoulos, V., Jose, J.M.: On social networks and collaborative recommendation. In: SIGIR, pp. 195–202 (2009)
17. Sun, Z., Han, L., Huang, W., Wang, X., Zeng, X., Wang, M., Yan, H.: Recommender systems based on social networks. J. Syst. Softw. **99**, 109–119 (2015)
18. Guo, G., Zhang, J., Thalmann, D.: Merging trust in collaborative filtering to alleviate data sparsity and cold start. Knowl. Based Syst. **57**, 57–68 (2014)
19. Papagelis, M., Plexousakis, D., Kutsuras, T.: Alleviating the sparsity problem of collaborative filtering using trust inferences. In: Herrmann, P., Issarny, V., Shiu, S.C.K. (eds.) iTrust 2005. LNCS, vol. 3477, pp. 224–239. Springer, Heidelberg (2005)
20. Wu, H., Yue, K., Pei, Y., Li, B., Zhao, Y., Dong, F.: Collaborative topic regression with social trust ensemble for recommendation in social media systems. Knowl. Based Syst. **97**, 111–122 (2016)
21. Nguyen, V.-D., Huynh, V.-N.: A community-based collaborative filtering system dealing with sparsity problem and data imperfections. In: Pham, D.-N., Park, S.-B. (eds.) PRICAI 2014. LNCS, vol. 8862, pp. 884–890. Springer, Heidelberg (2014)
22. Tang, L., Liu, H.: Community Detection and Mining in Social Media. Synthesis Lectures on Data Mining and Knowledge Discovery. Morgan & Claypool Publishers, San Rafael (2010)
23. Dempster, A.P.: Upper and lower probabilities induced by a multivalued mapping. Ann. Math. Stat. **38**, 325–339 (1967)
24. Shafer, G.: A Mathematical Theory of Evidence. Princeton University Press, Princeton (1976)
25. Smets, P.: Practical uses of belief functions. In: UAI 1999, pp. 612–621. Morgan Kaufmann Publishers Inc., San Mateo (1999)

26. Gregory, S.: A fast algorithm to find overlapping communities in networks. In: Daelemans, W., Goethals, B., Morik, K. (eds.) ECML PKDD 2008, Part I. LNCS (LNAI), vol. 5211, pp. 408–423. Springer, Heidelberg (2008)

27. Xie, J., Szymanski, B.K.: Towards linear time overlapping community detection in social networks. In: Tan, P.-N., Chawla, S., Ho, C.K., Bailey, J. (eds.) PAKDD 2012, Part II. LNCS, vol. 7302, pp. 25–36. Springer, Heidelberg (2012)

28. Kim, J., Lee, J.: Community detection in multi-layer graphs: a survey. ACM SIGMOD Rec. 44(3), 37–48 (2015)

29. Sun, P.G., Gao, L.: A framework of mapping undirected to directed graphs for community detection. Inf. Sci. 298, 330–343 (2015)

30. Herlocker, J.L., Konstan, J.A., Borchers, A., Riedl, J.: An algorithmic framework for performing collaborative filtering. In: SIGIR 1999, pp. 230–237. ACM (1999)

31. Bloch, I.: Some aspects of Dempster-Shafer evidence theory for classification of multi-modality medical images taking partial volume effect into account. Pattern Recogn. Lett. 17(8), 905–919 (1996)

32. Hewawasam, K.K.R., Premaratne, K., Shyu, M.: Rule mining and classification in a situation assessment application: A belief-theoretic approach for handling data imperfections. IEEE Trans. Syst., Man, Cybern., Syst., Part B, 37(6), 1446–1459 (2007)

Exploiting Social Relations to Recommend Scientific Publications

Tin Huynh, Trac-Thuc Nguyen[✉], and Hung-Nghiep Tran

University of Information Technology - Vietnam,
Km 20, Hanoi Highway, Linh Trung Ward, Thu Duc District,
Ho Chi Minh City, Vietnam
{tinhn,thucnt,nghiepth}@uit.edu.vn

Abstract. To study about the state of the art for a research project, researchers must conduct a literature survey by searching for, collecting, and reading related scientific publications. By using popular search systems, online digital libraries, and Web of Science (WoS) sources such as IEEE Explorer, ACM, SpringerLink, and Google Scholar, researchers could easily search for necessary publications related to their research interest. However, the rapidly increasing number of research papers published each year is a major challenge for researchers in searching for relevant information. Therefore, the aim of this study is to develop new methods for recommending scientific publications for researchers automatically. The proposed ones are based on exploiting explicit and implicit relations in the academic field. Experiments are conducted on a dataset crawled from Microsoft Academic Search [1]. The experimental results show that our proposed methods are very potential in recommending publications that are meet with research interest of researchers.

Keywords: Publication recommendation · Academic social networks · Researcher profile · Academic trust relationship

1 Introduction

To start a research project, senior researchers who have strong domain knowledge may already know which related publications should read for their research. However, the rapidly increasing number of research papers published each year is a major challenge for researchers in searching for publications related to their research interest because of information overload. Besides, students or junior researchers are less experience in finding relevant publications. They don't know which related publications should read for their research and they often need advices from their supervisor or experienced colleagues. Therefore, a publication recommender system that can automatically suggest a ranked list of relevant publications should be a very useful tool for both junior and senior researchers.

Content-based approach is one of the most successful approaches widely applied to develop new methods for recommending prospective scientific publications automatically. Content-based methods often study how to model profile

© Springer International Publishing Switzerland 2016
H.T. Nguyen and V. Snasel (Eds.): CSoNet 2016, LNCS 9795, pp. 182–192, 2016.
DOI: 10.1007/978-3-319-42345-6_16

describing research interest of researchers. After that, the researcher's profile is matched with content of existing publications to filter out $TopN$ of relevant ones for recommendation. However, content-based methods do not concern in academic social relations while these relationships are key factors affecting to research interest and decision of researchers. Thus, the aim of this study is to exploit implicit and explicit academic relationships to develop new methods for recommending publications.

The key contributions of this study are summarized as follows:

– Proposing ASN (Academic Social Networks) model for modeling relationships in the academic domain.
– Exploiting academic social relations in the ASN model to develop new methods for recommending scientific publications.

The remainder of this paper is organized as follows. In Sect. 2, we present a formal definition of the problem. Section 3 is brief survey of related works. Section 4 describes our approach in detail. The dataset, experimental results, and discussion are provided in Sect. 5. We conclude the paper and suggest future research in Sect. 6.

2   Problem Definition

The aim of this problem is to identify a utility function $f\colon R \times P \to \mathbb{R}$, to estimate how a paper $p_j \in P$ is useful to a specified researcher $r_i \in R$.

Where,

– $R = \{r\}$: set of all researchers.
– $P = \{p\}$: set of all publications.
– $P_c \subset P$: set of publications have been cited by other researchers.
– $Existed\_Rating = \{v_{r_i,p_j}\}$. Where, v_{r_i,p_j} presents a rating degree of $r_i \in R$ with $p_j \in P_c$ based on citation of r_i to p_j in the past.

For each given researcher $r_i \in R$, the paper recommender system need to generate a ranked list $(TopN)$ of prospective publications, $P_{TopN} = <p_1, p_2, ..., p_{TopN}>$, that are considered the best useful to r_i to do recommendation. $TopN$ of potential publications, $P_{TopN} = <p_1, p_2, ..., p_{TopN}>$, are selected according to the following constraints:

(i) $\forall p_k \in P_{TopN}, v(r_i, p_k) \notin Existed\_Rating$. It means that the system should recommend publications that r_i don't know yet.
(ii) $\forall p_k \in P_{TopN}, f(r_i, p_k) \geq f(r_i, p_{k+1})$, where $1 \leq k \leq TopN - 1$. It means that P_{TopN} is a ordered set (a ranked list). Therefore, the higher rank a publication have in the list, the higher priority it is recommended.
(iii) $\forall p_k \in P_{TopN}, \forall p_{no\_rec} \in P \backslash P_{TopN}, f(r_i, p_k) \geq f(r_i, p_{no\_rec})$. Utility value of recommended publication, which is computed by f function, must be greater than utility value of no-recommended publications.

3   Related Work

Relating to recommendation of scientific publications, there are several different sub-problems that the recent studies have been concerning on. For example, He et al. [3,4], Huang et al. [5] proposed new methods for recommending cited papers when researcher is writing a paper. Most of these studies aimed at developing a model for mapping between sentences in the writing paper with cited publications in references section. On the other hand, Lawrence et al. [8], Huynh et al. [6], has carried out research to develop new algorithms to recommend similar publications when users browse a publication in digital libraries.

In other studies, Sugiyama et al. proposed new methods to recommend scientific publications matching with the interest of researchers [9,10]. They proposed a method to build researcher's profile based on aggregating her/his publications with her references and citations from citation network. They collected 597 publications from the ACL conference (Association of Computational Linguistics) and consulted 28 researchers. These researchers have to read a list of 597 papers and label which publication related or not related to their research interests. After that, the authors has used this dataset to build the ground truth. The nature of the citation network is very sparse. Therefore, Sugiyama et al. has tried to reduce the sparse data by collaborative filtering to explore potential cited publications and used this result to refine the profile of candidates. Experimental results show that the exploitation of potential cited publications improved accuracy of recommendation [11].

In another study, Jianshan Sun et al. have proposed a new method for the recommendations of scientific publications by combining content of publications and social relations of researchers [12]. They have extracted a list of related publications and social relationships of researchers from the CiteULike website to build empirical dataset including ground truth, training set and testing set. Experimental result showed that their hybrid method outperform than content based approach.

Joeran Beel et al. conducted a survey of more than 170 papers, patents, web pages are published in this field [2]. This survey has shown that until now there are no benchmarks as well as methods/metrics to evaluate various approaches for this problem. Thus, it is very difficult to know strengths and weaknesses of the existing methods.

Currently, the study of Sugiyama et al. [9,10], Sun et al. [12], is the most similar to the problem which we are studying in this paper. However, most of these studies were not really interested in academic implicit social relationships, especially various types of trust relationship when making recommendations of scientific publications. Therefore, the aim of this study is to exploit implicit and explicit academic relationships for developing new methods to recommend publications. The next section presents details of our approach.

4   Our Approach

In order to take academic relationships into developing new methods, firstly these potential relations (both implicit and explicit ones) should be recognized from the collection of scientific publications. This study has proposed a model, ASN Model (Academic Social Networks), used for modeling implicit and explicit relationships from the collection (Fig. 1). The next section presents key components of the ASN model.

4.1   Modeling Academic Social Networks

The ASN model is a set of key components as following:

$$ASN = (CoNet, CiNet\_Author, CiNet\_Paper, AffNet, M) \qquad (1)$$

where,

- $CoNet$: Coauthor network.
- $CiNet\_Author$: Citation network among researchers.
- $CiNet\_Paper$: Citation network among publications.
- $AffNet$: Collaborative network among research institutes.
- M: set of computing methods used to estimate how strong relationships are.

When choosing a publication for reading, researchers are not only interested in content but also quality of publication. There are many different factors affecting to the quality of a publication such as: reputation of journal, conference,

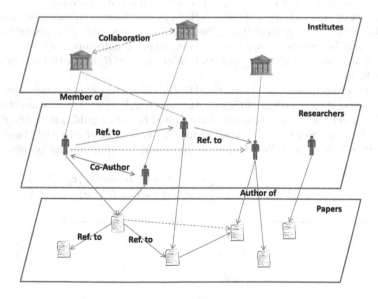

Fig. 1. Implicit and explicit relationships extracted from the collection of scientific publications (dashed and solid lines)

authors. In fact, researchers often have trust in some specific experts in relevant research topic and they tend to choose publications of these experts for their reading. For junior researchers, they often need advices, suggestions of their supervisor or experienced colleagues. Therefore, next section presents our proposed methods which combine content-based profile of researchers and academic relationships in ASN model. For these academic relations, this paper have specifically focused on trust relationships which are defined, computed by using CoNet and CiNet_Author in the ASN model.

4.2   Exploiting Explicit Relationships in CoNet and CiNet_Author

Assuming that the trust of a researcher for a paper depends on the trust level of her/his own and the trust of her/his co-author. Details of our method can be summarized as following:

Method: CB-TrendTrust1

Input:

- Set of all researchers $R = \{r\}$
- Set of all available publications $P = \{p\}$

Output: $\forall r \in R$, return a ranked list of Top-N publications $p \in P$ based on values predicted by the utility function.

1: **Step 1**: Building the coauthor network, $CoNet(R, E_1)$.
   - R: set of vertices of $CoNet(R, E_1)$. Each vertex is a researcher.
   - E_1: set of directed edges of $CoNet(R, E_1)$ presenting coauthor-ships. The direction from r_x to r_y present that r_x has coauthor-ships with r_y in r_x's publications. In the other words, r_x has trusted in r_y when r_x collaborate with r_y.
2: **Step 2**: Building the citation network $CiNet\_Author(R, E_2)$ with two key components R, E_2.
   - R: set of vertices of $CiNet\_Author(R, E_2)$. Each vertex is a researcher.
   - E_2: set of directed edges of $CiNet\_Author(R, E_2)$ presenting citation-ships. The direction from r_x to r_y presents that r_x cited r_y in r_x's publications ($r_x, r_y \in R$). In the other words, r_x has trusted in r_y. The weight of edges or the degree of trust relations in $CiNet\_Author(R, E_2)$ could be computed as follows:

$$w_{cite}(r_x, r_y, t_0) = \sum_{t_i=t_0}^{t_c} \frac{NumCitation(r_x, r_y, t_i)}{e^{(t_c - t_i)} * TotalCitation(r_x, t_i)} \quad (2)$$

3: Where,
   - $NumCitation(r_x, r_y, t_i)$: number of times that r_x cited r_y in year t_i.
   - $TotalCitation(r_x, t_i)$: Total number citations of r_x in year t_i.
   - t_c: the current year.
   - t_0: year to start considering the trend factor.
4: **Step 3:** Aggregating citation-ships r_x with citations of r_y's coauthors to compute degree of trust between r_x and r_y from t_0, $w_{trust}(r_x, r_y, t_0)$.

$$w_{trust}(r_x, r_y, t_0) = w_{cite}(r_x, r_y, t_0) +$$

$$\frac{\sum_{r_u \in CoAuthor(r_x)} w_{coauthor}(r_x, r_u, t_0) * w_{cite}(r_u, r_y, t_0)}{|CoAuthor(r_x)|} \quad (3)$$

$$W_{coauthor}(r_x, r_y, t_0) = \frac{f_{Trend}(r_x, r_y, t_0)}{\sum_{\forall r_u \in CoAuthor(r_x)} f_{Trend}(r_x, r_u, t_0)}, \quad (4)$$

$$f_{Trend}(r_x, r_y, t_0) = \sum_{t_i = t_0}^{t_c} n(r_x, r_y, t_i) * \frac{1}{e^{(t_c - t_i)}} \quad (5)$$

Where,
   - $CoAuthor(r_x)$: set of researchers who had coauthor-ship with r_x.
   - $n(r_x, r_y, t_i)$: total of papers which r_x had coauthor-ship with r_y in year t_i.
   - t_0: year to start considering the trend factor.
   - t_c: current year.
5: **Step 4:** Computing the trust weight of a researcher r_x to a publication p_j.

$$w_{trust}(r_x, p_j, t_0) = MAX(w_{trust}(r_x, r_y, t_0)) \quad (6)$$

(With $r_y \in R_{p_j}$: set of authors of publication p_j)
6: **Step 5:** Linear combination of trend-based content similarity and the degree of trust. $\forall r_x \in R, p_j \in P : RatingValue(r_x, p_j) = 0$,

$$RatingValue(r_x, p_j, t_0) = \alpha * w_{trust}(r_x, p_j, t_0) + (1 - \alpha) * Sim_{CB}(r_x, p_j) \quad (7)$$

7: **Step 6:** Recommendation. $\forall r_x \in R$,
   - Selecting Top-N of publications based on value of $RatingValue(r_x, p_j)$ to recommend for r_x.

Analyzing the complexity of CB-TrendTrust1:

- Steps 1 and 2 are considered as preprocessing steps.
- Step 3: This step computes the trust weight for all pairs of researchers. For each researcher $r \in R$, this step also computes for all her/his coauthors, $|CoAuthor(r)|$. Therefore, the complexity of this step is $\mathcal{O}(|R|^2 k)$ (k: average number of coauthors of a researcher).
- Step 4: This step computes the trust weight of all researchers to all publications. For each $p \in P$, this step has worked for $|Authors(p)|$ authors. Therefore, the complexity of this step is $\mathcal{O}(|R||P|l)$ (l: average number of authors of a publication p).

– In summary, because of $|P| >> |R|$, so the complexity of this method is $\mathcal{O}(|R||P|l)$ (where $|R|$: total number of researchers, $|P|$: total number of publications, l: average number of authors in one publication).

4.3   Exploiting Implicit Relationships in the CiNet_Author

In fact, one researcher typically follows citations in references section of reading papers to identify potential relevant publications. That is a implicit citing relation based on bridging relations in the CiNet_Author. From trust perspective, one researcher can put her/his trust in other researchers based on bridging trust relationships. Details of our method for mining potential trust relationships based on implicit citations can be summarized as follows:

Method: CB-TrendTrust2

Input:

– Set of all researchers $R = \{r\}$
– Set of all available publications $P = \{p\}$

Output: $\forall r \in R$, return a ranked list of Top-N publications $p \in P$ based on values predicted by the utility function.

1: **Step 1:** Similar to the CB-TrendTrust1 method
2: **Step 2:** Aggregating citation-ships of r_x with citations of researchers who are cited by r_x to compute the trust weight between r_x and r_y from t_0, $w_{trust}(r_x, r_y, t_0)$.
   $w_{trust}(r_x, r_y, t_0) = w_{cite}(r_x, r_y, t_0) +$

$$\frac{\sum\limits_{r_u \in CitedAuthor(r_x)} w_{cite}(r_x, r_u, t_0) * w_{cite}(r_u, r_y, t_0)}{|CitedAuthor(r_x)|} \tag{8}$$

3: **Step 3:** Similar to steps 4, 5, 6 of the CBTrendTrust1 method.

Analyzing the complexity of CB-TrendTrust2: similar to the CBTrendTrust1 method, the complexity of the CB-TrendTrust2 is $\mathcal{O}(|R||P|l)$ (where $|R|$: total number of researchers, $|P|$: total number of publications, l: average number of authors in one publication).

5   Experiments and Evaluation

5.1   Dataset

Joeran Beel et al. showed that until now there are no benchmarks as well as methods, metrics to evaluate various approaches for this problem [2]. Thus, it is very difficult to know strengths and weaknesses of existing methods. In this

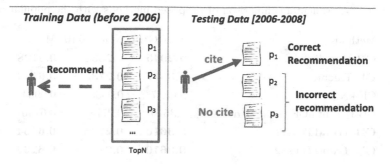

Fig. 2. Method to evaluate recommended publications

study, we have collected scientific publications at Microsoft Academic Search website to construct our empirical dataset. In order to contribute our dataset to research community, we have published our empirical dataset at https://sites.google.com/site/tinhuynhuit/dataset.

In our experiment, we randomly choose 1,000 researchers that published papers before 2006 and after 2006 as input data. The publications were published before 2006 selected as the training data. The papers that were cited by 1.000 researchers from 2006 to 2008 were chosen as ground truth to verify quality of recommendation methods. We assume that the result of recommendation is correct when recommended publication is cited in the ground truth by the researcher; otherwise it is incorrect (Fig. 2). The ground truth consists of 52.254 publications that were cited by 1.000 researchers from 2006 to 2008. Dividing the dataset into the past data (training data) and the future data (ground truth) to qualify recommendation methods commonly applied in other studies such as: Tang et al. [13], Sugiyama et al. [9,10], Sun et al. [12].

5.2 Experimental Results

We use NDCG [7], and MRR [14], which are used popularly in other studies to evaluate accuracy of recommendation. The higher these metrics are, the better the method is. The Table 1 shows the experimental results of our proposed methods compare with other popular methods such as content-based method (CB-Baseline), Collaborative Filtering (CF-kNN), CB-Recent that is proposed by Sugiyama et al. [9]. The experimental result shows that our propose methods outperforms the others (Fig. 3).

Basing on experimental results, we could bring out claims as following:

- The content-based approach is suitable for recommending scientific publications while the CF approach does not work for this problem.
- Modeling researcher's profile based on trend improved accuracy of recommendation.
- Combining trend-based content similarity with trust degree of a researcher to a publication has contributed to improve accuracy of recommendation, but the archived result is not really significant yet.

Table 1. Experimental comparison of proposed methods with the popular ones

Methods	NDCG@5	NDCG@10	MRR
CB-Baseline	**0.2945**	**0.2334**	**0.5128**
CB-Recent	**0.3577**	**0.2735**	**0.6142**
CF-kNN	0.0357	0.0330	0.0934
CB-Recent+CF (Best at $\alpha = 0.9$)	0.3570	0.2728	0.6140
CB-TrendTrust1	**0.3610**	**0.2778**	**0.6164**
CB-TrendTrust2	**0.3610**	**0.2778**	**0.6164**

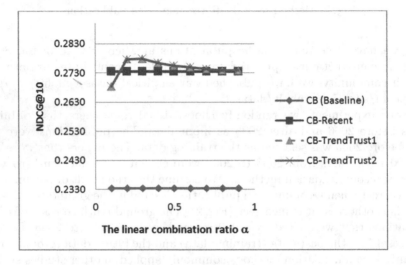

Fig. 3. Selecting the ratio α for linear combination of trend-based content similarity and the degree of trust relation

6   Conclusion and Future Work

The aim of this study is to develop new methods for recommending scientific publications for researchers. The key contributions of this study includes: proposing ASN (Academic Social Networks) model for modeling explicit and implicit social relationships; exploiting academic social relationships, especially trust relationships in the ASN model to develop new methods. The proposed methods are a linear combination of content-based similarity and trust degree computed by using social relations. The combination helps to improve the accuracy of recommendation, but the archived result is not really significant yet. Our next step in near future is to conduct experiment for different hybrid methods. We will also consider learning a model to predict potential publications for researchers by using multi features (content-based, link-based, time-aware features).

Acknowledgments. This research is funded by Vietnam National University HoChiMinh City (VNU-HCM) under grant number C2014-26-03

References

1. Microsoft Academic Search. http://academic.research.microsoft.com
2. Beel, J., Langer, S., Genzmehr, M., Gipp, B., Breitinger, C., Nürnberger, A.: Research paper recommender system evaluation: a quantitative literature survey. In: Proceedings of the International Workshop on Reproducibility and Replication in Recommender Systems Evaluation, RepSys 2013, pp. 15–22. ACM, New York (2013). http://doi.acm.org/10.1145/2532508.2532512
3. He, Q., Kifer, D., Pei, J., Mitra, P., Giles, C.L.: Citation recommendation without author supervision. In: Proceedings of the Fourth ACM International Conference on Web Search and Data Mining, WSDM 2011, pp. 755–764. ACM, New York (2011). http://doi.acm.org/10.1145/1935826.1935926
4. He, Q., Pei, J., Kifer, D., Mitra, P., Giles, L.: Context-aware citation recommendation. In: Proceedings of the 19th International Conference on World Wide Web, WWW 2010, pp. 421–430. ACM, New York (2010). http://doi.acm.org/10.1145/1772690.1772734
5. Huang, W., Kataria, S., Caragea, C., Mitra, P., Giles, C.L., Rokach, L.: Recommending citations: translating papers into references. In: Proceedings of the 21st ACM International Conference on Information and Knowledge Management, CIKM 2012, pp. 1910–1914. ACM, New York (2012). http://doi.acm.org/10.1145/2396761.2398542
6. Huynh, T., Luong, H., Hoang, K., Gauch, S., Do, L., Tran, H.: Scientific publication recommendations based on collaborative citation networks. In: Proceedings of the 3rd International Workshop on Adaptive Collaboration (AC 2012) as Part of The 2012 International Conference on Collaboration Technologies and Systems (CTS 2012), Denver, Colorado, USA, 21–25 May 2012, pp. 316–321 (2012)
7. Järvelin, K., Kekäläinen, J.: IR evaluation methods for retrieving highly relevant documents. In: Proceedings of the 23rd Annual International ACM SIGIR Conference on Research and Development in Information Retrieval, SIGIR 2000, pp. 41–48. ACM, New York (2000). http://doi.acm.org/10.1145/345508.345545
8. Lawrence, S., Giles, C.L., Bollacker, K.: Digital libraries and autonomous citation indexing. Computer **32**, 67–71 (1999)
9. Sugiyama, K., Kan, M.Y.: Scholarly paper recommendation via user's recent research interests. In: Proceedings of the 10th Annual Joint Conference on Digital Libraries, JCDL 2010, pp. 29–38. ACM, New York (2010). http://doi.acm.org/10.1145/1816123.1816129
10. Sugiyama, K., Kan, M.Y.: Serendipitous recommendation for scholarly papers considering relations among researchers. In: Proceedings of the 11th Annual International ACM/IEEE Joint Conference on Digital Libraries, JCDL 2011, pp. 307–310. ACM, New York (2011). http://doi.acm.org/10.1145/1998076.1998133
11. Sugiyama, K., Kan, M.Y.: Exploiting potential citation papers in scholarly paper recommendation. In: Proceedings of the 13th ACM/IEEE-CS Joint Conference on Digital Libraries, JCDL 2013, pp. 153–162. ACM, New York (2013). http://doi.acm.org/10.1145/2467696.2467701
12. Sun, J., Ma, J., Liu, Z., Miao, Y.: Leveraging content and connections for scientific article recommendation in social computing contexts. Comput. J. bxt086 (2013)

13. Tang, J., Wu, S., Sun, J., Su, H.: Cross-domain collaboration recommendation. In: Proceedings of the 18th ACM SIGKDD International Conference on Knowledge Discovery and Data Mining, KDD 2012, pp. 1285–1293. ACM, New York (2012)
14. Voorhees, E.M.: The TREC-8 question answering track report. In: TREC (1999)

Privacy-Preserving Ridesharing Recommendation in Geosocial Networks

Chengcheng Dai[1], Xingliang Yuan[1,2], and Cong Wang[1,2(✉)]

[1] Department of Computer Science, City University of Hong Kong,
Hong Kong, China
{cc.dai,xl.y}@my.cityu.edu.hk
[2] City University of Hong Kong Shenzhen Research Institute, Shenzhen, China
congwang@cityu.edu.hk

Abstract. Geosocial networks have received a lot of attentions recently and enabled many promising applications, especially the on-demand transportation services that are increasingly embraced by millions of mobile users. Despite the well understood benefits, such services also raise unique security and privacy issues that are currently not very well investigated. In this paper, we focus on the trending ridesharing recommendation service in geosocial networks, and propose a new privacy-preserving framework with salient features to both users and recommendation service providers. In particular, the proposed framework is able to recommend whether and where the users should wait to rideshare in given geosocial networks, while preserving user privacy. Meanwhile, it also protects the proprietary data of recommendation service providers from any unauthorised access, such as data breach incidents. These privacy-preserving features make the proposed framework especially suitable when the recommendation service backend is to be outsourced at public cloud for improved service scalability. On the technical front, we first use kernel density estimation to model destination distributions of taxi trips for each cluster of the underlying road network, denoted as cluster arrival patterns. Then we utilize searchable encryption to carefully protect all the proprietary data so as to allow authorised users to retrieve encrypted patterns with secure requests. Given retrieved patterns, the user can safely compute the potential of ridesharing by investigating the probabilities of possible destinations from ridesharing requirements. Experimental results show both the effectiveness of the proposed recommendation algorithm comparing to the naive "wait-at-where-you-are" strategy, and the efficiency of the utilized privacy-preserving techniques.

1 Introduction

With the proliferation of smartphones, geosocial networks are gaining increasing popularity for utilizing user-provided location data to match users with a place, event, or person relevant to their interests, and to enable further socialization activities based on such information [1]. Based on geosocial networks, many

© Springer International Publishing Switzerland 2016
H.T. Nguyen and V. Snasel (Eds.): CSoNet 2016, LNCS 9795, pp. 193–205, 2016.
DOI: 10.1007/978-3-319-42345-6_17

promising on-demand transportation services have been enabled and increasingly embraced by millions of mobile users. Among them, the trending ridesharing recommendation service has received lots of attentions, where users can submit their sources and destinations to find other users that match their trips. Ridesharing is particularly important for new on-demand transportation services such as Uber and DiDi. Besides saving money for users, ridesharing also brings potential to assuage traffic congestion and save energy consumption.

Despite the well understood benefits, such new services also raise unique security and privacy issues that are currently not very well investigated. Take the ridesharing recommendation service for example. Its recommendation mechanism depends on whether another passenger with the similar source and destination can appear in time. Thus, it would inevitably demand constant access to the users' current locations and/or their intended destinations. Such personal location data usually contain sensitive information, and should be always protected, as well recognised in the literature [11]. Besides, from the ridesharing service provider's perspective, the proprietary datasets, such as the recommendation algorithms, and valuable score functions learned from mass of data with data mining or machine learning techniques, are extremely valuable as well. They are crucial digital assets to the recommendation services and should also be strictly protected against any unauthorised access, especially given the rising concerns of recent data breach incidents [3]. Considering more and more emerging geosocial applications are directly hosted at commercial public cloud that are not necessarily within the trust domain of service providers, the security threats on unauthorised access of such proprietary information are further exacerbated.

In light of these observations, in this paper we propose a privacy-preserving framework for secure ridesharing recommendation with salient features to both users and recommendation service providers. In particular, the proposed framework is able to recommend whether and where the users should wait to rideshare in given geosocial networks, while preserving user privacy. The user privacy assurance hinges on the fact that all the data that leave from and arrive at the user's mobile devices are encrypted. Meanwhile, it also ensures that all the proprietary data from the recommendation service provider will always be encrypted during the service operations, and will never be exposed to any unauthorised users. These privacy-preserving features make the proposed framework resilient to data breach incidents at the service backend. They are also attractive when the recommendation service backend is to be outsourced at public cloud for improved service scalability, e.g., to satisfy the throughput and response time requirement for real-time queries.

In our framework, to deliver good ridesharing recommendation services, we exploit the fact that pick-ups and drop-offs of users' daily trips usually follow certain patterns. With the observation that user Alice may walk to some place nearby or change her destination to a Point-of-Interest (POI) nearby to increase her chance to rideshare, we investigate the probability for taxis to depart from somewhere near Alice's source towards somewhere near her destination. In particular, we first fragment the underlying road network into a number of road clusters, and then model destination distributions of taxi trips for each cluster, denoted as cluster arrival patterns, with kernel density estimation fused with departure probabilities for expected higher user satisfaction, as explained in Sect. 3. Based on these cluster arrival patterns, then we utilize off-the-shelf searchable encryption technique to carefully protect all the proprietary data so as to allow authorised users to retrieve encrypted patterns with secure requests. These patterns are always encrypted and stored on the cloud server while answering for authorised on-demand encrypted requests from mobile users.

The operation of our proposed framework starts from the client application on the user's smartphone. Given possible waiting places and destinations of a user, a secure query will be generated at the user client application, and then submitted to the cloud server. Subsequently, the server securely searches over encrypted patterns without decryption and returns encrypted result patterns. During this procedure, the privacy of both patterns and requested cluster ids (i.e., user source and destination information) are well-preserved. After decryption, the client application computes the ridesharing probability based on the patterns. If the potential to rideshare with others is not high enough for all nearby clusters, Alice is recommended to take a taxi directly. Otherwise, the client application highlights where to wait on the map for Alice. Thus in either cases, Alice can save either time or money.

The main contributions are summarized as follows: 1. We design a privacy-preserving recommendation framework to securely help users decide whether and where to wait for ridesharing. It also protects service provider's proprietary data from unauthorised users during operations. 2. Experimental results show the efficiency of the privacy-preserving techniques, and the effectiveness of the recommendation comparing to the naive "wait-at-where-you-are" strategy.

The rest of this paper is organized as follows. Section 2 states the system architecture, and Sect. 3 delineates the proposed privacy-preserving recommendation scheme. Section 4 gives the security analysis of the proposed scheme. Section 5 analyzes the performance. Section 6 discusses the related work. Finally, Sect. 7 concludes this paper.

2   System Model

As shown in Fig. 1, the architecture consists of three different parties: the *service provider*, the *user* and the *cloud server*. Service provider learns patterns with data mining or machine learning techniques, and encrypts these patterns before outsourcing them to the cloud. Users generate encrypted queries for certain patterns according to their ridesharing requests. Cloud server sends encrypted patterns to users in an "on-demand" manner. To enable search over encrypted pattern, searchable symmetric encryption (*SSE*) is utilized to securely index encrypted patterns. A secure pattern index will be uploaded as well.

Users. We consider authorised users with registration as prior work [5,18]. There is no malicious user that either shares her key with others or generates unnecessary queries to steal information from the server. As a client application on a

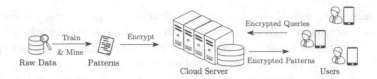

Fig. 1. Our proposed system model

user's smartphone, the city map is in the storage of the client application. When a user submits a query, the client application on the user's smartphone generates a secure search request to the cloud server. After receiving the encrypted patterns from the cloud server, the client application computes the ridesharing potential after decryption. If the user is recommended to rideshare, it will highlight the corresponding road for each recommended cluster on the map.

Besides, the user specifies her willingness in the preference setting of the client application, namely the maximum walking distance d_s to a new place from her source, the maximum walking distance d_e after she leaves the taxi to her own destination and maximum waiting time t_w at the new place for ridesharing.

Cloud Server. Sensitive patterns are encrypted and indexed before storing on a cloud server. The server is deployed in the cloud to provide the privacy-preserving recommendation service for a large number of real-time queries. In this paper, we consider a "honest-but-curious" cloud server, i.e., the server acts in an "honest" fashion, but is "curious" to infer and analyze the message flow to learn additional information on the user request and the pattern information.

Problem Definition. In our recommendation application, the user specifies her query as $Q = (ID, timestamp, l_s, l_d)$, where id is user id, $timestamp$ is when the query is submitted, l_s and l_d are respectively the source and the destination of the user. Given a query, we compute the potential of ridesharing and where the user should wait based on ridesharing requirements. Alice can rideshare with another passenger Bob if (i) the source of Bob is within her maximum walking distance d_s from her source l_s (ii) the destination of Bob is somewhere within her maximum walking distance d_e from her destination l_d (iii) Bob submits his request within waiting time t_w. Recall that d_s, d_e and t_w are set as their ridesharing willingness in the client application. For example, Alice can increase her chance of ridesharing by increasing her waiting time t_w.

When a user submits a query, the client application generates a search request according to ridesharing conditions. The server returns encrypted patterns to the client application. To allow an authorised group of users to search through the patterns and prevent unauthorised access, the server cannot infer any sensitive information of patterns from the encrypted storage before search and can only learn the limited information about the requested patterns and the results.

3   Our Proposed Design

In this section, we discuss how to perform privacy-preserving ridesharing recommendation with a third-party cloud server. To initialize the service, the service provider distributes search request generation keys to authorised users. We here assume the authorisation between the client application and the user is appropriately done on the smartphone. We discuss how to learn patterns (discussed in Sect. 3.1) and make recommendation (discussed in Sect. 3.2) with machine learning techniques, and perform ridesharing recommendation in a privacy-preserving way (discussed in Sect. 3.3).

3.1 Learning Patterns

Road Segment Clustering. Since modeling destinations of trips based on single road is too dynamic, the underlying road network is grouped into road clusters by applying k-means to the mid-points of road segments [16]. The cluster of the mid-point is the cluster that the road segment belongs to, recorded as $c_i = \{r_1, \ldots, r_N\}$. A grid index structure is built on the underlying road network. Given a location (lon, lat), we can find out the road on which the location is located and further get the cluster it belongs to. We category trip records into groups according to which clusters their sources belong to. Each cluster makes use of the corresponding group of records as samples to derive the kernel density estimator about the probability for other passengers (or taxis) to depart from somewhere in the cluster and have a given location (lon, lat) as their destination.

Kernel Density Estimation. For each cluster, given a trip record $(t_i, source_i, destination_i)$, we describe a training sample in the format $\mathbf{x_i} = (lon_i, lat_i, t_i)^T$, which is a 3-dimensional column vector with the longitude (lon_i) and latitude (lat_i) and the time (t_i), indicating a trip from somewhere in the cluster to $desination_i$ (lon_i, lat_i) happens at time t_i[1]. Intuitively ridesharing is related to when the query is submitted since traffic directions in modern cities depend on time. For example, traffics are likely to be from residencies to companies in the morning and right way around in the evening. Thus pick-up time t_i is taken into consideration.

Let $X_c = \langle \mathbf{x_1}, \mathbf{x_2}, \ldots, \mathbf{x_n} \rangle$ be the samples for a certain cluster c that follows an unknown density p. As described in [13], its kernel density estimator over X_c for a new sample $\mathbf{x}$ is given by: $p(\mathbf{x}) = \frac{1}{nh^d}\sum_{i=1}^{n} K(\frac{\mathbf{x}-\mathbf{x}_i}{h})$. $K(\mathbf{x}) = \frac{1}{\sqrt{(2\pi)^d}}e^{-\frac{1}{2}\|\mathbf{x}\|^2}$. h is a smoothing parameter and $K(.)$ is the widely used Gaussian kernel. With $d = 3$, we can obtain the probability to have a taxi that departs from a certain cluster c towards new location (lon, lat) at certain time as follows:

$$p(\mathbf{x}_{new}|X_c = \{\mathbf{x_1}, \mathbf{x_2}, \ldots, \mathbf{x_n}\}) = \sum_{i=1}^{n} \frac{1}{n}\frac{1}{(\sqrt{2\pi}h)^3}e^{-\frac{1}{2h^2}\|\mathbf{x}_{new}-\mathbf{x}_i\|^2} \qquad (1)$$

Equation 1 is equivalent to $\sum_i^n \frac{1}{n}\mathcal{N}(\mathbf{x}_{new}|\mathbf{x}_i, h^2\mathbf{I})$. The optimal smoothing parameter h [13] is $0.969n^{-\frac{1}{7}}\sqrt{\frac{1}{3}\sigma^T\sigma}$, where σ is the marginal standard deviation

---

[1] Instead of transforming the original pick-up time t_i into discrete values between 1 and 48 [6], we transform t_i to continuous values to keep more details about the time domain. Please refer to the experiment section for more details.

vector of values in X_c. Both samples $\mathbf{x}_1, \mathbf{x}_2, \ldots, \mathbf{x}_n$ in X_c and the smoothing parameter h are required to describe the kernel density estimator of each cluster.

Fusion with Departure Probability. For each cluster, the kernel density estimator describes the probability for taxis to depart from somewhere in it and have a given location (lon, lat) as their destination. However, it didn't consider the departure probabilities of different roads in the cluster. We distinguish each road with the departure probability, i.e., the probability to have a taxi departing from certain road, for more accurate prediction. Given $\mathbf{x}_{new} = (lon, lat, t)^T$, the probability $P(r_j \to \mathbf{x}_{new})$ to have a taxi departing at time t from road r_j in cluster c_i and towards destination (lon, lat) is:

$$P(r_j \to \mathbf{x}_{new}) = p(A_{r_j}) * p(\mathbf{x}_{new}) \tag{2}$$

where $p(A_{r_j})$ is the probability to have a taxi pick up a passenger at road r_j, i.e.,

$$p(A_{r_j}) = \frac{N_{r_j}}{\sum_{r_x \in c_i} N_{r_x}} \tag{3}$$

where N_{r_j} denotes the number of pick-ups on road r_j. Noted that $p(A_{r_j})$ does not incur any additional computation since it is obtained when we category records into groups according to which clusters their sources belong to. $p(\mathbf{x}_{new})$ is computed by Eq. 1.

3.2   Recommendation with Patterns

After learning patterns, and before talking about how to perform ridesharing recommendation in a privacy-preserving way, we describe how the client application can compute the ridesharing potential based on patterns to make recommendation. Specifically, given a ridesharing query $Q = (ID,\ timestamp,\ l_s,\ l_d)$, the client application performs network expansion technique [12] to get all the roads $R = \{r_1, \ldots, r_d\}$ that are within distance d_s from l_s. With the road clustering information in the client application, i.e., which road belongs to which cluster, we can easily group roads in R to candidate clusters. Denote the candidate set as $L = \{C_1, C_2, \ldots, C_x\}$ and $C_i = \{r_y | r_y \in R \wedge r_y \in c_i\}$. Cluster information, i.e., the kernel density estimator and departure probabilities of roads, are required to compute the ridesharing potential for each $C_i \in L$.

For roads that are reachable within d_e from l_d, which are obtained with the same technique, we generate $\mathbf{x}_{new_k} = (r_k^{lon}, r_k^{lat}, t)^T$ where (r_k^{lon}, r_k^{lat}) is the midpoint of road r_j and t is the time by transforming $timestamp$ in the query Q in the same way as records. Denote the sample set as D' and the probability to have taxis departing from roads in C_i toward roads in D as $P(C_i \to D)$, we have $P(C_i \to D) = \sum_{\mathbf{x}_{new_k} \in D'} P(C_i \to \mathbf{x}_{new_k})$, where $P(C_i \to \mathbf{x}_{new_k})$ is the probability for taxis to depart from any road $r_j \in C_i$, i.e., within distance d_s from l_s at time t towards somewhere on road r_k. By combining with Eq. 2, we have $P(C_i \to \mathbf{x}_{new_k}) = \sum_{r_j \in C_i} P(r_j \to \mathbf{x}_{new_k}) = p(\mathbf{x}_{new_k}) * \sum_{r_j \in C_i} p(A_{r_j})^2$.

---

[2] All roads in C_i share the same kernel density estimator and thus the same $P(\mathbf{x}_{new_k})$.

Since $\sum_{r_j \in C_i} P(A_{r_j})$ is not relative to $\mathbf{x}_{new_k}$, we further have

$$P(C_i \rightarrow D) = \sum_{\mathbf{x}_{new_k} \in D'} p(\mathbf{x}_{new_k}) * \sum_{r_j \in C_i} p(A_{r_j}) = \sum_{r_j \in C_i} p(A_{r_j}) * \sum_{\mathbf{x}_{new_k} \in D'} p(\mathbf{x}_{new_k}).$$

Noted that if no pick-ups exist on any the road in C_i, i.e., $\sum_{r_j \in C_i} P(A_{r_j}) = 0$, there is no need to further compute $p(\mathbf{x}_{new_k})$ by plugging in different $\mathbf{x}_{new}$ in Eq. 1. We display the computation of $P(C_i \rightarrow D)$ in Algorithm 1.

Algorithm 1. Ridesharing potential of cluster C_i

---

1 **foreach** $r_j \in C_i$ **do**
2 $\quad \llcorner$ $P_1 \mathrel{+}= p(A_{r_j})$;//Compute $p(A_{r_j})$ with Eq. 3;
3 **if** P_1 *is not 0* **then**
4 $\quad \mid$ $P_2 = 0$;
5 $\quad \mid$ **foreach** $r_k \in D$ **do**
6 $\quad \mid \quad \mid$ $\mathbf{x}_{new_k} = (r_j^{lon}, r_j^{lat}, t)$;
7 $\quad \mid \quad \llcorner$ $P_2 \mathrel{+}= p(\mathbf{x}_{new_k})$;//Compute $p(\mathbf{x}_{new_k})$ with Eq. 1;
8 $\quad \llcorner$ Return $P_1 * P_2$;

Only candidates with ridesharing potential $P(C_i \rightarrow D)$ greater than a threshold are considered valid for recommendation. If no valid cluster exists for Q, the user will be suggested not to wait for ridesharing and take taxi directly. In case many road clusters satisfy the condition, we return top-k clusters according to the probabilities. The client application will highlight the corresponding roads $r_j \in R$ for each recommended cluster on the map, i.e., roads that are within distance d_s from l_s.

3.3   Privacy-Preserving Ridesharing Recommendation

Given the discussion on learning patterns and making recommendation with patterns, we describe the overall privacy-preserving ridesharing recommendation. We organize the patterns as follows. For $Pattern_i$ of each cluster on the cloud server, it records samples and the smoothing parameter of the kernel density estimator, and the departure probabilities[3]. SSE is utilized on the server side to keep sensitive patterns confidential, while resuming the ability to selectively retrieve encrypted patterns. A SSE scheme is a collection of four polynomial-time algorithms (Kengen, BuildIndex, Trapdoor, Search) such that: (i) Keygen(1^k): outputs symmetric key K. (ii) BuildIndex($K, \mathcal{D}$): outputs a secure index I built on encrypted patterns $\mathcal{D}$ that helps the server to search without decryption. (iii) Trapdoor(K, w): outputs a trapdoor T_w. Cluster id w is associated with a trapdoor which enables server to search while keeping w hidden. (iv) Search(I, T_w):

---

[3] Departure probabilities are ordered according to road ids in ascending order. Suppose in $Pattern_1$ of cluster c_1 there are three probabilities 0.3, 0, 0.7 and $c_1 = \{r_1, r_3, r_4\}$. We get p(A_{r_1}) = 0.3, p(A_{r_3}) = 0 and p(A_{r_4}) = 0.7.

outputs the identifier of the pattern of cluster w. Noted that Kengen, BuildIndex, Trapdoor are run by the user, while Search is run by the server.

Let $Enc_s(\cdot)$, $Dec_s(\cdot)$ be semantic secure encryption and decryption functions based on symmetric key s. In addition, we make use of one pseudo-random function (PRF) $f : \{0,1\}^* \times key \rightarrow \{0,1\}^l$ and two pseudo-random permutations (PRP) $\pi : \{0,1\}^* \times key \rightarrow \{0,1\}^*$ and $\psi : \{0,1\}^* \times key \rightarrow \{0,1\}^*$. We are now ready for the details of the privacy-preserving ridesharing recommendation.

Generating Key. Generate random keys x, y and z where $x, y, z \xleftarrow{R} \{0,1\}^{k'}$ and output $K = (x, y, z, s)$.

Building a Secure Index. The secure index I is a look-up table, which contains information that enables one to locate the pattern of certain cluster c_i. Each entry corresponds to a cluster c_i and consists of a pair $\langle address, addr(Pattern_i) \oplus f_y(id_i) \rangle$. id_i is the id of cluster c_i. $Pattern_i$ is the pattern of cluster c_i. The address of $Pattern_i$, i.e., $addr(Pattern_i)$, is set to $\psi_x(id_i)$, which means that the location of a pattern is permutated and protected. $addr(Pattern_i) \oplus f_y(id_i)$ indicates that the address of $Pattern_i$ is encrypted using the output of a PRF $f_y(.)$. The other field, $address^4$, is used to locate an entry in the look-up table. We set $I[\pi_z(id_i)] = \langle addr(Pattern_i) \oplus f_y(id_i) \rangle$.

After building the index I, $Enc_s(Pattern_i)$ is performed for each pattern. Both the secure index and encrypted patterns are outsourced to the cloud server. Noted that we pad $Enc_s(Pattern_i)$ to the same length to prevent leaking the length information. Table 1 indicates the storage on the cloud server.

Table 1. Encrypted storage on the cloud server

Address	Key		Samples	h	Departure Prob.
$\pi_z(id_k)$	$addr(Pattern_k) \oplus f_y(id_k)$	$\rightarrow$	$Enc_s(\mathbf{x}_a), \forall x_a$	$Enc_s(h_k)$	$Enc_s(p(A_{r_d})), \forall r_d$
...	...	$\rightarrow$	...	...	...
$\pi_z(id_n)$	$addr(Pattern_n) \oplus f_y(id_n)$	$\rightarrow$	$Enc_s(\mathbf{x}_b), \forall x_b$	$Enc_s(h_n)$	$Enc_s(p(A_{r_e})), \forall r_e$
...	...	$\rightarrow$	...	...	...
$\pi_z(id_1)$	$addr(Pattern_1) \oplus f_y(id_1)$	$\rightarrow$	$Enc_s(\mathbf{x}_c), \forall x_c$	$Enc_s(h_1)$	$Enc_s(p(A_{r_f})), \forall r_f$

Trapdoor Construction. Output $T_w = (address, key)$, where $address = \pi_z(id_i)$, $key = f_y(id_i)$ and id_i is the requested cluster id.

Searching. With the trapdoor $T_w = (address, key)$, the server retrieves $\theta = I[address]$ and uses key to decrypt θ. Let $\langle addr(Pattern_i) \rangle = \theta \oplus key$. With the address of $Pattern_i$, the server sends the encrypted pattern of cluster c_i to the user. For each received pattern, the user performs $Dec_s(Pattern_i)$.

Multi-user SSE. To allow an arbitrary group of users submit queries to search the data on the cloud server, we combine single-user SSE with broadcast encryption to achieve multi-user SSE [5]. We assume the pre-sharing of the trapdoor

---

[4] We manage $address$ with indirect addressing [5] to provide efficient storage and access of sparse tables.

generation key between the service provider and users. Adding/revoking users can be properly done via broadcast encryption. An authorised user applies a PRP ϕ (keyed with a secret key r) on a regular single-user trapdoor T_w. Upon receiving $\phi_r(T_w)$, the server recovers the trapdoor by computing $\phi_r^{-1}(\phi_r(T_w))$. Unauthorised users cannot get the valid r to yield a valid trapdoor for searching.

On behalf of an authorised user, the client application generates a search request T_w for each required cluster via a certain one-way function with the trapdoor generation key. We have $T_w = (\phi_r(T_{w_1}), \phi_r(T_{w_2}), \ldots, \phi_r(T_{w_x}))$, where $T_{w_i} = (\pi_z(id_i), f_y(id_i))$ and id_i is the id of cluster c_i. After construction, T_w is submitted to the cloud server. Given T_w, the server recovers $T_{w_i} = (\pi_z(id_i), f_y(id_i))$ with key r and preforms searching. In this way, the server searches over the stored data without decryption, and sends back required cluster patterns, i.e., $Pattern_1, Pattern_2, \ldots, Pattern_x$. Noted that the server is not aware of which cluster is requested. After receiving the required patterns, the client application decrypts them on behalf of an authorised user and computes the ridesharing potential of each cluster C_i as shown in Algorithm 1. Recommendations about whether and where the user should wait for ridesharing are made as shown in Sect. 3.2.

4   Security Analysis

We follow the security definition in searchable symmetric encryption [5] that nothing beyond the encrypted outcome and the repeated search queries should be leaked from the remote storage. We adapt the simulation-based security model of [5] and prove non-adaptive semantic security is guaranteed. We first introduce notions used in [5]: 1. *History:* an interaction between the user and the cloud server, determined by a collection $\mathcal{C}$ of cluster patterns and a set of cluster ids searched by the user, denoted as $H = (\mathcal{C}, id_1, id_2, \ldots, id_x)$. 2. *View:* what the cloud server can see given a history H, denoted as $V(H)$, including the index I of $\mathcal{C}$, the trapdoors of the queried cluster ids $\{T_{id_1}, T_{id_2}, \ldots, T_{id_x}\}$, and the encrypted collection of $\mathcal{C}$. 3. *Trace:* what the cloud server can capture, denoted as $Tr(H)$, including the size of the encrypted patterns, the outcome of each search (i.e., the patterns $Pattern_i$) and whether two searches were performed for the same cluster id or not.

Given two histories with the identical trace, the cloud server cannot distinguish the views of the two histories. Our mechanism is secure since the cloud server cannot extract additional knowledge beyond the trace, which we are willing to leak. We can describe a simulator S such that, given trace $Tr(H)$, it can simulate a view V^* (composed of encrypted patterns, index and trapdoors) indistinguishable from the cloud server's view $V(H)$ [5]. In particular, the simulated encrypted pattern is indistinguishable due to the semantic security of the symmetric encryption. The indistinguishability of index and trapdoors is based on the indistinguishability of pseudo-random function output and a random string.

5 Experiments

Dataset. We make use of the Uber trip data of NYC[5]. Each record is in the format $(t, source, destination)$, where t is the pick-up time, *source* and *destination* are respectively pick-up location and drop-off location, described as (lon, lat). We transform the pick-up time from the original format $hh:mm:ss$ to $(hh*3600 + mm*60 + ss)/(24*3600)$ in preprocessing. We randomly select 1,000 records as ridesharing queries. The time, source and destination of the trip are treated as $timestamp$, l_s and l_d in the queries.

Effectiveness Evaluation. We compare our ridesharing recommendation (RR) with the naive strategy to "wait at where your are" (WW). In WW users wait for ridesharing at where they are, i.e., the cluster that l_s is in. To evaluate the effectiveness of recommendation, we category ridesharing recommendation into two types, namely *to-rideshare* and *not-to-rideshare*. An accurate *to-rideshare* means that users can rideshare with others at the recommended locations. An accurate *not-to-rideshare* indicates that users are recommended not to wait for ridesharing and there are indeed no others to satisfy the ridesharing requirements. We consider the following measurements. (i) Ridesharing successful ratio (*RSRatio*). We measure the ratio of successful ridesharing of both RR and WW by *RSRatio*, defined as $RSRatio = \frac{\#\ accurate\ to\text{-}rideshare}{\#\ to\text{-}rideshare}$. (ii) Prediction accuracy (*Accuracy*). We measure the accuracy of predicting whether the user should wait for ridesharing or not by *accuracy*, defined as $Accuracy = \frac{\#\ accurate\ not\text{-}to\text{-}rideshare + \#\ accurate\ to\text{-}rideshare}{\#\ queries}$. (iii) Recommendation quality[6]. To find out how many clusters that actually have others to rideshare a query are discovered by our framework, we employ standard metrics, i.e., *precision* and *recall*: $precision = \frac{\#\ discovered\ clusters}{k}$, $recall = \frac{\#\ discovered\ clusters}{\#\ positive\ clusters}$. Positive clusters are clusters with others to rideshare a query. Discovered clusters are the positive clusters in the recommended clusters. *Precision* and *recall* are averaged over all queries.

Table 2. Effect of waiting time t_w (seconds)

Metrics	RR					WW				
	150	300	450	600	750	150	300	450	600	750
RSRatio	0.218	0.255	0.309	0.327	0.364	0.061	0.078	0.082	0.091	0.124
Accuracy	0.535	0.543	0.567	0.574	0.588	0.061	0.078	0.082	0.091	0.124
Precision	0.475	0.478	0.544	0.556	0.586	-	-	-	-	-
Recall	0.618	0.658	0.686	0.694	0.712	-	-	-	-	-

---

[5] https://github.com/fivethirtyeight/uber-tlc-foil-response. Destinations are generated based on a check-in dataset of Foursquare from http://download.csdn.net.

[6] We didn't study precision and recall of WW since users wait at where they are.

Table 3. Execution time

Learning patterns	enc.pattern	SSE.index	SSE.token	SSE.search	Recommendation
263.37 s	14.12 s	40.34 ms	117.95 μs	24.60 μs	0.269 s

Table 2 shows the influence of waiting time t_w on the performances of RR and WW. The default values of maximum walking distances d_s and d_e are set to 500 m. The number of recommended clusters k is set to 5. As t_w increases, users can wait for ridesharing for a longer time. *RSRatio* and *Accuracy* increase for RR and WW. As more passengers appear during a longer t_w, the potential to rideshare increases. Both the number of discovered clusters and the number of positive clusters increase, leading to the increase in *precision* and *recall*.

Performance Evaluation. Experiments were performed on a Windows machine with Intel i7-3770 CPU and 16 GB RAM. We measure the time of learning patterns (Learning patterns), the time to make recommendation after obtaining patterns from the server (Recommendation). To study the efficiency of *SSE*, we measure the average execution time to generate indexes (SSE.index), generate trapdoors per query (SSE.token), and execute search operations per query (SSE.search), as well as the time to encrypt pattern (enc.pattern) before outsourcing to a server. We have 1,000 patterns with the size of 106MB. As shown in Table 3, the major cost is introduced in the learning pattern stage and the encryption stage, which is acceptable because it is a one-time setup. The introduced security overhead (enc.pattern + SSE.index) is around 5 % to the cost of learning patterns which is also required in plaintext applications. Meanwhile, the time cost for processing each secure query (SSE.token + SSE.search) is small, and obtaining the potential for ridesharing from encrypted patterns is also efficient.

6 Related Works

Private Searching in Cloud. Public-key searchable encryption is usually adapted to deal with third-party data [2], where anyone with the public key can write to the data stored on the server but only users with the private key can search. For user-owned data, symmetrically encryption is widely adapted and client uploads additional encrypted data structures to help search, including oblivious RAMs [7] and searchable symmetric encryption (*SSE*) [5,8]. *SSE* is applicable for any forms of private retrieval based on keywords [4,15,17], varying from exact matching [5,8] to similarity search on textual files [15] or images [4]. Built on locality-sensitive hashing, similarity search is transformed to keyword search to handle millions of encrypted records [17]. Yet, the above studies do not focus on achieving privacy-preserving recommendation in geosocial networks.

Privacy-Preserving Recommender Systems. Privacy-preserving tests for proximity is proposed to test whether a friend is close to her without revealing location information of any of them [11]. To perform online behavioral advertising without compromising user privacy, an efficient cryptographic billing system is proposed so that the correct advertiser is billed without knowing which advertisement is displayed to the user [14]. Secure image-centric social discovery [18] is modeled as secure similarity retrieval of encrypted high-dimensional image profiles. Social media sites can recommend friends or social groups for users from public cloud without disclosing the encrypted image content.

Ridesharing. Current studies about ridesharing can be categorized into two scenarios, either drivers change their routes to pick up and drop off passengers [10] or users walk certain distance to get on a taxi and to their own destinations after getting off [9]. We consider ridesharing in geosocial networks as the second scenario to recommend users with the places that are most likely to have other users with similar trips. Unlike previous work [6] that only considers destinations of trips to model the pattern of each cluster, we fusion with departure probabilities of each road to distinguish each road with their pick-up probabilities for more accurate prediction. Noted that none of existing works [6,9,10] consider privacy issues in ridesharing, we preserve user privacy as well as protect the proprietary data of service providers from any unauthorised access.

7 Conclusion

In this paper, we proposed a privacy-preserving framework to recommend whether and where the users should wait to rideshare in geosocial networks. The privacy of both users and recommendation service providers are well protected. As future work, we plan to study how to enable the server to directly compute the results in the encrypted domain.

Acknowledgement. This work was supported in part by the Research Grants Council of Hong Kong (Project No. CityU 138513), and the Natural Science Foundation of China (Project No. 61572412).

References

1. Bao, J., Zheng, Y., Mokbel, M.F.: Location-based and preference-aware recommendation using sparse geo-social networking data. In: SIGSPATIAL, pp. 199–208 (2012)
2. Boneh, D., Di Crescenzo, G., Ostrovsky, R., Persiano, G.: Public key encryption with keyword search. In: Cachin, C., Camenisch, J.L. (eds.) EUROCRYPT 2004. LNCS, vol. 3027, pp. 506–522. Springer, Heidelberg (2004)
3. Bost, R., Popa, R.A., Tu, S., Goldwasser, S.: Machine learning classification over encrypted data. In: NDSS (2015)
4. Cui, H., Yuan, X., Wang, C.: Harnessing encrypted data in cloud for secure and efficient image sharing from mobile devices. In: INFOCOM, pp. 2659–2667 (2015)
5. Curtmola, R., Garay, J.A., Kamara, S., Ostrovsky, R.: Searchable symmetric encryption: improved definitions and efficient constructions. In: CCS, pp. 79–88 (2006)

6. Dai, C.: Ridesharing recommendation: whether and where should i wait? In: Cui, B., Zhang, N., Xu, J., Lian, X., Liu, D. (eds.) WAIM 2016. LNCS, vol. 9658, pp. 151–163. Springer, Heidelberg (2016). doi:10.1007/978-3-319-39937-9_12

7. Goldreich, O., Ostrovsky, R.: Software protection and simulation on oblivious RAMs. J. ACM **43**(3), 431–473 (1996)

8. Kamara, S., Papamanthou, C., Roeder, T.: Dynamic searchable symmetric encryption. In: CCS, pp. 965–976 (2012)

9. Ma, S., Wolfson, O.: Analysis and evaluation of the slugging form of ridesharing. In: SIGSPATIAL, pp. 64–73 (2013)

10. Ma, S., Zheng, Y., Wolfson, O.: Real-time city-scale taxi ridesharing. IEEE Trans. Knowl. Data Eng. **27**(7), 1782–1795 (2015)

11. Narayanan, A., Thiagarajan, N., Lakhani, M., Hamburg, M., Boneh, D.: Location privacy via private proximity testing. In: NDSS (2011)

12. Papadias, D., Zhang, J., Mamoulis, N., Tao, Y.: Query processing in spatial network databases. In: VLDB, pp. 802–813 (2003)

13. Silverman, B.W.: Density Estimation for Statistics and Data Analysis, vol. 26. CRC Press, Boca Raton (1986)

14. Toubiana, V., Narayanan, A., Boneh, D., Nissenbaum, H., Barocas, S.: Adnostic: privacy preserving targeted advertising. In: NDSS (2010)

15. Wang, C., Ren, K., Yu, S., Urs, K.M.R.: Achieving usable and privacy-assured similarity search over outsourced cloud data. In: INFOCOM, pp. 451–459 (2012)

16. Wang, R., Chow, C., Lyu, Y., Lee, V.C.S., Kwong, S., Li, Y., Zeng, J.: TaxiRec: recommending road clusters to taxi drivers using ranking-based extreme learning machines. In: SIGSPATIAL, pp. 53:1–53:4 (2015)

17. Yuan, X., Cui, H., Wang, X., Wang, C.: Enabling privacy-assured similarity retrieval over millions of encrypted records. In: Pernul, G., Y A Ryan, P., Weippl, E. (eds.) ESORICS. LNCS, vol. 9327, pp. 40–60. Springer, Heidelberg (2015). doi: 10.1007/978-3-319-24177-7_3

18. Yuan, X., Wang, X., Wang, C., Squicciarini, A.C., Ren, K.: Enabling privacy-preserving image-centric social discovery. In: ICDCS, pp. 198–207 (2014)

Complex Network Approach for Power Grids Vulnerability and Large Area Blackout

An T. Le[✉] and Ravi Sankar

University of South Florida, Tampa, USA
atle2@mail.usf.edu, sankar@usf.edu

Abstract. In a power grid, a device protection trip could cause a number of consequent or cascading trips. These consequent trips could be transferred and redistributed and cause a large area blackout. The development of complex network theory provides a new approach for power system study especially on vulnerability and large blackout prevention. By mapping the power grid to a graph, analyze the degree and link properties, and utilizing the small-world power grid to study the behavior of blackout.

Keywords: Protection and control · SCADA · Complex network · BES · Smart-Grid · Power grid · N-1-1 contingency · Cascading failure · NERC

1 Introduction

In a power grid, a trip is an interrupting operation to isolate a failure or overload or fault component out of the power grid. A consequent or a cascade trip associating with the original trip could occur if:

- The tripping device fails to trip. For this condition the upper stream protective devices will be requested to trip. This is typically called transfer trip [1]. The trip transferring chain will not stop until the failure device is totally isolated from the power grid.
- The power distribution (flow) changes after the first successive trip, results in other components becoming overloaded and as a consequence they also trip. The new trips will initiate other transfer trips, and so on.

The failure of a device or a line trip could lead to a bus (node) trip. Figure 1 illustrates a system one line diagram. For example, if the line between bus 3 and 5 needs to be tripped (isolated), and if the breaker on bus 5 fails to trip, the entire bus 5 will be requested to trip. That means all breakers that connect to bus 5 will trip, including the generator breaker. As a result the whole system will lose power. The more the cascade domino trip can go to the upper stream, the more extensive the power lost from the trip could be. In Power operation system [2], there will be an alarm if the cascade trip domino effect goes too far. The Operator could manually trip other devices or the automatic load shedding system could be activated to reduce the impact of an upper stream trip. The upper stream trip could cause a large area blackout.

© Springer International Publishing Switzerland 2016
H.T. Nguyen and V. Snasel (Eds.): CSoNet 2016, LNCS 9795, pp. 206–213, 2016.
DOI: 10.1007/978-3-319-42345-6_18

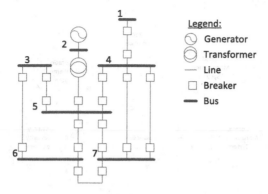

Fig. 1. A simple power grid system one line diagram

In a legacy power grid protection and control system the power flow and the short circuit program will estimate the power grid status after a trip. In most of the case, the Operator will use the real-time measurement values to make a decision. Human error can also contribute to the blackout risk.

With a complex network, we are looking at a different approach for automatic transfer trip which will request a system wide constraint before executing a trip. The constraint will include but not limited to the network behavior prediction and the possibility of initiating and executing a new cascade trip. We will also do a careful study of the power grid during the development and planning phase before design and construction. As part of this study, a new connection should be developed to minimize the cascading failure effect.

2 Power Grid Topology, Complex Power Network's Vertex Degree and Connection Properties

2.1 Complex Power Grid Networks

A complex power grid network could be represented as shown in Fig. 2. Other discipline networks such as SCADA (Supervisory Control and Data Acquisition), System Protection, and Energy Management System (EMS) etc. are not truly independent since each has a number of connections to the complex power network and other discipline networks.

A power grid consists of nodes and lines. A power bus is a voltage node. The power grid could be classified under nominal voltage, i.e. 69 kV, 138 kV 230 kV grids, etc. These grids also tie together via power transformers. Using the concept of power flow all power grids could be combined into one. Power flows from node to node via power lines. Power flow comes in a node via the source line and comes out of the load via the load line. Blackout results from losing all power source lines.

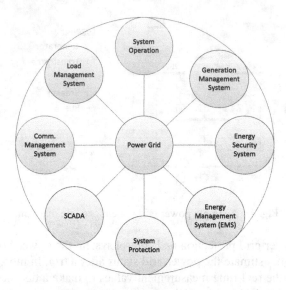

Fig. 2. A complex power network

2.2 Degree of a Vertex and Degree Properties

The simplest format of power grid modelling under complex network is to call a bus a node or a vertex and a line an edge. Local characteristic of a vertex is its degree k: the total number of the edges or number of connection attached to a vertex [3]. The network graph of the one line diagram on Fig. 1 is shown as Fig. 3 and the degree matrix is shown below:

$$
\begin{bmatrix}
0 & 0 & 0 & 1 & 0 & 0 & 0 \\
0 & 0 & 0 & 0 & 1 & 0 & 0 \\
0 & 0 & 0 & 0 & 1 & 1 & 0 \\
1 & 0 & 0 & 0 & 1 & 0 & 2 \\
0 & 1 & 1 & 1 & 0 & 1 & 1 \\
0 & 0 & 1 & 0 & 1 & 0 & 1 \\
0 & 0 & 0 & 2 & 1 & 1 & 0
\end{bmatrix}
\tag{1}
$$

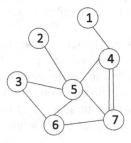

Fig. 3. A power network graph

It is obvious in the degree matrix that kij = kji.

As we mentioned above, each connection has a direction going in or going out of the vertex. Therefore we will have in-degree k_i and out-degree k_o. Another way to define vertex degree is the number of nearest neighbor of a vertex. That means there is maximum one edge or connection from one vertex to another. We call this definition a single connection graph. In the power grid we could combine all lines between two nodes into one. This definition is recommended for complex power network since this definition will simplify the degree 1 vertex.

The in-degree 1 is a dead-end load or a distribution only vertex. The out-degree 1 is the generator. The network graph of the above mentioned one line diagram is shown as Fig. 4 and the degree matrix becomes:

$$\begin{bmatrix} 0 & 0 & 0 & 1 & 0 & 0 & 0 \\ 0 & 0 & 0 & 0 & 1 & 0 & 0 \\ 0 & 0 & 0 & 0 & 1 & 1 & 0 \\ 1 & 0 & 0 & 0 & 1 & 0 & 1 \\ 0 & 1 & 1 & 1 & 0 & 1 & 1 \\ 0 & 0 & 1 & 0 & 1 & 0 & 1 \\ 0 & 0 & 0 & 1 & 1 & 1 & 0 \end{bmatrix} \tag{2}$$

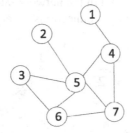

Fig. 4. Power network graph with single connection

Complex Power Network Degree Property: In a complex power network with $k \geq 2$, the vertex exist if and only if $k_i \neq 0$ and $k_o \neq 0$. This "lemma" could be proven by Kirchhoff's circuit laws. In other words, we can say if there is no power flows in a node then there will be no power flows out of a node and vice versa. In a power network, it is possible a connection could be in or out.

2.3 Edge (Connection) Properties

The most important property of a connection is total power flow and the direction. The degree matrix of Fig. 4 could be written as follows:

$$\begin{bmatrix} 0 & 0 & 0 & -1 & 0 & 0 & 0 \\ 0 & 0 & 0 & 0 & 1 & 0 & 0 \\ 0 & 0 & 0 & 0 & -1 & 1 & 0 \\ 1 & 0 & 0 & 0 & -1 & 0 & -1 \\ 0 & -1 & 1 & 1 & 0 & 1 & 1 \\ 0 & 0 & -1 & 0 & -1 & 0 & 1 \\ 0 & 0 & 0 & 1 & -1 & -1 & 0 \end{bmatrix} \quad (3)$$

In this case vertices degree is $k = |ki| + |ko|$.

3   Cascade Failure and Mitigation

3.1   Redundancy Network

In our simple example, since vertex 2 is only the source, the path from vertex 2 to any other vertices should go through vertex 5. Failure of vertex 5 causes a cascade failure of the entire network. The first rule of cascade failure mitigation is, if possible, the in-degree of a vertex should be greater or equal to 2. Using the same example network, and adding two connections as shown in Fig. 5 the network has a chance not to blackout if it loses any one connection. We can turn the network into higher reliability or a redundancy network by providing alternate connections for any vertices.

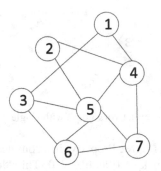

Fig. 5. Full redundancy network

3.2   Cascading Failure for Breaker Failure

In power operation a N-1-1 contingency is a sequence of events consisting of initial loss of single generator or transmission component (Primary Contingency), followed by system adjustments, and followed by another loss of a single generator or transmission component (Secondary Contingency). Typically, an Optimal Power Flow (OPF) or Security-Constrained Optimal Power Flow (SCOPF) may be used with contingency analysis to help forecast system adjustment [4, 5]. The adjustment includes but not limited to redirect power flow. If the real-time OPF or SCOPF is not fast enough, a stochastic analysis [6–8] for cascading failure could help speed-up the calculation process.

Focusing on cascading breaker failure in a network such as Fig. 5, any loss of single generator or transmission component may not create a blackout. However, as we mentioned in the introduction a failure occurring on a connection (line) could lead to the loss of two adjacent vertices.

- *Case 1:* A fault on line 4–5 will request the breakers on both node 4 and 5 to trip. However if they fail to trip, all breakers that are tied to node 4 and 5 should be tripped. Loss of both 4 and 5 vertices (nodes) will cause the loss of vertex 2 and create a blackout. This blackout is caused by a first stage of cascade failure.
- *Lemma 1:* To eliminate the possibility of the blackout of a node caused by a first stage of cascade failure, at least one shortest path from one of its neighbor to another neighbor consisting of greater than or equal to one vertices should be found.

Figure 6 illustrates a modified version of the previous network example. We have replaced the connection from vertex 2 to 4 by a connection from vertex 2 to 1. In this graph, 1 and 5 are neighbors of 2 and the shortest path from 1 and 5 is via node 3 or 4 vertices (hop). The first stage of losing any vertex or connection that is not directly tied to a vertex will not cause this vertex to isolate from the network.

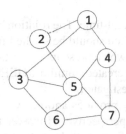

Fig. 6. Modified network to eliminate blackout from a first stage cascade failure

- *Lemma 2:* Moving forward, to eliminate a second stage of cascade failure, at least one shortest path from one of its neighbor to another neighbor's vertex consisting of greater than or equal to two vertices should be found.

3.3   From a Node to a Zone

With a large area network, a small-world power grid is preferred [9]. A group of BUS [10, 11] could be used. Each group could become a node. Figure 7 is an example of how we could group a power zone to a node. Obviously, each zone always have the same property: Total power coming in should equal total power coming out a zone. Note that in all our graphs, all distribution lines are not shown, although they are there. The strategy for cascade blackout prevention then will be the same. In this case we can have five vertices instead of 34. This approach could be used for regional forecasting.

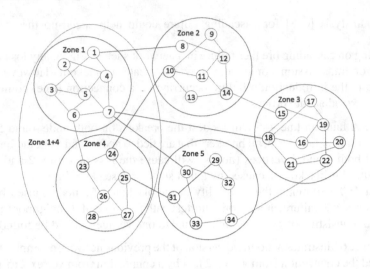

Fig. 7. Power network with zones

4   Future Studies

The complex power network are real-time. The relation between graph and linear values and statistical and empirical method should be studied further in future. Most of power network data are non-disclosure which presents an obstacle for detailed study. A close-to-real-world database should be created. Similar to IEEE-30 [10, 11] real and complex situations or other should be investigated.

The complex network theory provides another visual look of large area blackout prevention on the connectivity perspective. However there should also be a system security and stability perspectives as well. In order to take complex network to real-world power network, all power operators and regulators should work together as stake-holders to promote further studies on blackout prevention techniques.

References

1. Blackburn, L., Domin, T.J.: Protective Relaying: Principles and Applications, 4th edn. CRC Press, Boca Raton (2014)
2. Miller, R., Malinowski, J.: Power System Operation, 3rd edn. McGraw-Hill Professional, New York (1994)
3. Dorogovtsev, S.N., Mendes, J.F.F.: Evolution of networks. Adv. Phys. **51**, 1079 (2002)
4. Dahman, S.R.: N-1-1 Contingency Analysis using PowerWorld Simulator. http://www.powerworld.com/files/SimulatorN-1-1.pdf
5. NERC: Reliability Concepts. http://www.nerc.com/files/concepts_v1.0.2.pdf
6. Wang, Z., Scaglione, A., Thomas, R.J.: A Markov-transition model for cascading failures in power grids. In: 45th Hawaii International Conference on System Science, pp. 2115–2124 (2012)

7. Rahnamay-Naeini, M., Wang, Z., Mammoli, A., Hayat, M.M.: A probabilistic model for the dynamics of cascading failures and blackouts in power grids. In: IEEE Power and Energy Society General Meeting, pp. 1–8 (2012)
8. Rahnamay-Naeini, M., Wang, Z., Ghani, N., Mammoli, A., Hayat, M.M.: Stochastic analysis of cascading-failure dynamics in power grids. IEEE Trans. Power Syst. **29**, 1767–1779 (2014)
9. Sun, K.: Complex networks theory: a new method of research in power grid. In: IEEE/PES Transmission and Distribution Conference and Exhibition: Asia and Pacific, pp. 1–6 (2005)
10. Kachore, P., Palandurkar, M.V.: TTC and CBM calculation of IEEE-30 bus system. In: 2nd International Conference on Emerging Trends in Engineering and Technology (ICETET), pp. 539–542 (2009)
11. Divya, B., Devarapalli, R.: Estimation of sensitive node for IEEE-30 bus system by load variation. In: International Conference on Green Computing Communication and Electrical Engineering (ICGCCEE), pp. 1–4 (2014)

A Hybrid Trust Management Framework for Vehicular Social Networks

Rasheed Hussain[1]([✉]), Waqas Nawaz[1], JooYoung Lee[1],
Junggab Son[2], and Jung Taek Seo[3]

[1] Institute of Informatics, Innopolis University, Innopolis, Russia
{r.hussain,w.nawaz,j.lee}@innopolis.ru
[2] Department of Mathematics and Physics,
North Carolina Central University, Durham, USA
json@nccu.edu
[3] College of Information Security, Soonchunhyang University,
Asan, South Korea
seojt@sch.ac.kr

Abstract. This paper addresses the trust management problem in the emerging Vehicular Social Network (VSN). VSN is an evolutionary integration of Vehicular Ad hoc NETwork (VANET) and Online Social Networks (OSN). The application domain of VSN inherits the features of its parental VANET and OSN, providing value-added services and applications to its consumers, i.e. passengers and drivers. However, the immature infrastructure of VSN is vulnerable to security and privacy threats while information sharing, and hard to realize in the mass of vehicles. Therefore, in this paper, we particularly advocate for communication trust establishment and management during information exchange in VSN. First, we establish functional architectural frameworks for VSN that are based on the underlying applications. Second, based on these frameworks, we propose two trust establishment and management solutions, i.e. email-based social trust and social networks-based trust, to target different sets of applications. Third, we discuss the contemporary research challenges in VSN. Our proposed scheme is a stepping stone towards the secure and trustworthy realization of this technology.

Keywords: VANET · Social networks · Vehicular Social Network (VSN) · Trust · Reputation · Security · Privacy

1 Introduction

Vehicular Ad hoc NETwork (VANET) is poised to offer the drivers and passengers with a safe, at least fail safe, reliable, and infotainment-rich driving environment. From the research results in the field of vehicular networks (semi-autonomous) and driverless (autonomous) cars, it can be easily speculated that intelligent transportation system (ITS) technologies, which are realized through VANET, will be soon pervading our highways. There are few challenging issues

© Springer International Publishing Switzerland 2016
H.T. Nguyen and V. Snasel (Eds.): CSoNet 2016, LNCS 9795, pp. 214–225, 2016.
DOI: 10.1007/978-3-319-42345-6_19

that are keeping the stakeholders and investors at bay from deploying these technologies on a mass scale. These issues include security, privacy, trust, framework design, initial deployment, data and user privacy, to name a few [17].

The mobility patterns, based on space and time, are predictable in VANET and linked to online social networks (OSNs). For example, the traffic tends to be dense during rush hours because people are going to office in the morning and coming back home in the evening, which is not the case for non-peak hours. This phenomena develops a unique social relationships among neighbors who tend to share same interests and/or likely schedule. The recent developments in OSNs gives rise to the concept of VSN [19] by providing a preferred mean of sharing social activities among VANET users. Consequently, many VSN applications are developed for this purpose such as Tweeting car[1], SocialDrive [9], Social-based navigation (NaviTweet) [14], CliqueTrip [5], and GeoVanet. Beside the technological advancements, it is essential to look at the social perspective of VANET [3,4].

The credibility of both the stakeholders and the information shared through OSN in VANET infrastructure using VSN applications is a challenging task. The former is achieved through tools and methods from cryptography and public-key infrastructure (PKI), and the later cannot be guaranteed with the first line of defense, i.e. traditional PKI-based approach. The credibility of information can be indirectly measured through trust evaluation and management. Recently, a number of studies were conducted to look into the possibility of merging VANET with social networks and harvest the features of both technologies to enrich the application space of ITS [19]. A plethora of techniques proposed various solutions for trust establishment in VANET [1,2,6,10,11,13,15,16,18,20]. However, there is a significant gap between stakeholder and information trust. Specifically, the data level trust is overlooked by existing studies. To overcome these issues, we proposed architectural frameworks for VSN. Further, we establish two trust methods, namely email-based and social network-based trust, to guarantee the credibility of information in VSN.

The structure of the rest of this paper is organized as follows: Sect. 3 describes functional architectural frameworks for VSN. Our proposed trust management scheme is outlined in Sect. 4. We discuss the unique VSN research challenges in Sect. 5 followed by concluding remarks and future directions in Sect. 6.

2 Related Work

Trust is one of the many challenges in VANET. A number of studies have proposed various solutions for trust establishment in VANET. Node/entity trust is achieved in VANET through well-established cryptographic solutions. The cryptographic mechanisms help to prove the legitimacy of the source of communication. In other words, secure and efficient authentication mechanisms guarantee node trust in VANET [6,16,18,20]. Furthermore, trust management schemes

---
[1] http://www.engin.umich.edu/college/about/news/stories/2010/may/caravan-track-hits-the-road.

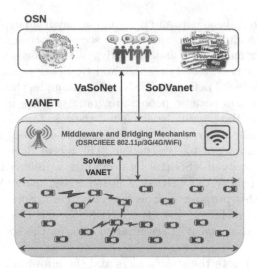

Fig. 1. VSN network and communication model

have also been implemented to build trust among the VANET users for information exchange [1,2]. In [15], the authors consider both data trust and node trust, and propose an attack-resistant trust management solution for vehicular networks. They achieve data trust through data collection from multiple sources (vehicles) and node trust through functional approach and recommendation approach. Moreover, a trust quantification mechanism is also proposed in [13]. Another email-based social trust establishment scheme has been proposed by Huang et al. [11]. Our email-based trust management in VSN is inspired by [11]. Huang et al. proposed a situation-aware trust framework in [10]. It includes an attribute-based policy control model for highly sporadic VANET, which is a proactive trust model to build trust among VANET nodes, and an email-based social network trust system to enhance trust among users. It is worth noting that the research community has focused on node/entity trust in VANET where the sender is judged based on the confidence of trust. A very small attention has been given to the data trust. In this paper we try to minimize the gap between node trust and data trust.

3   Functional and Architectural Frameworks of VSN

This section comprehensively discusses the network and communication model, taxonomy of VSN application areas, followed by potential architectural frameworks for VSN.

3.1   Network and Communication Model

The network and communication model for our proposed scheme is shown in Fig. 1. Social networks have their own setup and they run both desktops and mobile versions. On the other hand, VANET is based on the dedicated short-range communication (DSRC) which mandates V2V and V2I communication. Vehicular nodes and roadside units are equipped with on-board units (OBUs) and tamper-resistant hardware (TRH). TRH is responsible for storing the security-related keys and other cryptographic material. OBUs send out different kinds of messages that include frequent beacon messages, service requests, key update requests, warning messages, and so forth. In order to bridge vehicular networks with the OSN, we have a number of options and roadside units (RSUs) is one of them. Today's high-end 3/4G network capable cars can also send and/or receive data to/from OSN to VANET. For instance, mobile devices with social network applications can connect to vehicle through WiFi and Bluetooth protocols.

3.2 Taxonomy

There are many application domains that benefit from VSN either directly or indirectly. Some of these application domains include entertainment, information exchange, diagnostic/control, health-care, platooning, cooperative cruise control, crowdsourcing, cooperative navigation, content delivery, social behavior, clustering-based communication, and vehicular clouds [19]. The communication among vehicles is the first entry point to the social networking paradigm, because both follow the same baseline principle of real world communication. Therefore, the information exchange is rendered as social interaction among vehicles. In order to understand the aforementioned application domains, we outline a detailed taxonomy of these applications based on varying architectures of VSN. We divide VSN into three functional architectural frameworks namely Social Data-driven vehicular networks (SoDVanet), Social VANET (SoVanet), and Vanet data-driven Social Networks (VaSoNet). Figure 2 outlines the taxonomy of VSN applications based on the underlying framework. These frameworks encompass the potential application domains from vehicular communications to user behavioral perspective.

3.3 Social Data-Driven Vehicular Networks (SoDVanet)

In SoDVanet framework, the existing vanet infrastructure uses social data obtained from users. Therefore, this framework broadens the application space and offers more services to pure vanet users. The SoDVanet architecture assumes that both vanet and social networks are established and there is a bridging mechanism to integrate these two in a seamless fashion. To be precise, vanet uses data from the available social network for its specific class of applications and provides the required services to the users. The integration is user-centric because every vanet user rely on its social network direct (friends) or multi-hop contacts (friends of friends) depending upon the required degree of connectivity. For instance, a comprehensive social data-driven information system would help the vanet users to be updated for certain events on the road, city, and/or across the

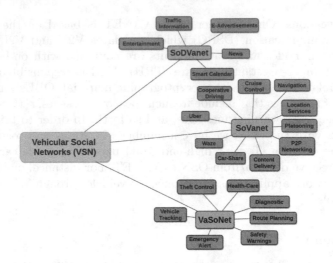

Fig. 2. Taxonomy of VSN applications

country. We can achieve this through a pull-based strategy at vanet infrastructure where the information shared by users in social networks is collected by the car and present it to the user based on his/her preferences.

3.4   Social VANET (SoVanet)

Vehicular nodes communicate with each other through different types of messages, for instance beacons and warning messages. Beacons are shared with neighbors and with infrastructure for cooperative awareness. The nature of these messages determine the social behavior among vehicles at communication and application level. Eventually, this results in realization of a number of applications through vehicular social communications. Besides beacons, vehicle also share critical information such as warning message as a result of some designated incident on the road, black ice on the pavement, ambulance approaching warning, traffic jam warning. Additionally, vehicle users can also share their experience related to daily social activities, e.g. experiences about a restaurant, availability of parking lots, new movies in the theatre. We call SoVanet as infrastructureless social network, because vehicles use existing vanet infrastructure and use social parameters for information exchange.

3.5   Vanet Data-Driven Social Networks (VaSoNet)

OSN users leverage the data obtained from vehicular communications in VaSoNet framework. The data obtained from VANET infrastructure is related to transportation. There are different variations for this communication paradigm; however, the most interesting one is the inquiry of transportation related information ubiquitously through VANET infrastructure. A certain query is executed either in a centralized (at the server), or distributed fashion (by the nodes in the area of interest). For instance, before leaving home on a busy national holiday, one would like to know the current traffic situations on the road. The communication model of this framework is based on an efficient and secure bridging mechanism between OSN and VANET. A well defined mechanism is required at first place to authenticate the data sources in VANET and to preserve both user and location privacy. This paradigm comes with a unique challenge to stimulate the VANET nodes to share their experience and/or data, e.g. pictures-on-demand, real time traffic information, in correspondence with OSN queries.

4   Proposed Hybrid Trust Management Scheme

In this section we outline our proposed hybrid email-based and social network-based trust management scheme for the aforementioned VSN frameworks.

4.1   Baseline Overview

The health of information shared among nodes in VSN, is of paramount importance and cannot be achieved through traditional cryptographic techniques. Therefore we employ a trust management mechanism to make sure that the exchanged information is healthy and trustworthy. In the light of the fact that most of the users use email as a mean of communication, therefore we use the frequency of email interactions for trust calculation. In order to calculate peer trust, users look into the email interactions with neighbors. If the node is in the trusted list of the receiver node, then the information is likely to be trusted, otherwise there are a number of other options, for instance generating a query about the trust value of the sender node and/or looking into the social relations of the sender, and so forth. We particularly focus on the 2-hop trust propagation where the trust query propagates to friends and friends of friends. The nodes also maintain their peer trust based on personal interactions with the neighbors and if needed, share this trust information with neighbors. The users also calculate trust based on their social interactions through OSN and depending on application, they calculate the resultant trust from intermediate trust values.

4.2   System Initialization

In order to enable email-based trust, department of motor vehicles (DMV) initializes the OBU by performing a number of operations. First the user enters its anonymized email address and DMV registers it against the user. DMV also issues a number of pseudonyms $\{Ps_1^u, Ps_2^u, Ps_3^u, \ldots Ps_l^u\}$ to the user u. Furthermore, DMV issues public private key pair to each pseudonym $<PK_{Ps_i}^u, SK_{Ps_i}^u>$ and another master secret key SK_u which is derived from the email ID, E_u. The email ID E_u of user u serves as public key which it shares with the neighbors. DMV shares the trapdoor of the pseudonyms with revocation authorities (RAs) as well which will be used in revocation process. Due to space limitation, we refer the readers to [12]. Each user also runs a local email agent that connects to email service provider. The user maintains its contacts in different groups such as family (fm), friends (fr), acquaintances (aq), and work (wr). In email-based trust, certain groups such as family, is static changes to family group are less likely while others will be dynamic. An absolute confidence value c_i is assigned to each group where i is the group. In the dynamic groups, the nodes can earn the privilege to upgrade to a different group with higher c_i. It is also to be noted that, $c_{fm} > c_{fr} > c_{wr} > c_{aq}$ which defines the preferences. There is also a baseline unknown confidence c_U which is assigned to the contacts who are either first timers or unknown.

4.3   Hybrid Trust Management in VSN

Our proposed trust management is composed of two modules, email-based trust calculation and social network-based trust calculation module. Based on application, these modules will work in adaptive and robust manner. For instance email-based trust can be ideal for the applications in SoDVanet and social network-based trust can be used for applications in VaSoNet. Trust mechanism is divided into three processes, trust bootstrap, trust calculation and evaluation, and trust query. We describe these processes in detail.

Trust Bootstrap. Bootstrapping of trust information is the first step before the real communication among users. Every user maintains three lists namely Known friends list (KFL), anonymous friends list (AFL), and random encounter list (REL). KFL contains the information about the friends that are known either through social networks or emails. The real distinction among friends is done through the aforementioned confidence value c_x where x is degree of closeness. AFL contains information about acquaintances developed through either vehicular communications, social networks, or emails. Lastly the lowest level of relation is the random encounter through vehicular communications or emails. The definition of random encounter is debatable; however, for the sake of understanding we argue that communications carried out with neighbors for less than a defined threshold t_σ, will be placed in REL, where σ is the lowest threshold for which the nodes must be communicating with each other to get the level of AFL. At the system initialization every node populates KFL and AFL (based on immediate previous experience) and an empty REL. These lists store the information against anonymous pseudonyms instead of real identities, whereas in case of KFL, a certain degree of node information is also known. Pseudonyms become handy in case of AFL where the receiving node is not sure about the real identity and the trust level is in its infancy. Moreover the pseudonyms serve other purposes like preserving conditional privacy and revocation when needed.

Trust Calculation and Evaluation. There are two kinds of trust evaluations, local trust evaluation through received messages and the recommendation from neighbors as a result of mutual communication. The trust calculation mechanism can be either sender-centric or receiver-centric. Local trust is receiver-centric whereas the recommendation can be both sender-centric or receiver-centric. In case of sender-centric, the receivers of the message calculate the trust value for the sender and in case of receiver-centric, the receiving node calculates trust value for the source of the message. For our proposed scheme, we consider sender-centric trust where a node waits for confidence values that it accumulates for itself from the neighbors.

Each node i calculates the trust value for its neighbor j based on two factors: (i) encounters (number of beacons η_b to be more precise) that i had with j. (ii) endorsement for j by its neighbors as a result of event message generated/broadcast by j. The net trust is calculated as follows:

$$\tau_j^i = \alpha \times \eta_b + (1 - \alpha) \times \sum_{i=1}^{n} \tau_e \times T_j^e$$

τ_j^i is the trust value calculated for node j by node i. α is the priority factor (weight) for the means of trust calculation (value between 0 and 1). In this case, we argue that the direct encounters carry more weight for the trust calculation than the endorsement of the neighbors. τ_e is the endorser's own trust value perceived from neighbors and T_j^e is the trust value endorsed by endorser e for the node j. It is worth noting that these values are obtained through group query-response process. The direct communication with the nodes will give more confidence to node i to calculate local trust value for the neighbors. Therefore the condition $\alpha > (1-\alpha)$ must hold. For the nodes in KFL, the trust calculator signs their certificates and pseudonyms with highest confidence. Whereas for the AFL nodes, the trust calculator signs the certificates with confidence $c_{AFL} < c_{KFL}$. The value of c_{AFL} will vary depending on the current neighborhood status of the node. For the nodes in REL, the certificates will be signed with a value of baseline confidence c_{REL}. The relation $c_{KFL} > c_{AFL} > c_{REL}$ must hold during the trust calculation.

In the email-based trust evaluation, every node assigns the trust to the nodes based on which list they currently belong to. For nodes in KFL, the trust calculator node assigns the fully-trusted status. In other words, if $n_i \in \{KFL\}$ and the contact frequency is above a threshold (certain emails in a specified amount of time) then $\tau_i = FullyTrusted$. On the other hand, for AFL, the trust calculator assigns the trust value based on heuristics from the previous trust value that was possessed by the node in question. There is a base trust value for AFL denoted by τ_{base}. If $n_i \in \{AFL\}$ then $\tau_i = \tau_{prev} + \tau_{cur}$, and $\tau_{prev} = \tau_{base} + \tau_{cur}$. This calculation is recursive and the only limit is the upper-bound of the AFL and REL. It is worth noting that the value of $\tau_{previous}$ will be between the base value for AFL and the base value for KFL. In other words, the maximum trust value of the nodes in AFL cannot exceed the base value of KFL. Similarly the social network-based trust calculation is same except for the lists management where only family and best friends are fully trusted while the trust of other nodes will depend on the frequency of communications. It is to be noted that if the credentials of a node are legitimate then the trust calculator will sign the credential; however, the trust value will be calculated according to the aforementioned mechanism. Need for efficient interaction among social network, vehicular network and email service is essential for the trust management solution. The provision of intermediary service among these networks is out of scope of this paper.

Transitive Global Trust (Trust Query). In order to get a geographically-controlled global view of the neighbor's trust, a cooperative approach is employed where a node generates trust query to its immediate neighbors (including RSU). The query contains the email ID of the query originator, its pseudonym, and other credentials that will prove the legitimacy of the node. This query is broadcasted over DSRC channel. When the neighbors receive such query, first of all, they check for the validity of the cryptographic material in the query that includes certificate verification and validation. This is done by applying public key of the DMV/trusted party to the certificate. Then the receiver also checks for the validity of the pseudonym and certificate which can be checked through pseudonym revocation list (PRL) and/or certificate revocation list (CRL). We assume that an efficient PRL and CRL mechanisms are already in place [7,8]. After credential verification, the node traverses through its lists to determine the trust value for the node in the query. If the node is found, its trust value is sent back to the query originator along with the confidence value and signature of the responder. The query originator accumulates all the replies and updates the trust status of the node in the query. More precisely the query originator combines the trust values from the neighbors and calculates the net trust value for the node in the query. However if the node is not in the list of the responder, and the responder also received message from the node in query, then the responder seconds the query and show the interest to know the trust value of the node in query as well.

5   Research Challenges in VSN and Open Questions

5.1   Deployment

VANET is on the verge of deployment whereas OSN is fully deployed with unbelievably huge number of users and still growing. There are a number of problems that have caused the impeded momentum in VANET deployment. Few prominent issues include security, privacy, hardware, and lack of infrastructure. The deployment of VSN will face additional problems that are unique. For instance, investors will be reluctant to put their huge investment at stake. Therefore, at the first deployment stage, the traditional off-the-shelf hardware will become handy. More insight is needed to counter these issues at the very beginning of VSN deployment.

5.2   Security of Information Exchange

The security of information exchange is important in traditional VANET and OSN; however, in case of VSN the information exchange may violate user privacy.

On the other hand, the level of user privacy may be different in different applications. Therefore, the context information must be taken into account before preserving user and location privacy in VSN. It is also worth noting that the revocation mechanisms will vary from application to application in VSN.

5.3   Cross-Platform Conditional Privacy

The level of privacy is hard to generalize and seems application dependent in VSN. Moreover, the semantics of user privacy are different in OSN and VANET. Therefore, the cross-platform applications must take the privacy requirements into account while using data and preserve the user and/or location privacy accordingly. This phenomenon is going to be a daunting challenge in VSN and will require a thorough investigation.

5.4   Audit and Incentives

Most of the VSN applications are cooperative in nature where the data is collected through cooperation among nodes. However, selfish behavior from legitimate nodes is still not out of question. Therefore, a secure, efficient, and privacy-aware incentives mechanism is essential to stimulate active participation of the nodes.

5.5   Information Update/Decay

With the passage of time, the size of lists and their trust values will grow exponentially. Deep insight is required to decide on the frequency of the updates, to the lists, and the trust values. In order to find optimum frequency, the traffic scenario, spatial and temporal statistics must be taken into account. Moreover, the calculated trust values are not permanent and subject to change depending on the behavior of the neighbors. Therefore, the lifetime parameter of trust value is of paramount importance to guarantee the scalability of trust management scheme. The trust value should be valid for a certain amount of time after which the nodes will need to re-establish the trust. Determining the optimal time during is also an open problem.

5.6   Mobility vs Social Factors

In VANET, the mobility of vehicles is restricted to the road networks that will likely exhibit in VSN as well. Whereas in traditional OSN, there is no such restriction (although the behavior of users is still predictable). The data shared between VANET and OSN will definitely help the application to grow and provide the consumers with better services, but may also impact the social values of the users in both networks. For instance, profilation, user behavior, and social interests are prone to be abused as a result of such integration. Therefore, clear distinction is necessary between sensitive users' data and application data.

6   Conclusions and Future Work

In this paper, we aimed at a new paradigm shift referred to as vehicular social network (VSN) and proposed application-based architectural frameworks. First we proposed the application taxonomy of VSN and then three architectural frameworks namely Social Data-driven vehicular networks (SoDVanet), Social VANET (SoVanet), and Vanet data-driven Social Networks (VaSoNet). Furthermore, we proposed trust management system for VSN which leverages two approaches, email-based and social network-based trust management. In email-based trust management, the nodes calculate the trust values for neighbors based on the frequency of their email communication. The nodes also leverage social distinction among neighbors in terms of family, friends, work, and acquaintances. We also proposed social network-based trust management scheme for VSN. When nodes calculate the trust values for the neighbors, they consider the possibility of social relation with those neighbors through online social networks. Based on the nature of relation, respective trust value is calculated for the neighbor. In the proposed system, a node can also query trust status from its neighbors. We also outlined the research issues and open questions in VSN. We aim to implement the reputation system based on the real-world data and work on the optimization of scheme selection to incorporate trust scalability in VSN.

References

1. Abumansoor, O., Boukerche, A.: Towards a secure trust model for vehicular ad hoc networks services. In: Global Telecommunications Conference (GLOBECOM2011), pp. 1–5. IEEE, December 2011
2. Alriyami, Q., Adnane, A., Smith, A.K.: Evaluation criterias for trust management in vehicular ad-hoc networks (VANETs). In: 2014 International Conference on Connected Vehicles and Expo (ICCVE), pp. 118–123, November 2014
3. Cunha, F.D., Vianna, A.C., Mini, R.A.F., Loureiro, A.A.F.: How effective is to look at a vehicular network under a social perception? In: 2013 IEEE 9th International Conference on Wireless and Mobile Computing, Networking and Communications (WiMob), pp. 154–159, October 2013
4. Cunha, F.D., Maia, G.G., Viana, A.C., Mini, R.A., Villas, L.A., Loureiro, A.A.: Socially inspired data dissemination for vehicular ad hoc networks. In: Proceedings of the 17th ACM International Conference on Modeling, Analysis and Simulation of Wireless and Mobile Systems, MSWiM 2014, NY, USA, pp. 81–85 (2014). http://doi.acm.org/10.1145/2641798.2641834
5. Ekler, P., Balogh, T., Ujj, T., Charaf, H., Lengyel, L.: Social driving in connected car environment. Proc. Eur. Wirel. Conf. **2015**, 1–6 (2015)
6. Feiri, M., Pielage, R., Petit, J., Zannone, N., Kargl, F.: Pre-distribution of certificates for pseudonymous broadcast authentication in VANET. In: 2015 IEEE 81st Vehicular Technology Conference (VTC Spring), pp. 1–5, May 2015
7. Ganan, C., Munoz, J.L., Esparza, O., Mata-Diaz, J., Alins, J., Silva-Cardenas, C., Bartra-Gardini, G.: RAR: risk aware revocation mechanism for vehicular networks. In: 2012 IEEE 75th Vehicular Technology Conference (VTC Spring), pp. 1–5, May 2012

8. Haas, J.J., Hu, Y.C., Laberteaux, K.P.: Efficient certificate revocation list organization and distribution. IEEE J. Sel. Areas Commun. **29**(3), 595–604 (2011)
9. Hu, X., Leung, V.C., Li, K.G., Kong, E., Zhang, H., Surendrakumar, N.S., TalebiFard, P.: Social drive: a crowdsourcing-based vehicular social networking system for green transportation. In: Proceedings of the Third ACM International Symposium on Design and Analysis of Intelligent Vehicular Networks and Applications, DIVANet 2013, NY, USA, pp. 85–92 (2013). http://doi.acm.org/10.1145/2512921.2512924
10. Huang, D., Hong, X., Gerla, M.: Situation-aware trust architecture for vehicular networks. IEEE Commun. Mag. **48**(11), 128–135 (2010)
11. Huang, D., Zhou, Z., Hong, X., Gerla, M.: Establishing email-based social network trust for vehicular networks. In: 2010 7th IEEE Consumer Communications and Networking Conference, pp. 1–5, January 2010
12. Hussain, R., Kim, D., Tokuta, A.O., Melikyan, H.M., Oh, H.: Covert communication based privacy preservation in mobile vehicular networks. In: Military Communications Conference, MILCOM 2015, pp. 55–60. IEEE, October 2015
13. Kim, Y., Kim, I., Shim, C.Y.: Towards a trust management for vanets. In: The International Conference on Information Networking (ICOIN 2014), pp. 583–587, February 2014
14. Lequerica, I., Longaron, M.G., Ruiz, P.M.: Drive and share: efficient provisioning of social networks in vehicular scenarios. IEEE Commun. Mag. **48**(11), 90–97 (2010)
15. Li, W., Song, H.: Art: an attack-resistant trust management scheme for securing vehicular ad hoc networks. IEEE Trans. Intell. Transp. Syst. **17**(4), 960–969 (2016)
16. Lo, N.W., Tsai, J.L.: An efficient conditional privacy-preserving authentication scheme for vehicular sensor networks without pairings. IEEE Trans. Intell. Transp. Syst. **PP**(99), 1–10 (2016)
17. Qu, F., Wu, Z., Wang, F.Y., Cho, W.: A security and privacy review of VANETs. IEEE Trans. Intell. Transp. Syst. **16**(6), 2985–2996 (2015)
18. Shao, J., Lin, X., Lu, R., Zuo, C.: A threshold anonymous authentication protocol for VANETs. IEEE Trans. Veh. Technol. **65**(3), 1711–1720 (2016)
19. Vegni, A.M., Loscr, V.: A survey on vehicular social networks. IEEE Commun. Surv. Tutorials **17**(4), 2397–2419 (2015)
20. Wang, F., Xu, Y., Zhang, H., Zhang, Y., Zhu, L.: 2FLIP: a two-factor lightweight privacy-preserving authentication scheme for VANET. IEEE Trans. Veh. Technol. **65**(2), 896–911 (2016)

Distributed and Domain-Independent Identity Management for User Profiles in the SONIC Online Social Network Federation

Sebastian Göndör(✉), Felix Beierle, Senan Sharhan, and Axel Küpper

Telekom Innovation Laboratories, TU Berlin, Berlin, Germany
{sebastian.goendoer,beierle,axel.kuepper}@tu-berlin.de,
senan.mh.sharhan@campus.tu-berlin.de

Abstract. As of today, communication habits are shifting towards Online Social Network (OSN) services such as *WhatsApp* or *Facebook*. Still, OSN platforms are mostly built in a closed, proprietary manner that disallows users from communicating seamlessly between different OSN services. These lock-in effects are used to discourage users to migrate to other services. To overcome the obvious drawbacks of proprietary protocols and service architectures, SONIC proposes a holistic approach that facilitates seamless connectivity between different OSN platforms and allows user accounts to be migrated between OSN platforms without losing data or connections to other user profiles. Thus, SONIC builds the foundation for an open and heterogeneous *Online Social Network Federation* (OSNF). In this paper, we present a distributed and domain-independent ID management architecture for the SONIC OSNF, which allows user identifiers (GlobalID) to remain unchanged even when a profile is migrated to a different OSN platform. In order to resolve a given GlobalID to the actual URL of a social profile the Global Social Lookup System (GSLS), a distributed directory service built on peer to peer technology is introduced. Datasets called *Social Records*, which comprise all information required to look up a certain profile, are stored and published by the GSLS. Following this approach, social profiles can be migrated between OSN platforms without changing the user identifier, or losing connections to other users' social profiles.

1 Introduction

As of today, a strong trend can be observed that shows that communication habits are shifting towards Instant Messaging (IM) and Online Social Networks (OSN). While old fashioned communication habits such as voice calls are declining, usage of OSN and IM services is steadily rising [1,2]. OSN platforms allow their users to communicate via text, audio, and video, share content, or just stay in contact with friends and relatives. While a large number of competing OSN platforms with a broad variety of features exist as of today, *Facebook*, which was founded in 2004, managed to overcome its predecessors and competitors by far in terms of number of users and popularity [3], and continues to be the world

H.T. Nguyen and V. Snasel (Eds.): CSoNet 2016, LNCS 9795, pp. 226–238, 2016.
DOI: 10.1007/978-3-319-42345-6_20

leader in terms of users accessing the service [4]. Competitors were forced out of the market or had to focus on niche markets such as modeling relations to business partners (e.g., *LinkedIn* and *Xing*) or to address different aspects of social interactivity (e.g., communication via *WhatsApp*) or activities (e.g., publishing images via *Instagram*). Most OSN designs promote a closed, proprietary architecture that disallows users from communicating seamlessly between different OSN services. The well-calculated lock-in effects of proprietary platforms are used to bind users to the service, as migrating to another OSN platform would result in a loss of connections to friends and the data one has accumulated as part of his social profile [5]. Alternative OSN architectures propose a federation of servers or make use of peer to peer technology to distribute control over the social graph and associated data [6,7]. Still, communication between different OSN platforms is mostly not possible, or just enabled via special plugins or services, which are used to replicate data between different accounts of the same user on different OSN platforms [8]. To overcome the obvious drawbacks of proprietary protocols and architectures in the area of OSN services, SONIC [9] proposes a holistic approach that facilitates seamless connectivity between different OSN platforms and allows user accounts to be migrated between OSN platforms without losing data or connections to other user profiles. Following the *Interop* theory [10], the vision of SONIC proposes an open and heterogeneous *Online Social Network Federation* (OSNF), in which social profiles are managed independently from the platform they are hosted on [11]. To allow seamless and transparent communication between different OSN platforms, identification of user profiles as well as resolving identifiers to a profile's actual location is a crucial task. As profiles may be migrated at any time, identifiers that are bound to a domain name of a OSN platform cannot be employed. Hence, identifiers in SONIC need to be domain agnostic and created in a distributed fashion. This allows users to keep their identifier even after migrating to a new OSN platform on a different domain. Anyhow, introducing domain agnostic global identifiers requires for means of resolving an identifier to the current network location of the respective social profile. For this reason, SONIC introduces the Global Social Lookup System (GSLS), a distributed directory service built on peer to peer technology using distributed hash tables (DHT).

In this work, we present an identification architecture for decentralized OSN ecosystems. The architecture features GlobalIDs as domain agnostic, globally unique identifiers, which can be generated in a distributed fashion without the need for a central authority. The architecture introduces a distributed directory service, the GSLS, which is utilized to resolve GlobalIDs to a user profile's actual location. The GSLS manages a digitally signed dataset, the *Social Record*, which comprises information about the social profile identified by the GlobalID. Following this paradigm, social user profiles can be identified independently of the OSN platform's operator. Furthermore, users can change the location of their profile at any time without losing connections between other social profiles [11]. The architectural requirements for the SONIC federation have been defined in [12], comprising a decentralized architecture, the use of open protocols and formats, the option for users to migrate their social accounts, seamless communication, the use of a single social profile, and global user identification. The remainder of this paper is organized as follows: The following chapter provides an overview about existing approaches, protocols, and standards in the area of identity management. Section 3 gives an overview of the concept of the SONIC federation, followed by a description of the identity management architecture in Sect. 4. Section 5 describes the implementation of the proposed solution, which is evaluated in Sect. 6. Section 7 concludes the paper.

2   Related Work

Services that manage multiple users or objects require a measure of identification to distinguish between individual users or objects. For this purpose, an identifier is assigned to each entity, where an identifier is a name that usually is a sequence of letters, numbers, or symbols, with the usual intent of being unique in a certain domain. This assures that each user or object can be uniquely addressed via its identifier, and two equal entities can be distinguished. In applications and services that are used by multiple users, each user is traditionally assigned a user name, which is unique in the domain of the application or service. A well known example is the Linux operating system, where each user gets to chose a unique user name and a serial number (uid). The uid is used by the system to identify users, while the actual user name is mostly used for authentication and displaying purposes. Social applications and services also usually identify users by a numerical user identifier, which in most cases has to be unique within the domain of this service or application. In addition, most services allow their users to pick a display name, which is shown to other users. This display name is then not necessarily used as an identifier, but as a normal name. As of this, the display name is not necessarily unique and functions similar to a given name.

While issuing and resolving user identifiers within the same domain is comparatively easy, identifying entities across different domains is a more complex task. Here, usually composed identifiers are used that comprise a local identifier, which is unique in its issuing domain, and a domain identifier that uniquely identifies the domain. This way, a local user name "Marc" can exist in two separate domains at the same time, while only the domain name is required to be unique. This kind of composed identifiers is used by most Internet-based services or applications, where the domain identifier is the full qualified domain name (FQDN) of the service. One example are email addresses [13] or jabber-ids (JID) as employed by XMPP in the format `local-id@domain-id` [14]. Resolving this kind of composed identifiers depends on the Domain Name Service (DNS) [15], which is required to resolve the domain part of the identifier, while the local user name is resolved by the service itself.

Similar to this identifier format, Unified Resource Identifiers (URI) or International Resource Identifiers (IRI) [16] can be used to uniquely identify an entity or person. Here, a path can be specified to further describe categories or a classes of identities, e.g., http://company.com/berlin/employees/alice. By utilizing actual URLs as identifiers, users and services can easily resolve an identifier to e.g., a document, which provides further information about the linked entity. This approach is employed by e.g., WebID [17], where a URI is resolved to a profile document using the DNS. The protocol WebID+SSL [18] further involves exchange and verification of encryption keys to establish a trusted and secure connection between two individuals. Also the authentication protocol OpenID employs URIs as user identifiers [19]. While the advantage of these kinds of composed identifiers are that services can freely assign user names for identification purposes, identifiers created in this fashion are bound to the domain they were created in, and hence cannot be migrated to another domain.

In scenarios, where entities need to be identified independently of a fixed domain or service, different approaches have to be applied. To avoid collision of identifiers created without coordination of the id generating services, randomness can be used to make a collision unlikely. Following this approach, cryptographic hash functions are used to create a random number from a combination of deterministic or random input values. Universally Unique Identifiers (UUID) - also known as Globally Unique Identifiers (GUID) - are 128 bit identifiers created by using hash algorithms [20]. The UUID standard defines 4 types of identifiers. Depending on the type of the UUID, different data is used for its creation. For example, a version 1 UUID uses the machine's MAC address and datetime of creation, while version 5 uses SHA1 with a namespace part. The uniqueness of UUIDs is based on the assumption that generating the same UUID twice is very unlikely. In 2003, the OASIS group introduced eXtensible Resource Identifiers (XRI) as an identifier scheme for abstract identifiers [21]. XRIs are designed to be domain-, location-, and platform-independent and can be resolved to an eXtensible Resource Descriptor Sequence (XRDS) document via HTTP(S). Work on the XRI 2.0 specification was discontinued in 2008 by the XRI Technical Committee at OASIS. Twitter Snowflake [22] is an identifier schema based on hashing a timestamp, a preconfigured machine number, and a sequence number. Twitter Snowflake was built for fast and distributed id generation without the need for the machines generating the ids to coordinate with each other. Snowflake was discontinued in 2010, but other implementations of the approach exist, e.g. PIIP Cruftflake [23]. Boundary Flake, which follows a similar approach as Twitter Snowflake, is a "decentralized, k-ordered id generation service" [24]. Here, the machine's MAC address, a UNIX timestamp, and a 12 bit sequence number are hashed to create a 128-bit identifier. In comparison to composed identifiers, distributed identifiers can be generated in a distributed fashion, i.e., without a central control entity. Anyhow, verification of an entity's identity might be problematic, as any entity can assume any ID. To circumvent this, distributed entity's need to be resolvable in a trusted and secure manner.

To verify an identity, identifiers are usually resolved to a data record or a document comprising further information about the identified entity. Usually, such data records are maintained in a network-based database and made accessible to authorized clients by a directory service. In directory services, data records (entries) are organized in a hierarchical structure, where each entry has a parent entry. Each entry is identified by a distinguished name (DN), which is not necessarily unique. Therefore, each entry is uniquely identified by it's path from the root entry, the relative distinguished name (RDN). As entries might be shifted to another branch or level in the tree-like structure, it's RDN is not guaranteed to remain stable. An existing and widely used standard for directory services is the Lightweight Directory Access Protocol (LDAP) [25–27] based on the ITUT standard X.500 [28]. One of the most used and well known directory services is the Domain Name System (DNS) [15,29]. The DNS is a hierarchically and decentrally organized directory service, that allows users and services to resolve human readable domain names into IP addresses, therefore mapping a name to a location. The data is stored in resource records (RR), which are replicated throughout the system. Still, both LDAP and the DNS build on a hierarchical design, which requires one organization or company to maintain control. To circumvent certain drawbacks and security issues in the DNS, Distributed Hash Tables (DHT) have been adopted for the use of directory services. In [30], Ramasubramanian and Sirer propose a DHT-based alternative for the DNS. This approach provides equal performance as the traditional hierarchical DNS, but showed a far better resilience against attacks [31].

3   The SONIC OSN Federation

As today's OSN platforms are mostly closed solutions that keep users from freely communicating and connecting with each other, several alternative solutions and architectures have been proposed over the last years. Here, either alternative centralized OSN platform solutions were built or ones relying on federated or completely decentralized peer-to-peer architectures [6]. Anyhow, all proposed alternatives require a user to sign up for a new user account within the new platform, while seamless interaction with other OSN platforms is not possible. Hence, there is no real incentive for users to abandon one service for another closed solution. In contrast to the proposed alternatives discussed above, SONIC follows a different approach. Here, a common protocol is used to allow different kinds of OSN platforms to interact directly by implementing a common API and using common data formats. This would allow to exchange social information across platform borders in a entirely transparent manner. This way, users are able to freely choose an OSN platform of their liking while staying seamlessly connected to all friends using other platforms. As of this, it becomes irrelevant which of your friends are using the same or a different OSN service. The resulting ecosystem is called *Online Social Network Federation* (OSNF) defined as a *heterogeneous network of loosely coupled OSN platforms using a common set of protocols and data formats in order to allow seamless communication between different platforms* [12]. Prerequisites for the OSNF comprise a *decentralized architecture*, the use of *open protocols and formats*, *seamless communication* between platforms, *migration* of user accounts to other OSN platforms [11], and a *single profile* policy with *global user identification* [12].

4   User Identification

In the SONIC OSNF, every user and every platform is identified by a globally unique identifier, the GlobalID. GlobalIDs are domain and platform independent and remain unchanged even when a user account is moved to a new
domain. This way, a user account can be addressed regardless of where it is
actually hosted. Furthermore, migration of user profiles is made possible without losing connectivity between social user accounts - even when the location
of a profile is changed frequently [11]. A user's GlobalID is derived from an
PKCS#8-formatted RSA public key and a salt of 8 bytes (16 characters) length
using the key derivation function PBKDF#2 with settings SHA256, 10000 iterations, 256bit output length. The result is converted to base36 (A-Z0-9), resulting
in a length of up to 50 characters length (see Fig. 1). An example of a GlobalID is
2UZCAI2GM45T16OMDN44OIQ8GKN5GGCKO96LC9ZOQCAEVAURA8. Each entity in the
SONIC ecosystem maintains two RSA key pairs, the *PersonalKeyPair* and the
AccountKeyPair. While the *PersonalKeyPair* is used to derive the GlobalID,
the *AccountKeyPair* is used to sign and verify all communication payload data
within SONIC. As a result, the *PersonalKeyPair* can never be changed while
AccountKeyPairs can be revoked and exchanged with a new key pair. GlobalIDs
are registered in a global directory service, the Global Social Lookup System
(GSLS). By resolving a GlobalID via the GSLS, the actual network location
(URL) of a user's account can be determined. Information about the actual profile's location, as well as other information required for verification of authenticity
and integrity are stored in a dataset called *Social Record.*

4.1   Global Social Lookup System

Following the idea of a fully decentralized OSN ecosystem that does not depend
on any entity or service controlled by a single corporation or group, the GSLS
was designed as a directory service built on DHT technology. Similar to the DNS,
any participant in the SONIC ecosystem is able to host a GSLS server that is
automatically integrated into the DHT, forming a dynamic, heavily distributed
directory service. The GSLS operates as a global directory service with a REST-
based interface for read and write operations as described in Table 1. As data in
the GSLS is public and may be overwritten by unauthorized entities, the data
is digitally signed using the user's *PersonalKeyPair*. As the GlobalID is derived

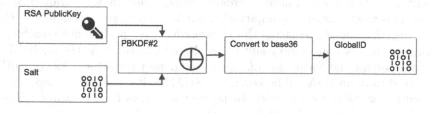

Fig. 1. Creation of a GlobalID.

Table 1. GSLS REST interface

Method	Path	Description
GET	/	Request a status message from the GSLS node
GET	/:globalID	Retrieves the *Social Record* as a signed JWT for the specified GlobalID.
POST	/	Sends a new *Social Record* as a signed JWT to be stored in the DHT.
PUT	/	Sends a new version of an already existing *Social Record* as a signed JWT to the DHT. The already existing version will be overwritten

directly from the enclosed public key and the salt, unauthorized changes in the payload would result in either an invalid digital signature or - in case the key pair is exchanged - an altered GlobalID.

4.2   The Social Record

The GlobalID and associated information is published in a dataset called *Social Record*, which comprises information that is required to resolve the GlobalID to the profile's location. The actual contents of the *Social Record* are described in Table 2. For security reasons, the GSLS API requires data to be formatted as a signed JSON Web Token (JWT [32]) using RS512 to digitally sign the payload using the owner's *PersonalKeyPair*. The *Social Record* data itself is a private claim named socialRecord and has to be a serialized, Base64URL-encoded JSON object. The digital signature of the JWT is created using the *Personal-PrivateKey*, so the signature can be verified by everyone using the *PersonalPublicKey*, which is included in the signed dataset. In case that the *AccountKeyPair* should be revoked, a key revocation certificate is created. Similar to [33], this certificate comprises the revoked public key, date and time of the revocation, a numerical indication of the reason for the revocation, and a digital signature. All revocation certificates are published in the *Social Record*, while the outdated *AccountKeyPair* is replaced with a new one.

   GlobalIDs are generated in a distributed fashion. Hence, without a central authority controlling the process, attacks are possible that aim at taking over a SONIC identity. As GlobalIDs are derived directly from the *PersonalKeyPair* and salt, an attacker would need to create a valid signature for a crafted *Social Record*. This would mean that an attacker would need to get access to the *PersonalPrivateKey* itself. Replacing the *PersonalKeyPair* itself is not possible, as exchanging the key pair would result in an altered GlobalID. As the GlobalID is used for resolving the *Social Record*, this would deflect the attack. As GlobalIDs are derived from an RSA public key using SHA256, it's uniqueness depends on the security of the key generation. To prevent attackers from creating rainbow tables for all available GlobalIDs in order to find a collision for a random *Social Record*, a cryptographic salt has been introduced. This salt is used by PBKDF#2 in

Table 2. Contents of the *Social Record*

Attribute	Description
type	Type of the *Social Record*
globalID	The identifier for the user profile
platformGID	GlobalID of the associated OSN platform
displayName	Human-readable username for on screen display
salt	Cryptographic salt of 16 characters length
accountPublicKey	RSA public key
personalPublicKey	RSA public key
datetime	XSD DateTime timestamp
keyRevocationList	List of revoked account key pairs
active	Flag that describes the current status of the *Social Record*

the generation process of the GlobalID. The usage of the salt, which is randomly created for each *Social Record*, aggravates brute force attacks as a new key cannot be checked against multiple *Social Records* for a collision, but needs to be hashed again for each GlobalID. Anyhow, as generating an RSA key pair is the most time consuming task in creating a GlobalID, an attacker might chose a key and just alter the salt in oder to find a collision. To limit the possibility of this attack to succeed, the length of the salt has been fixed to 8 bytes. By limiting the length of the salt, only 4.2×10^9 possible salts can be used, thus effectively eliminating the chance of creating a collision through manipulation of the salt. Using the birthday bound, an attacker would need to create 4.8×10^{37} key pairs and salts for a 1 % chance of a collision, thus rendering an attack extremely unlikely.

5 Implementation

This section describes the implementation details of the GSLS. It has been implemented as a Java server daemon based on Eclipse Jetty, a lightweight application server capable of handling REST requests. The application is run via Jsrv to run as a server daemon. The GSLS exposes a REST-based interface on port 5002 that allows clients to commit and request *Social Records*. The interface features operations for retrieving and writing *Social Records* as described in Table 1. For storage of the *Social Records*, the GSLS implements TomP2P, a Kademlia-based DHT implementation written in Java [34]. Kademlia is based on on a reactive key-based routing protocol, which uses other node's search queries to update and stabilize the routing tables. As a result, Kademlia-based DHTs are very robust and performant, as separate stabilization mechanisms are not necessary [35].

To prevent manipulation of the dataset by malicious participants, the dataset is stored as a signed JSON Web Token (JWT). The token is signed using RS512. For compatibility reasons, the dataset is encoded using Base64URL and stored in the JWT as a private claim named `data`. The token is then signed with the private key matching the enclosed public key. This way, the integrity of the dataset can always be verified. *Social Record* datasets sent to the GSLS will are validated by the service regarding integrity and format to ensure that no faulty datasets are managed or delivered by the GSLS. The API allows no `DELETE` requests, as a hard delete would allow a previously occupied GlobalID to be reused by a new *Social Record* with a matching GlobalID. Even though being unlikely, identity theft would be made possible this way. As of this, the GSLS only supports a soft delete, where the `active` flag of the *Social Record* is set to 0 to mark the dataset as inactive.

5.1   SONIC SDK

In order to ease the integration of the SONIC protocol into both existing OSN platforms as well as to support the development of new OSN projects, the SONIC SDK has been implemented. The SDK features a set of classes that provide functionality for formatting, parsing, signing, and validating SONIC data formats, as well as handling requests to and from other SONIC compliant platforms. For resolving GlobalIDs, the SONIC SDK incorporates an API to retrieve *Social Records* from the GSLS, as well as for creating, publishing, and updating *Social Records*. The SONIC SDK automatically resolves GlobalIDs in the background, including an automated integrity verification process. The SDK has been integrated in a proof of concept implementation of the SONIC project as well as in the well-known OSN platform Friendi.ca.

5.2   SONIC App

To ease the management of the *Social Record* and the associated key pairs for the user, the SONIC App has been implemented as a mobile application based on Android (see Fig. 2). The SONIC App allows a user to create both the *Social Record* and associated key pairs and is able to synchronize the data with the GSLS, as well as with the user's SONIC platform. While the *AccountKeyPair* needs to be accessible by the platform in order to sign data as well as requests, the *PersonalKeyPair* is only used to sign the *Social Record*. Using the SONIC App, the *PersonalKeyPair* is managed by a device owned by the user and is not made available to the platform or any other - possibly untrusted - third party. To ease the setup process when creating a user account on a SONIC platform, the SONIC App automates the creation of keys. Here, the platform displays a QRCode encoding the login credentials of the platform, which can be scanned by the SONIC App. After creating a new *Social Record* and the associated keys, the SONIC App uploads the necessary data to the platform. Besides creating a new *Social Record* or editing an existing one, the SONIC App also automates the process of migrating a social profile to a new platform as described in [11]. Here, a new user account is created at the target platform and all profile information is copied to the target location. As part of this migration protocol, the SONIC App will automate the process of updating the *Social Record* in the GSLS. Further

(a) Editing a *Social Record*   (b) Scanning a QR code

Fig. 2. User interface of the SONIC App.

features of the SONIC App include exporting a *Social Record* to a text file, importing a *Social Record* from a text file, and scanning a QRCode encoding the GlobalID of another user in order to directly send a friend request to him.

6   Evaluation

For the evaluation of the GSLS, a testbed with 3 virtual machines has been set up. Each node was configured to use 1 virtual CPU and 1 GB of RAM, running Debian Linux "Wheezy". To perform the evaluation of the writing performance of the system, 50,000 unique *Social Records* were created by a script and directly pushed to the GSLS. For each *Social Record* dataset being sent to the GSLS, the total duration of the request to complete was measured and logged to a database for later analysis (see Fig. 3). Each write request comprised a payload of approximately 4 KB depending on the *Social Record's* contents.

Analysis of the logged data showed that most requests were fully processed in approximately one second, with a minimum of 0.956 s and an average of 2.312 s (median value 1.032 s). While 30.8 % of all requests were processed in less than a second, 89.6 % of all requests were processed in less that 2 s. Only 4.9 % of the requests took more than 3 s and 3.6 % of the requests took more than 6 s. Even though the overall writing performance of the GSLS can be considered good, a small fraction of requests took a - partly significant - longer amount of time to complete. As no request timeout was configured on both server and client side during the test, the client waited until a response was received. Here, response times of up to 227.548 s were measured. To perform an evaluation of the reading performance of the GSLS, 10,000 requests for randomly chosen GlobalIDs for existing *Social Records* were sent to the one of the nodes. Again, all requests were answered successfully. The average response time for the requests was found to

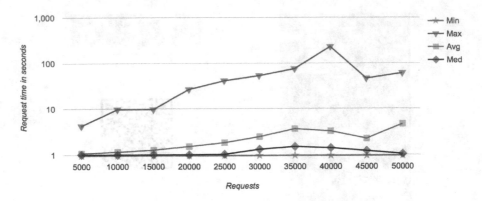

Fig. 3. GSLS writing performance for 50,000 consecutive write requests.

be 0.034 s with a minimum of 0.009 s and a maximum of 4.085 s. The median time to answer a request took 0.014 s. While the reading performance of the GSLS while accessing stored *Social Records* showed to be stable and fast, writing new datasets to the DHT showed a slower performance. Still, the median response time for a successful request was 1.032 s, with few requests that took longer to complete.

7   Conclusion

To overcome the obvious drawbacks of proprietary protocols and service architectures, SONIC proposes a holistic approach that facilitates seamless connectivity between different OSN platforms and allows user accounts to be migrated between OSN platforms without losing data or connections to other user profiles. Thus, SONIC builds the foundation for an open and heterogeneous *Online Social Network Federation.* In this paper, we presented a distributed and domain-independent ID management architecture for the SONIC OSNF, which allows user identifiers to remain unchanged even when a profile is migrated to a different OSN platform. These so called GlobalIDs are derived from a public key pair using PBKDF#2 and are therefore domain-agnostic. In order to resolve a given GlobalID to the actual URL of a social profile the GSLS, a distributed directory service built on DHT technology has been introduced. Datasets called *Social Records*, which comprise all information required to look up a certain profile, are stored and published by the GSLS. For security reasons, *Social Records* are digitally signed using the user's private key. For easing key management and exchanging of GlobalIDs, a mobile SONIC app has been implemented based on Google Android. This application allows to create, edit, import, and export *Social Records*, exchange GlobalIDs, and directly send friend requests using the SONIC protocol [9].

Acknowledgment. The work described in this paper is based on results of ongoing research and has received funding from the projects SONIC (http://sonic-project. net) and reThink (http://rethink-project.eu). SONIC (grant no. 01IS12056) is funded as part of the *SoftwareCampus* initiative by the *German Federal Ministry of Education and Research (BMBF)* in cooperation with *EIT ICT Labs Germany GmbH* and *Deutsches Luft- und Raumfahrtzentrum (DLR)*. reThink (grant no. 645342) is funded as part of the European Union's research and innovation program *Horizon 2020*.

References

1. Perrin, A.: Social Media Usage 2005–2015 (2015). http://www.pewinternet.org/ files/2015/10/PI_2015-10-08_Social-Networking-Usage-2005-2015_FINAL.pdf
2. Ofcom: Communications Market Report 2012 (2012). http://stakeholders.ofcom. org.uk/binaries/research/cmr/cmr12/CMR_UK_2012.pdf
3. Ugander, J., Karrer, B., Backstrom, L., Marlow, C.: The Anatomy of the Facebook Social Graph. arXiv preprint arXiv:1111.4503 (2011)
4. Cosenza, V.: World map of social networks (2016). http://vincos.it/ world-map-of-social-networks/
5. Yeung, C., Liccardi, I., Lu, K., Seneviratne, O., Berners-Lee, T.: Decentralization: the future of online social networking. In: W3C Workshop on the Future of Social Networking Position Papers, vol. 2 (2009)
6. Paul, T., Famulari, A., Strufe, T.: A survey on decentralized online social networks. Comput. Netw. **75**(Part A), 437–452 (2014)
7. Heidemann, J.: Online social networks - Ein sozialer und technischer Überblick. Informatik-Spektrum **33**(3), 262–271 (2010)
8. Hu, P., Fan, Q., Lau, W.C.: SNSAPI: A Cross-Platform Middleware for Rapid Deployment of Decentralized Social Networks. arXiv preprint arXiv:1403.4482 (2014)
9. Göndör, S., Beierle, F., Sharhan, S., Hebbo, H., Küçükbayraktar, E., Küpper, A.: SONIC: bridging the gap between different online social network platforms. In: 2015 IEEE 8th International Conference on Social Computing and Networking (SocialCom). IEEE (2015)
10. Palfrey, J.G., Gasser, U.: Interop: The Promise and Perils of Highly Interconnected Systems. Basic Books, New York (2012)
11. Göndör, S., Beierle, F., Küçükbayraktar, E., Hebbo, H., Sharhan, S., Küpper, A.: Towards migration of user profiles in the SONIC online social network federation. In: ICCGI, IARIA, pp. 1–2 (2015)
12. Göndör, S., Hebbo, H.: SONIC: towards seamless interaction in heterogeneous distributed OSN ecosystems. In: 2014 IEEE 10th International Conference on Wireless and Mobile Computing, Networking and Communications (WiMob), pp. 407–412. IEEE (2014)
13. Resnick, P.: Internet Message Format (2008). https://tools.ietf.org/html/rfc5322
14. Saint-Andre, P.: Extensible Messaging and Presence Protocol (XMPP): Instant Messagingand Presence (2004). http://tools.ietf.org/html/rfc3921
15. Mockapetris, P.: Domain Names - Concepts and Facilities (1987). https://tools. ietf.org/html/rfc1034
16. Berners-Lee, T., Fielding, R., Masinter, L.: Uniform Resource Identifier (URI): Generic Syntax (2005). https://www.tools.ietf.org/html/rfc3986
17. W3C: WebID 1.0 Web Identity and Discovery (2013). http://dvcs.w3.org/hg/ WebID/raw-file/tip/spec/identity-respec.html

18. Story, H., Harbulot, B., Jacobi, I., Jones, M.: FOAF+SSL: restful authentication for the social web. In: Proceedings of the First Workshop on Trust and Privacy on the Social and Semantic Web (SPOT 2009) (2009)
19. Recordon, D., Reed, D.: OpenID 2.0: a platform for user-centric identity management. In: Proceedings of the Second ACM Workshop on Digital Identity Management, DIM 2006, pp. 11–16. ACM (2006)
20. Leach, P., Mealling, M., Salz, R.: A Universally Unique IDentifier (UUID) URN Namespace (2005). http://tools.ietf.org/html/rfc4122
21. Reed, D., McAlpin, D.: Extensible Resource Identifier (XRI) Syntax V2.0 (2005). https://www.oasis-open.org/committees/download.php/15377
22. Demir, B.: Twitter Snowflake (2010). https://github.com/twitter/snowflake
23. Gardner, D., Vasconcelos, L.: Cruftflake (2015). https://github.com/davegardnerisme/cruftflake
24. Featherston, D., Debnath, S., Nyman, T., Veres-Szentkirlyi, A., Countryman, M.: Boundaryflake (2015). https://github.com/boundary/flake
25. Wahl, M., Howes, T., Kille, S.: Lightweight Directory Access Protocol (v3) RFC 2251 (1997). http://www.ietf.org/rfc/rfc2251.txt
26. Zeilenga, K.: Lightweight Directory Access Protocol (LDAP) Transactions RFC 5805 (2010). http://tools.ietf.org/html/rfc5805
27. Sermersheim, J.: Lightweight Directory Access Protocol (LDAP) The Protocol RFC 4511 (2006). http://tools.ietf.org/html/rfc4511
28. International Telecommunication Union (ITU-T): X.500: Information technology - Open Systems Interconnection - TheDirectory: Overview of concepts, models and services (2012). http://www.itu.int/rec/T-REC-X.500/en
29. Mockapetris, P.: Domain Names - Implementation and Specification (1987). https://tools.ietf.org/html/rfc1035
30. Ramasubramanian, V., Sirer, E.G.: The design and implementation of a next generation name service for the internet. ACM SIGCOMM Comput. Commun. Rev. 34(4), 331–342 (2004)
31. Massey, D.: A Comparative Study of the DNS Design with DHT-Based Alternatives. In: Proceedings of IEEE INFOCOM (2006)
32. Jones, M., Bradley, J., Sakimura, N.: JSON Web Token (JWT). Technical report, IETF (2015). http://tools.ietf.org/html/rfc7519
33. Cooper, D., Santesson, S., Farrell, S., Boeyen, S., Housley, R., Polkk, W.: Internet X.509 Public Key Infrastructure Certificate and Certificate Revocation List (CRL) Profile (2008). https://tools.ietf.org/html/rfc5280
34. Bocek, T.: TomP2P, a P2P-based high performance key-value pair storage library (2012). http://tomp2p.net
35. Maymounkov, P., Mazières, D.: Kademlia: a peer-to-peer information system based on the XOR metric. In: Druschel, P., Kaashoek, M.F., Rowstron, A. (eds.) IPTPS 2002. LNCS, vol. 2429, pp. 53–65. Springer, Heidelberg (2002)

Proposal of a New Social Signal for Excluding Common Web Pages in Multiple Social Networking Services

Hiroyuki Hisamatsu[1](✉) and Tomoaki Tsugawa[2]

[1] Osaka Electro-Communication University,
1130-70 Kiyotaki, Shijonawate, Osaka, Japan
`hisamatu@osakac.ac.jp`
[2] CubeSoft Inc., 21-9 Machikaneyama, Toyonaka, Osaka, Japan
`tsugawa@cube-soft.jp`
`http://sotome.org`

Abstract. In recent years, a social signal that is given by a social network service (SNS) on the World Wide Web where a huge quantity of information exists has attracted attention. A social signal is an index that measures how much a web page is a hot topic among users on the SNS. For instance, the number of "retweets" on Twitter and the number of "likes" on Facebook are social signals. Generally speaking, a social signal is evaluated by the administrator of the SNS and displayed on a web page and can be read by anyone. By utilizing a social signal, acquiring current hot-topic web pages efficiently is expected. However, if web pages are simply chosen on the basis of the magnitude of the social signal of an SNS that has many users, most of web pages will be the common web pages among SNSs, and the characteristic web pages that can only be seen when using a certain SNS is buried in the common web pages. Therefore, in this paper, we propose a new social signal that assesses the degree to which a certain web page is a hot-topic web only in an SNS by combining the social signals of SNSs. As a result of a performance evaluation, we show that by acquiring web pages on the basis of the magnitude of the proposed new social signal, hot-topic web pages in multiple SNSs are excludable.

Keywords: Social signal · Social Networking Service (SNS) · Content curation

1 Introduction

In recent years, with the explosive deployment of the Internet, the information resources on the World Wide Web continue increasing every day. The types of information resources on the World Wide Web vary such as a text, photographs, and video. Methods have been researched and developed to efficiently acquire the necessary information from these information resources for many years; consequently, many websites and web services have been created. The utilization

© Springer International Publishing Switzerland 2016
H.T. Nguyen and V. Snasel (Eds.): CSoNet 2016, LNCS 9795, pp. 239–248, 2016.
DOI: 10.1007/978-3-319-42345-6_21

of a search engine, which is a web service provided by Google [3] and Bing [1] and so on, is one of the most widely used methods for efficiently acquiring the necessary information from the World Wide Web. A search engine returns web pages that have highest estimated relevance from a given user input; then, the user finds the information that he/she needs from the web pages. Thus, the search engine significantly shortens time until a user arrives at the information that he/she needs. The search engines are expected to continue occupying an important position as a method of collecting necessary information efficiently.

When utilizing a search engine, a user needs to input the appropriate keyword(s) that takes the user to the information that he/she needs. However, if the information that the user needs is vague, he/she may be unable to input a specific keyword—for instance, current news that attracts concern socially or the newest information on some topic that the user is interested in. In order to meet these demands of users, a different method from the search engine is necessary.

The number of views (the number of accesses) by the users of web pages has been used as an evaluation index for efficiently selecting web pages from the World Wide Web without using a search engine. In recent years, as an evaluation index, a social signal has attracted attention. A social signal is a value that shows how much a web page is mentioned by users in a social network service (SNS). If a web page is a hot topic, the social signal of the web page is large. A social signal can be utilized as an index for measuring how much a web page receives attention from SNS users—for instance, the number of "retweets" on Twitter [8], or the number of "likes" on Facebook [2]. A social signal is expressed numerically by an SNS, and the social signal displayed on the World Wide Web in a form that anyone can read and utilize. Therefore, it is expected that a social signal will be utilized to select web pages instead of the number of view, which only the administrator of a website can acquire.

Some users dissatisfied when choosing web pages using the social signal since novel web pages such the beginning of the SNS is lost as time passes. With web services like an SNS, in particular, the deviation in the tastes of users who utilize the SNS at beginning of the SNS when there are few users is large. Consequently, at the beginning of an SNS, using the social signal of the SNS, we can select the characteristic web pages that cannot be a hot topic for other SNSs. However, when the popularity of an SNS increases and many users begin to utilize the SNS, the deviation in the tastes does not greatly vary, and the characteristic web page cannot be found using the social signal. Consequently, when accounting for the magnitude of the social signal simply for selecting hot-topic web pages, the problem is that most of web pages that are obtained are common web pages that everyone knows.

There have already been studies about SNSs and information recommendation systems that utilize SNSs [9,10]. In [10], the authors proposed a social-media recommendation framework focused on the reliability of information. In [9], the authors proposed a social signature that is the set of tokens that gives us the paraphrasing of web pages. It would be expected that we can use a social signature to rank search results, and organize content.

In this paper, we propose a new social signal that utilizes the tendency that the social signal of a web page that everyone knows is large in SNSs with many users. Specifically, utilizing the multiple social signals of multiple SNSs, we derive a new social signal that assesses the degree to which a certain web page is a hot topic only in an SNS. Furthermore, we evaluate the performance of our social signal and show that the web pages that are hot topic in by multiple SNSs can be avoided with a high accuracy when the web pages are acquired on the basis of the new social signal.

The construction of this paper is as follows. First, in Sect. 2, we explain a social signal and the features of an SNS that has many users such as Twitter and Facebook. Next, in Sect. 3, we investigate the distributions of the social signals of Twitter and Facebook for web pages and the degree of duplication of hot-topic web pages. In Sect. 4, we propose a new social signal. In Sect. 5, we present a performance evaluation of our new social signal. Finally, in Sect. 6, we conclude this paper and discuss our future works given the results presented this paper.

2   Social Signal

A social signal is an evaluation index that shows assesses the degree to which a certain web page is a hot topic among the users of the SNS. A social signal is expressed numerically by the SNS, and in many cases, is open to the public on the World Wide Web in the form of a button with a balloon. Figure 1 shows an image showing an example of appearance of social signals on a web page. Moreover, only the numerical value of a social signal is acquirable in many cases in the form of a web application program interface (API). Thus, it is not limited to the SNS company and SNS users, and anyone can know the value of a social signal easily.

The number of "retweets" on Twitter and the number of "likes" on Facebook are representative examples of social signals. For these social signals, a user's action to increase the social signal value differs. Specifically, for Twitter, if the URL of the web page is contained in the text posted by the user, this post is regarded as a retweet. Therefore, the number of "retweets" is increased by a post in which user inputs the URL, a reply post that quotes the original post that included the URL, and a retweeted post that is a post using a feature offered by Twitter for copying and re-posting the original text.

On the contrary, the number of "likes" on Facebook is increased only by pushing the "like" button displayed on the web page. Therefore, the action of the user that increases the number of "likes" is limited compared to the action of the user that increases the number of "retweets" on Twitter. The interface through which the numerical value of a social signal increases differs among SNSs. SNSs that offer a similar method to Facebook include Google+ [4], LinkedIn [5], Pinterest [6], and Pocket [7].

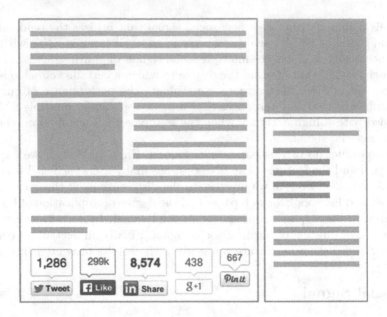

Fig. 1. Appearance of social signals in a Web page

3   Degree of Duplication of Hot-Topic Web Pages Between SNSs

In this section, for SNSs utilized by many users, we examine the degree of duplication of hot-topic web pages between SNSs. In this survey, we collected the web pages that obtained three or more social signals by Twitter and/or Facebook for one year from November 1, 2014 to October 31, 2015, and we used these web pages for the survey. In addition, the total number of the web pages collected during this period was 716,209 Web pages.

Figure 2 shows the cumulative distributions of social signals of web pages. For 90 % or more of the web pages, this figure shows that the social signals of Twitter and Facebook are less than 300. On the other hand, the results show that there are 295 web pages on Twitter and 1,254 Web pages on Facebook in which a social signal exceeds 10,000. The social signals for these web pages are large, such that many users on the SNSs take action on the web pages, and the web pages are displayed many times in the SNSs when these actions are taken. By simply selecting web pages using the value of the social signal, it is expected that a small number of web pages, which have large social signal, will be selected.

Figure 3 shows the duplication ratio of hot-topic web pages between Twitter and Facebook according to the day. This figure shows the top 20, 50, and 100 web pages with the value of the social signal. The figure shows that 20–30 % of the web pages overlap between Twitter and Facebook. As mentioned above, when simply selecting the web pages using the value of a social signal, it turns out that the same web pages are selected.

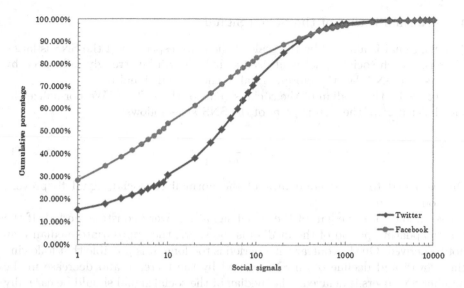

Fig. 2. Cumulative distribution of a social signal

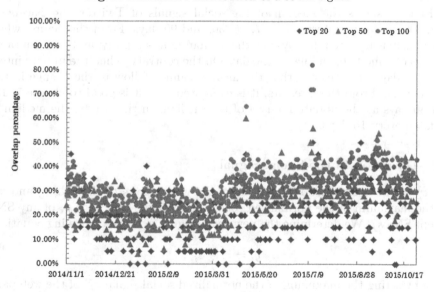

Fig. 3. Overlap ratio of hot-topic web pages on Twitter and Facebook

4   A New Social Signal

In this section, we propose a social signal that assesses the degree to which a certain web page is a hot topic only on an SNS. First, in Sect. 4.1, we explain the normalization of a social signal value, and in Sect. 4.2, we derive a new social signal.

4.1   Normalization of the Social Signal

A social signal is affected by the scale of the SNS, especially if the SNS is large. That is, if each social signal is utilized as it is, it will be strongly influenced by a large-scale SNS. We therefore normalize each social signal first.

Let m^j be the median of the social signals of the SNS s^j. We normalize the social signal z_i^j of the web page p_i of the SNS s^j as follows:

$$z_i^j = \frac{v_i^j}{m^j + 1}. \tag{1}$$

One is added to the denominator of the normalized social signal to prevent division by zero.

We update the median of the social signal for every constant period. If the target updating period of the median is too short, the appropriate median may not be derived. On the contrary, if a period is too long, it is possible that following the growth and decline of an SNS caused by the increase and decrease in the number SNS users is delayed. The median of the social signal should be carefully given.

Figure 4 shows the median of the social signals of Twitter and Facebook when the updating interval is 1, 7, 15, 30, and 90 days. From this figure, when the median is updated day-by-day, the median moves rapidly, and an unsuitable value is obtained depending on the day. On the contrary, when the update interval is 90 days, it turns out that the median cannot follow in the change in the social signal. From these results, it is estimated that it is good to choose 7, 15, and 30 days as the update interval of the median. In this study, the median is updated every 15 days.

4.2   Deviation in the Social Signal

Let z_i^j be the normalized social signal of the web page p_i of the SNS s_j and z_i^m be the maximum of the normalized social signal of the web page p_i of any SNS except SNS s_j. We determine the new social signal g_i by the following equation:

$$g_i = z_i^j - z_i^m. \tag{2}$$

By subtracting the maximum of the normalized social signal z_i^m of the web page of any SNS except the SNS s_j from the normalized social signal z_i^j of the web page of SNS s_j, our social signal can show that assesses the degree to which a certain web page is a hot topic only for the SNS s_j.

5   Performance Evaluation

In this section, we evaluate the performance of our new social signal.

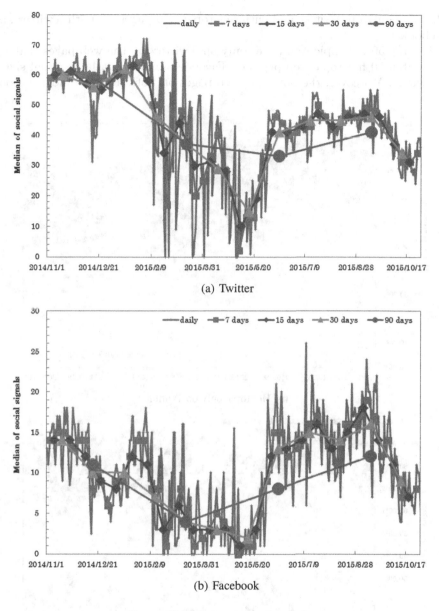

Fig. 4. Movement of social signals

5.1   Evaluation Environment

In the performance evaluation, we also use the web pages that we collected in Sect. 3. We then obtain the top 50 web pages day-by-day according to the value of our new social signal. Moreover, we calculate the coincidence ratio of the web pages that are obtained on the basis of our social signal with the top 50 hot-topic web pages only on Twitter and with hot-topic web pages on both Twitter and Facebook.

The top 50 hot-topic web pages only on Twitter are the web pages that are not the top 50 hot-topic web pages on Facebook and have a large social signal on Twitter. Moreover, the hot-topic web pages on both Twitter and Facebook

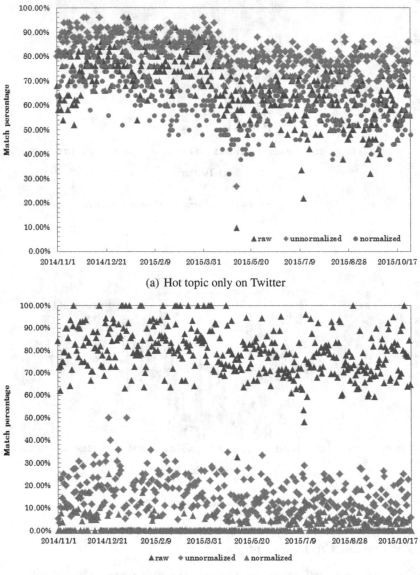

(a) Hot topic only on Twitter

(b) Hot topic on both Twitter and Facebook

Fig. 5. Evaluation results

are the web pages that are not the top 50 hot-topic web pages only on Twitter and have the social signal on Twitter of 50th place of "the top 50 hot-topic web pages only on Twitter" or more.

5.2   Evaluation Results

Figure 5 shows evaluation results. The horizontal axis of this figure expresses the time, and the vertical axis expresses the coincidence ratio. In this figure, "raw" means that hot-topic web pages are acquired only on the basis of the social signal of Twitter (raw), "unnormalized" means that hot-topic web pages are acquired on the basis of the difference between the unnormalized social signals of Twitter and Facebook, and "normalized" means that hot-topic web pages are acquired on the basis of our social signal.

From Fig. 5, when web pages are acquired only on the basis of the social signal of Twitter (raw), the coincidence ratio for hot-topic web pages on both Twitter and Facebook exceeds 60 % for almost all days. This result shows that there are many common web pages when we simply use the social signal for acquiring hot-topic web pages. Moreover, although the coincidence ratio for hot-topic web pages only on Twitter is high when the social signal that is not normalized (unnormalized) is used for acquiring hot-topic web pages, the coincidence ratio for hot-topic web pages on both Twitter and Facebook is about 20 %. That is, when we utilize a social signal without normalization, it turns out that we acquire common hot-topic web pages. On the other hand, when using our social signal (normalized), it turns out that the coincidence ratio for hot-topic web pages only on Twitter maintains a high value, and the coincidence ratio for hot-topic web pages on both Twitter and Facebook is almost zero. That is, our social signal can exclude the hot-topic web pages on both Twitter and Facebook, which means common hot topics, for acquiring web pages in the World Wide Web. The above result shows the effectiveness of our social signal.

6   Conclusion and Future Work

In this paper, we proposed a new social signal that is large if a web page is a hot topic only on one SNS and small if a web page is a hot topic on multiple SNSs by using the tendency that the social signal of a web page that everyone knows is large on multiple SNSs. By combining multiple social signals, our social signal that assesses the degree to which a certain web page is a hot topic for a specific SNS. As a result of a performance evaluation, we showed that we can exclude the common hot topic when we acquire the web pages on the basis of our social signal.

In this study, we only targeted on Twitter and Facebook; the performance evaluation of many SNSs such as Google+, LinkedIn, and more is planned for future work.

Acknowledgment. This work was partly supported by Dayz Inc.

References

1. Bing. https://www.bing.com/
2. Facebook. https://www.facebook.com/
3. Google. https://www.google.com/
4. Google+. https://plus.google.com/
5. LinkedIn. https://www.linkedin.com/
6. Pinterest. www.pinterest.com
7. Pocket. https://getpocket.com/
8. Twitter. https://twitter.com/
9. Alonso, O., Bannur, S., Khandelwal, K., Kalyanaraman, S.: The world conversation: web page metadata generation from social sources. In: Proceedings of the 24th International Conference on World Wide Web, WWW 2015 Companion, pp. 385–395 (2015)
10. Wu, J., Chen, L., Yu, Q., Han, P., Wu, Z.: Trust-aware media recommendation in heterogeneous social networks. World Wide Web **18**(1), 139–157 (2013)

Measuring Similarity for Short Texts on Social Media

Phuc H. Duong, Hien T. Nguyen[✉], and Ngoc-Tu Huynh

Faculty of Information Technology, Ton Duc Thang University, Ho Chi Minh City, Vietnam
{duonghuuphuc,hien,huynhngoctu}@tdt.edu.vn

Abstract. In this paper, we present a method for measuring semantic similarity between short texts by combining two different kinds of features: (1) distributed representation of word, (2) knowledge-based and corpus-based metrics. Then, we present experiments to evaluate our method on two popular datasets - Microsoft Research Paraphrase Corpus and SemEval-2015. The experimental results show that our method achieves state-of-the-art performance.

1 Introduction

Measuring semantic similarity between two short texts, e.g., news headlines, tweets or comments in public forums, plays an important role in social network analysis, sentiment and opinion analysis, summarization of posts/replies, information retrieval, etc. Since a short text is usually limited in the number of characters, context-poor, irregular or noisy, techniques in natural language processing proposed for short texts are not tailored to perform well on those tasks. Most of the proposed methods in literature exploit corpus-based or knowledge-based to compute the degree of similarity between given texts by measuring the word-to-word similarity [1]. Other approaches take the advantage of machine translation metrics [2], discourse information [3]. In [2], the authors implement a heuristic alignment algorithm to identify pairs of plagiarism sentences, then, pass them to a learning algorithm for training a classifier. The approach proposed in [3] divides sentences into elementary discourse units (EDUs), aligns EDUs, and computes the overall similarity between sentences based on aligned EDUs. Although some previous work focuses on the preprocessing phase, it still does not consider many factors, for example, the number of tokens constructs a meaning word. Hence, in this paper, we elaborately consider many aspects, as presented below, in measuring similarity between short texts and apply those to our preprocessing phase.

- One of the most challenge task in determining the similarity of words or concepts is that they usually do not share actual terms in common. Consider an example, in analyzing a text, the concepts *"Artificial Intelligence"* and *"AI"* are similar to each other in the context of computer science. In other example, *"The Pentagon"* and *"United States Department of Defense"*, the two terms are different, but similar in meaning. Therefore, our method performs named entity recognition and named entity co-reference resolution to isolate them from the texts for other steps.
- Beside named entities, the number of tokens constructing a meaning word is also importance. Much previous work considers each token as a meaning word; however,

© Springer International Publishing Switzerland 2016
H.T. Nguyen and V. Snasel (Eds.): CSoNet 2016, LNCS 9795, pp. 249–259, 2016.
DOI: 10.1007/978-3-319-42345-6_22

that is not always true. For instance, in English grammar, *"pull out"* is a phrasal verb and has the same meaning with *"extract"*. If separating *"out"* from *"pull"*, we lose the word *"pull out"* and lose the chance to capture similarity between *"pull out"* and *"extract"* when they occur in two given texts. In order to overcome this drawback, our proposed method includes a step, namely *tokenizer*, that preserves phrasal words like the case of *"pull out"*.

Furthermore, in order to make our proposed method becomes flexible, we design a model which is suitable for measuring similarity for both formal and informal texts. We also investigate three different kinds of features and show that our proposed method achieves state-of-the-art performance.

In summary, the contribution of this paper is two-fold as follow: First, we preserve phrasal words, take named entities and their co-reference relations among them into account, which were not exploited in literature; Second, we exploit two different similarity measures as features: (1) Word-embedding-based similarity, (2) Knowledge-based and corpus-based similarity; Finally, we conduct experiments to evaluate our method and show that word-embedding-based similarity superior contribution to the performance.

The rest of this paper is organized as follows. First, we present related work in Sect. 2. Section 3 presents our method and the two features for measuring similarity. Then, experimenting our method on the two popular datasets are described in Sect. 4. Finally, Sect. 5 concludes the paper.

2   Related Work

There have been many studies on scoring the similarity degree between two short texts. In [4], the authors propose a method which combines semantic and syntactic information in the given texts. For semantic information, this approach exploits knowledge-based and corpus-based to reflect both the meanings and the actual usage of words. For syntactic information, the method represents the given texts as word-order vector to measure a number of different words and word pairs in a different order. In [5], the authors use pointwise mutual information, latent semantic analysis and six knowledge-based methods [1] for measuring word-to-word similarity, then, conclude the degree of similarity between two texts. In [6], the authors present the discriminative term-weighting metric, known as TF-KLD, which is an improvement of traditional TF-IDF and WTMF [7]. Then, they form a feature vector from the latent representations of each text segment pair and input to SVM classification. In [8], the authors combine the longest common subsequence and skip *n*-gram with WordNet[1] similarity.

In [2], the authors re-examine 8 machine translation metrics for identifying para-phrase in two datasets, and method proposed in [9] gains the best performance. This study shows that a system only employs machine translation metrics can achieve prom-ising results. The approach in [3] takes advantage of the elementary discourse units (EDUs) to identify paraphrase. Method in [10] presents a probabilistic model which

---

[1] http://wordnet.princeton.edu.

combine semantic and syntactic using quasi-synchronous dependency grammars. In [11], the authors present an unsupervised recursive auto-encoders to learn the feature vectors which contain the similarity of single word and multi-word extracted from parse trees of two text segments. In [12], the authors present two components in modular functional architecture. For the sentence modeling component, they use convolutional neural network, for the similarity measurement component, they compare pairs of regions of the sentence representations by combining distance metrics. In [13], the authors propose a kernel function which takes the advantage of search engine (e.g., Google) and TF-IDF for computing query expansion, then applies kernel function to multiply the two query expansion of two given texts to conclude the degree of similarity.

Because the basic element in constructing a text is words (tokens), the degree of similarity between two text snippets depends on the similarity between pairs of word of two given texts. There have been many approaches [1], but in this paper, we roughly classify word-to-word similarity metrics into three major types: (1) knowledge-based, (2) corpus-based, and (3) vector space representation. Methods in knowledge-based approach [14] exploit the semantic information from structure knowledge sources (e.g., WordNet, Wikipedia[2]) to measure the similarity of two given concepts. Other field of study which rely on the statistical information of concepts in large corpus, well-known methods in this approach are information content, latent semantic analysis, hyperspace analogue to language, latent dirichlet allocation. In [15], the authors present a word embedding approach using two-layer neural network with continuous bag-of-words (CBOW) or continuous skip-gram architecture. In [16], the authors consider both the proximity and disambiguation problems on word embedding method, and then they propose Proximity-Ambiguity Sensitive model to tackle them.

3 Proposed Method

In this section, we present our method for computing the semantic similarity between two given snippets of texts. Figure 1 presents our model for measuring of similarity between two short texts. We explain in detail our proposed method in Sects. 3.1 and 3.2.

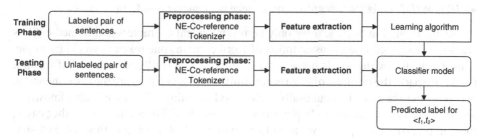

Fig. 1. Our proposed model of measuring similarity between short texts

---

[2] https://en.wikipedia.org/.

3.1   Preprocessing

Short texts (e.g., news title, message, tweet) often contain some special characters (e.g., dollar sign, colon, emoticon), but they do not contribute much semantic information for measurement. Therefore, we suggest to ignore those special characters in given texts but still preserve their structures.

In order to gain the best performance in computing similarity, we recognize named entities and then perform named entity co-reference resolution. A named entity often contains more than one word, e.g., *"United States"* is semantically different from *"United"* and *"States"*. To recognize named entities, we take the advantage of Wikipedia, which is an open encyclopedia contributed by a large community of users. Since Wikipedia contains named entities and common concepts (e.g., tree, data structures, algorithm), we treat those common concepts in Wikipedia as "named" entities. In reality, an entity may have more than one alias and an alias may corresponding to many entities in different context. For example, in Wikipedia, *"United States"* has up to four difference aliases {*United States of America, America, U.S., USA*}, that means, all of them are similar to each other. By practice, we found out that named entity often has four tokens, thus, we propose to set a sliding window of four to get a set of all candidate named entities from given text. Next, we detect the orthographic co-reference between those recognized named entities by using rules proposed in [17]. After perform co-reference resolution step, named entities which referent to each other are grouped in co-reference chains. Finally, with the co-reference entities, we assign them a unique identifier (*"ID#"*) to make them become similar entities. Let's consider the example below:

- Obama calls on tech industry at SXSW to help solve nation's problems.[3]
- Obama, at South by Southwest, calls for law enforcement access in encryption fight.[4]

By using *exact match* and *equivalent* rules to perform named entity co-reference resolution, there are two pairs of co-reference named entities, which are {*"Obama"*$_1$, *"Obama"*$_2$} and {*"SXSW"*$_1$, *"South by Southwest"*$_2$}. Therefore, we replace them to *"ID#"* format, the input sentences become:

- *ID*1 calls on tech industry at *ID*2 to help solve nation's problems.
- *ID*1, at *ID*2, calls for law enforcement access in encryption fight.

As mentioned in Sect. 1, if we only consider special characters and named entities are not enough, because the assumption of word contains one token is weak. Example, consider the following words in the same context, *"cut a rug"* and *"dance"*, if we split the white space, the meaning of them is not similar. However, they are the same meaning, because *"cut the rug"* is a culturally understood meaning of *"dance"*, also known as *idiom*. We can see that not only phrasal verbs, but also idioms and many other cases, thus, in preprocessing phase, we need to recognize all of them, and this task is a sub-task of tokenizer. To perform this task, we use Wiktionary[5], a free dictionary contributed

---

[3] http://usat.ly/1pla4oI.
[4] http://nyti.ms/1QS47Ga.
[5] http://en.wiktionary.org/.

by community members, contains 644,966[6] entries including 547,056 with gloss definitions. We apply longest matching algorithm, which find the first best matching between series of tokens and dictionary. Then, marking them with underscore symbol between tokens to group them together, for instance, *"look_after"* and *"take_care_of"*.

After perform named entity recognition and tokenizer, we have finished preprocessing phase, and two given texts are now ready for computing semantic similarity.

3.2 Feature Extraction

In this section, we systematically introduce two features in measuring semantic similarity for given short texts. They are (1) word-embedding-based similarity, (2) knowledge-based and corpus-based similarity.

Word-Embedding-Based Similarity (*Sim_word-embedding*). Before explain the method to score the similarity of given texts, we introduce an approach for measuring the degree of similarity between two words by learning distributed representation of words. The distributional hypothesis states that the words are similar meanings if they are in similar context. Therefore, we take advantage of the simplified neural network skip-gram model, which predicts surrounding words given the current word by sliding a context window along the text and uses back-propagation to train the network. Figure 2 shows the main idea of skip-gram model.

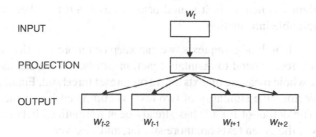

Fig. 2. Skip-gram model [18]

Given a sequence of words $\{w_1, w_2, ..., w_T\}$, the training objective of skip-gram model is maximizing the average log probability. In Eq. (1), c is the size of the training context. The larger the context size is; the higher accuracy the model will be. However, it does expenses more training time. Therefore, in order to overcome the time consuming problem but maintain the accuracy, we use negative-sampling as softmax function. Unlike hierarchical softmax function, instead of considering all context of w at each iteration, negative-sampling considers a few words by randomly chosen from context,

---

[6] This information is generated from the 03 March 2016 dump.

thus it can reduce training time. In experiment, we use the Google News dataset containing 100 billion words to train our skip-gram model.

$$\frac{1}{T} \sum_{t=1}^{T} \sum_{-c \leq j \leq c, j \neq 0} \log p\left(w_{t+j} | w_t\right) \tag{1}$$

After calculating similarity between words using word embedding, we present a metric to compute the similarity of two given texts. We have three sub-tasks in this phase: (1) create a joint word set, (2) create semantic vectors, (3) normalize and compute the distance between vectors. Let's consider the example below:

- I am studying Artificial Intelligence.
- I learn AI with my friends.

First, we create a joint word set W contains all distinct words in given texts, denoted by T_1 and T_2, as proposed in [4]. With the example above, after go through preprocessing phase, the W would be $W = \{$I, am, study, $ID1$, learn, $ID1$, with, my, friend$\}$. Because W is directly derived from given texts, we use it as standard semantic vector for comparing with T_i. Second, we represent T_1 and T_2 as semantic vectors, denoted by V_i. The V_i's length is equal W, and each element of V_i will be assigned as the following rules:

- **Rule 1:** if w_i appears in T, assign 1 to v_i position in V.
- **Rule 2:** unless, compute the similarity score s between w_i and each word in T. If s exceeds a preset threshold τ, then assign s to the considering position in V, otherwise, assign 0. When s is near to 0, it would better to assign 0 to v_i because it does not contribute valuable information.

Depend on the length of given texts, we can keep or ignore function words. In case of short texts, we recommend to maintain function words, but we can assure that they do not affect the whole meaning of texts due to our preset threshold. Finally, after having two semantic vectors, the similarity of two texts is computed by cosine coefficient of those vectors. The output of Eq. (2) has already been normalized between 0 and 1. As the value nears 1, the given texts are more similar, and vice versa.

$$Sim_{word-embedding}(T_1, T_2) = \cos(\theta) = \frac{V_1 \cdot V_2}{\|V_1\| \|V_2\|} \tag{2}$$

Knowledge-Based and Corpus-Based Similarity ($Sim_{knowledge-and-corpus}$). In previous section, we have presented an approach using neural language model to represent word as semantic vector. In this section, we present a method which exploits knowledge base and corpus. With knowledge-based method, we use a semantic graph structure (e.g., WordNet), in which words (also known as *concepts*) are organized as a hierarchy, to measure the relatedness between words. The meaning of relatedness is more general than similarity, for example, *"car"* and *"wheel"* are not similar, but between them exists *part-of* relationship. In WordNet, concepts are grouped to *synsets*, which means *sets of synonyms*, and represented as graph structure, together with six types of relationship: (1) synonymy, (2) antonymy, (3) hyponymy, (4) meronymy, (5) troponomy and

(6) entailment. In order to identify the relatedness, we take into account the path between two concepts, its length reflects the degree of relationship. However, only considering the path length may lose the generalization, we also consider the lowest common subsumer (LCS) [19] concept, which is the nearest to the compared concepts. Although we have looked for the LCS of two concepts, it does not reflect the contribution of both LCS and two concepts. Therefore, we combine the statistical technique on large corpus, e.g., Brown corpus[7]. As proposed in [20], first, we form a set of LCSs that subsume two concepts, then, we compute the probability that each element in LCSs set appears in the corpus and get the maximum probability. This metric denoted by Sim_{F2}, as Eq. (3).

Though WordNet is a good choice in many semantic metrics, it does not cover all up-to-date concepts. For instance, with the growth of social networks, there are many new concepts created in every day, e.g., *"selfie"*, *"emoji"*. Therefore, to overcome this drawback, when a concept not found in WordNet, we will find it in Wiktionary. However, the structure of Wiktionary is not well for finding LCS, we use another metric, called gloss-based. Each concept in Wiktionary comes with descriptions, called as *gloss texts*. The method proposed in [21] is based on the assumption that the level of overlapping between gloss texts of concepts is proportional to the level of similarity. After calculating similarity between words based on knowledge and corpus, we represent given short texts as vectors and compute the similarity between them using Eq. (3).

$$Sim_{F2}(c_1, c_2) = \begin{cases} \max_{c \in LCS(c_1, c_2)} \left[-\log p(c) \right], \{c_1, c_2\} \in \text{WordNet} \\ gloss(c_1) \cap gloss(c_2), \text{otherwise} \end{cases} \tag{3}$$

4 Experiments

4.1 Datasets

We conduct experiments on two datasets: (1) Microsoft research paraphrase corpus (MSRP) [22], and (2) SemEval-2015[8]. The MSRP is a well-known dataset for the problem of paraphrase identification, containing pairs of labeled sentences, if two sentences are paraphrase, the label will be 1 and vice versa. This dataset can be applied to supervised learning approaches, the training set contains 4,076 sentences (2,753 positive, ~67.5 %), and the test set contains 1,725 sentences (1,147 positive, ~66.5 %). The SemEval is series of evaluation of computational semantic analysis systems. The SemEval-2015 dataset also contains two parts, training and test set. Both sets are divided into five domains, in which, each pair of texts is manually semantic annotated by human, in range of [0,5]; the score is proportional to the similarity degree.

---

[7] https://en.wikipedia.org/wiki/Brown_Corpus.
[8] http://alt.qcri.org/semeval2015/task2/.

4.2 Experimental Results

In order to measure the performance of our proposed method, we train our model by using support vector machine learning algorithm on MSRP and SemEval-2015 training sets, and then, test the model on two datasets respectively. However, to show the contribution of the presented features, we perform independently two training and testing tasks: (1) only consider $Sim_{word\text{-}embedding}$, denoted by F_1; (2) combine $Sim_{word\text{-}embedding}$ with $Sim_{knowledge\text{-}and\text{-}corpus}$, denoted by $F_1 + F_2$.

In Table 1, we present the performance of our method by evaluating the contribution of the features on two testing sets, but with SemEval-2015 dataset, we only show the best result of all domains. Tables 2 and 3 present our experiment results on two datasets in comparison to other approaches. With MSRP dataset, we use the accuracy to present the performance of our system, with SemEval-2015 dataset, we use the Pearson correlation coefficient.

Table 1. Evaluate the combination of the presented features

Features	Datasets	
	MSRP *(accuracy)*	SemEval-2015 (ρ)
F_1	0.83	0.89
$F_1 + F_2$	0.82	0.87

Table 2. Experiment results on MSRP dataset

Method	Accuracy
Madnani *et al.* [2]	77.4 %
Ji and Eisenstein [6]	80.4 %
Milajevs *et al.* [25]	73.0 %
Nguyen *et al.* [23]	80.7 %
This paper	**83.6 %**

Table 3. Experiment results on SemEval-2015

Domain	Sultan *et al.* [24]	This paper	Feature
Answer-forums	0.73	**0.75**	F_1
Answers-students	0.78	**0.79**	$F_1 + F_2$
Belief	0.77	**0.76**	F_1
Headlines	0.84	**0.89**	F_1
Image captions	0.86	**0.87**	$F_1 + F_2$

With the experiment results in Table 1, we can see that the contribution of F_1 does yield the best performance on two datasets. When we combine F_1 with F_2, the results are not quite good, because WordNet does not contain all up-to-date concepts, thus we combine with gloss-based method on Wiktionary. By this combination, it may increase the noise in our model, as gloss-based method does not perform well when the gloss texts are short, and the part-of-speech of words may also affect the selection of appropriate gloss texts.

In Table 2, the experiment results on MSRP dataset shows that our method yields a better result than our proposed method in [23] when using $Sim\text{-}_{word\text{-}embedding}$ feature. The main difference between this method and the previous method is how to measure word-to-word similarity. In [23], Nguyen *et al.* use WordNet as knowledge base with information content metric, but WordNet can cover about 64.5 % words on MSRP dataset. On the other hand, in this study, we use the word embedding model to exploit the context surrounding words and combine with tokenizer in preprocessing phase to conclude the level of similarity, and this overcomes the previous drawback. In Table 3, with the results on SemEval-2015, our performance is slightly better than the method proposed in [24]. In [24], the authors gained the best experiment results when using S_1 method, which is quite similar to our method, but differs from the training set for word-similarity metric.

5   Conclusion

We have presented our method for measuring the semantic similarity between short texts on social media by independently evaluating and combining the two different kinds of features: (1) distributed representation of word, (2) knowledge-based and corpus-based metrics. The main contribution of our work can be summarized as follow:

- First, by performing the named entity co-reference resolution, we have increased the system performance because of removing the influence of them. Besides that, we have showed the assumption "each token is a meaning word" is weak, thus, we do tokenizer in our preprocessing phase.
- Second, using skip-gram model to represent word as semantic vector to measure the semantic similarity between words, instead of only relying on semantic graph structure (WordNet) and corpus (Brown Corpus).
- Third, by evaluating the contribution when combines the two features on MSRP and SemEval-2015 datasets, we realize that word embedding feature performs better than another feature, and also significantly improves the performance of our method.
- Finally, our proposed method is quite easy for re-implementing and evaluating other datasets, and can also apply to many applications of natural language processing with an acceptable performance.

References

1. Duong, P., Nguyen, H., Nguyen, V.: Evaluating semantic relatedness between concepts. In: IMCOM, pp. 20:1–20:8. ACM (2016)
2. Madnani, N., Tetreault, J., Chodorow, M.: Re-examining machine translation metrics for paraphrase identification. In: HLT-NAACL, pp. 182–190 (2012)
3. Bach, N., Nguyen, M., Shimazu, A.: Exploiting discourse information to identify paraphrases. Expert Syst. Appl. **41**(6), 2832–2841 (2014)

4. Li, Y., McLean, D., Bandar, Z., O'Shea, J., Crockett, K.: Sentence similarity based on semantic nets and corpus statistics. IEEE Trans. Knowl. Data Eng. **18**(8), 1138–1150 (2006)
5. Mihalcea, R., Corley, C., Strapparava, C.: Corpus-based and knowledge-based measures of text semantic similarity. In: AAAI, pp. 775–780 (2006)
6. Ji, Y., Eisenstein, J.: Discriminative improvements to distributional sentence similarity. In: EMNLP, pp. 891–896 (2013)
7. Guo, W., Diab, M.: Modeling sentences in the latent space. ACL **1**, 864–872 (2012)
8. Kozareva, Z., Montoyo, A.: Paraphrase identification on the basis of supervised machine learning techniques. In: Salakoski, T., Ginter, F., Pyysalo, S., Pahikkala, T. (eds.) FinTAL 2006. LNCS (LNAI), vol. 4139, pp. 524–533. Springer, Heidelberg (2006)
9. Snover, M., Madnani, N., Dorr, B., Schwartz, R.: TER-Plus: paraphrase, semantic, and alignment. Mach. Transl. **23**(2–3), 117–127 (2009)
10. Das, D., Smith, N.: Paraphrase identification as probabilistic quasi-synchronous recognition. In: Su, K.-Y., Su, J., Wiebe, J. (eds.) ACL/IJCNLP, pp. 468–476 (2009)
11. Socher, R., Huang, E., Pennington, J., Ng, A., Manning, C.: Dynamic pooling and unfolding recursive autoencoders for paraphrase detection. In: Shawe-Taylor, J., Zemel, R., Bartlett, P., Pereira, F., Weinberger, K. (eds.) NIPS, pp. 801–809 (2011)
12. He, H., Gimpel, K., Lin, J.: Multi-perspective sentence similarity modeling with convolutional neural networks. In: Lluís, M., Callison-Burch, C., Pighin, D., Marton, Y. (eds.) EMNLP, pp. 1576–1586 (2015)
13. Sahami, M., Heilman, T.: A web-based kernel function for measuring the similarity of short text snippets. In: Carr, L., Roure, D., Iyengar, A., Dahlin, M. (eds.) WWW, pp. 377–386 (2006)
14. Witten, I., Milne, D.: An effective, low-cost measure of semantic relatedness obtained from Wikipedia links. In: Proceeding of AAAI Workshop on Wikipedia and Artificial Intelligence: An Evolving Synergy, pp. 25–30. AAAI Press, Chicago (2008)
15. Mikolov, T., Chen, K., Corrado, G.: Efficient estimation of word representations in vector. In: Proceedings of International Conference of Learning Representations (2013)
16. Qiu, L., Cao, Y., Nie, Z., Yu, Y.: Learning word representation considering proximity and ambiguity. In: Brodley, C., Stone, P. (eds.) AAAI, pp. 1572–1578 (2014)
17. Bontcheva, K., Dimitrov, M., Maynard, D., Tablan, V., Cunningham, H.: Shallow methods for named entity coreference resolution. In: Chaînes de références et résolveurs d'anaphores, Workshop TALN (2002)
18. Mikolov, T., Sutskever, I., Chen, K., Dean, J.: Distributed representations of words and phrases and their compositionality. In: Advances in Neural Information Processing Systems, pp. 3111–3119 (2013)
19. Wu, Z., Palmer, M.: Verbs semantics and lexical selection. In: Proceedings of the 32nd Annual Meeting on Association for Computational Linguistics, pp. 133–138 (1994)
20. Resnik, P.: Using information content to evaluate semantic similarity in a taxonomy. In: IJCAI, pp. 448–453 (1995)
21. Lesk, M.: Automatic sense disambiguation using machine readable dictionaries: how to tell a pine cone from an ice cream cone. In: Proceedings of the 5th Annual International Conference on Systems Documentation, pp. 24–26 (1986)
22. Dolan, W., Brockett, C.: Automatically constructing a corpus of sentential paraphrases. In: Proceedings of IWP (2005)
23. Nguyen, H.T., Duong, P.H., Le, T.Q.: A multifaceted approach to sentence similarity. In: Huynh, V.-N., Inuiguchi, M., Demoeux, T. (eds.) IUKM 2015. LNCS, vol. 9376, pp. 303–314. Springer, Heidelberg (2015). doi:10.1007/978-3-319-25135-6_29

24. Sultan, M., Bethard, S., Sumner, T.: DLS@ CU: Sentence similarity from word alignment and semantic vector composition. In: Proceedings of the 9th International Workshop on Semantic Evaluation, pp. 148–153 (2015)
25. Milajevs, D., Kartsaklis, D., Sadrzadeh, M., Purver, M.: Evaluating neural word representations in tensor-based. In: EMNLP, pp. 708–719 (2014)

Fi-Senti: A Language-Independent Model for Figurative Sentiment Analysis

Hoang Long Nguyen[1], Trung Duc Nguyen[2], and Jason J. Jung[1(✉)]

[1] Department of Computer Engineering, Chung-Ang University, Seoul, Korea
longnh238@gmail.com, j2jung@gmail.com
[2] Faculty of Information Technology, Vietnam Maritime University,
Haiphong, Vietnam
duc.nguyentrung@gmail.com

Abstract. This paper focuses on identifying the polarity of figurative language in the very short text collected from Social Network Services. Although this topic is not new, most computer scientists have solved this issue by using natural language processing techniques. This seems difficult for non-native English speakers because they have to rely on heuristics in language. Therefore, our target in this work is to find a language-independent approach to solve the problem without using any semantic resources (e.g., dictionaries and ontologies). A statistical method based on two main features (i.e., (i) textual terms and (ii) sentimental patterns) is proposed to determine the sentiment degree of three popular types of figurative language (i.e., sarcasm, irony, and metaphor). We experimented on two Test sets with about 3,800 tweets and used Cosine similarity as the correlation measurement for evaluating the performance. The results show that our Fi-Senti model (Figurative Sentiment analysis model) well performs in determining the sentiment intensity of the figurative language with the best achievement is 0.8952 with sarcasm and 0.9011 with irony.

Keywords: Figurative sentiment analysis · Language-independent · Sarcasm · Irony · Metaphor

1 Introduction

Nowadays, every activities are conducted online. This leads to the exploration of information, especially on the Social Network Services (SNSs). Until now, there are more than 200 different SNSs all over the world with hundred million users. Due to this reason, researching about data on SNSs is a real challenging and excitement for researchers.

Sentiment analysis is one of the interesting topics [8,9] which aims to determine the polarity of specific documents without directly communication [12]. The results of sentiment analysis can be very helpful in many different fields (e.g., companies can understand about their customer opinion in order to have appropriate strategies for promptly adapting their demand, political parties are

© Springer International Publishing Switzerland 2016
H.T. Nguyen and V. Snasel (Eds.): CSoNet 2016, LNCS 9795, pp. 260–272, 2016.
DOI: 10.1007/978-3-319-42345-6_23

also to achieve lots of advantages by understanding citizens to make quick decision). Moreover, sentiment analysis is also used for reputation analysis [1] or election result prediction [18].

There are two types of languages which are literal and figurative language. In case of literal language, the meaning of a text is directly stated. However, the meaning is underlying with figurative language and we have to use our imagination to understand its real meaning. Due to this fact, precisely analyzing sentiment of figurative documents is still difficult for computers as well as human beings. Figurative language can be found everywhere from stories, musics to movies and SNSs is not an exception. By retrieving public tweets on Twitter, we obtained a lot of figurative tweets with different meanings on various topics. There are many types of figurative languages, however, this paper only focuses on identifying the polarity of sarcasm, irony, and metaphor. Both sarcasm and irony are used once someone wants to say something but the meaning can be opposite. While sarcasm is often used to hurt someone's feelings, irony can be seen in funny situations. On the other hand, use of metaphor is to make an implicit comparison between two entities. Many works have been done on these three types [5,20,24].

We select Twitter as a case study of SNSs for researching due to both its advantage and challenge: (i) the data on Twitter is easily collected because of the open API privacy which is more strict in other SNSs and (ii) the length of a text which is used to represent the status of a user is very short which makes the task of figurative sentiment analysis even more difficult. There are also many studies with various approaches which also work on analyzing figurative language on Twitter [2,13,16,17]. But they almost solve the problem by using their heuristics in language. By proposing a language-independent model, we believe that this is an effective approach for non-native English speakers and easy-extending for characterizing other types of figurative devices.

In this section, we give the overview of our motivation. In Sect. 2, we summarize related work. Some basic notions are addressed in Sect. 3. Section 4 explains the proposed method in detail. Further, the performance of our models will be presented in Sect. 5. To sum up, we conclude and discuss future works in Sect. 6.

2 Related Work

In this section, we survey previous works which study about figurative language. Very soon contribution was proposed by [23] to comprehend and create metaphors using similes. The set of similes were collected from a list of antonymous adjectives using Google API with two queries (i.e., "as ADJ as *" and "as * as NOUN") to achieve nearly 75,000 simile instance. Though this work manually generated similes but it opened an effective way for identifying figurative language. Another important contribution that we want to mention here is from [21] by creating a method for retrieving figurative language. In this work, the author defined a list of operators (i.e., neighborhood (?X), cultural stereotype (@X), and ad-hoc category (^X)) and the compound rules for expressing sentences. Using this methodology is able to create a lot of meaningful queries. By integrating these two techniques together and extending one more operator (antonym (-X)) from [21], the author gave a method for categorizing texts into straight simile and ironic simile by using strategies of ironic subversion [22]. In another research [19], unsupervised methods were used to find the associate from a small set of metaphorical expressions by verb and noun clustering. The knowledge which is extracted can be useful to detect similarity structure of metaphor in a larger domain with high precision (0.79). However, the recall and F-measure were still not mentioned in this work.

Using customer reviews which are collected from Amazon, Slashdot, TripAdvisor, the goal of [14] was to identify salient components of verbal irony for: (*i*) distinguishing between irony and non-irony data and (*ii*) automatically finding ironic data. With 6 categories of features (i.e., n-grams, POS-grams, funny profiling, positive/negative profiling, affective profiling, and pleasantness profiling), the model showed high classification results for the first task by using naive bayes, support vector machine, and decision tree as classifiers. However, results of the second task was not clearly mentioned in this paper. With the attempt to automatically detect irony, [15] proposed three conceptual layers including eight textual features to represent the core of irony. This paper showed interesting problems about irony in general.

Above researches only focused on understanding figurative text without proposing a method for analyzing its sentiment degree. Overcome this problem, [3] identified the polarity of tweets by using information retrieval techniques. In this work, the authors constructed a set of indexes based on the dependencies between terms in the Training set. Further, the given tweet was built as a set of queries with same procedure which is used for building the index to predict the intensity. This method can work well with well-formed documents, however, limit of performance with short texts like tweets. In another way, [7] solved the problem with linear support vector machine approach. This model included two steps which are preprocessing step to build featured dictionary from the set of features and classification step to calculate the sentiment of tweets.

3   Basic Notions

3.1   Tweet

Tweet is a message which is sent on Twitter. In this work, we focus on very short text tweets (i.e., 30 characters excluding hashtags and 40 characters including hashtags).

Table 1 shows the example of some tweets which are collected for the training purpose. We denote T as a set of tweets in the Training set with their properties is defined as follows

$$t = <c_t, s_t> \text{ with } s_t \in \mathbb{Z}, \ s_t \in [-5, 5] \tag{1}$$

where t is a tweet, c_t and s_t is the content and the score of that tweet respectively.

Table 1. The example of some tweets in the Training set

No	Content	Score
1	Happy Weekend G! I promise not to bother you again! lol! #not Cheers!	1
2	Tiago Splitter plays basketball about as gracefully as Cosmo Kramer would	−2
3	Breakdowns at the beginning of the day really make me happy. #sarcasm	−3
4	I currently feel like the biggest disappointment of the century. #greatfeeling #not	−4

3.2   Term

In this paper, we consider term as a single word which is extracted from tweets and make sense of sentiment. The pre-processing step is first conducted to increase the performance of term extraction task by:

- converting tweets to lowercase,
- removing any redundant information (e.g., stopwords, tagged people, URLs, and non-ascii characters),
- deleting unneeded characters (e.g., lmaoooo → lmaoo).

Especially, hashtags and emoticons has to be kept because it is the importance elements which represent the sentiment of tweets. We call W_T is the set of terms which is extracted from the set of tweets T.

4   Fi-Senti Model

Fi-Senti model basically uses statistics-based approach based on two main modules which are Textual term-based and Sentimental pattern-based module. The results of these two module are integer numbers in the range of 11 point sentiment scale (from −5 to 5, including 0) which is calculated by using a fuzzy equation

$$S = W_{TT} \times TT + W_{SP} \times SP \tag{2}$$

where S is the final score of the given tweet, TT is the score which is determined by Textual term-based module, SP is the score which is determined by Sentimental pattern-based module, W_{TT} and W_{SP} are weights which are identified by conducting the experiments, with $W_{TT} + W_{SP} = 1$.

4.1   Textual Term-Based Module

Basically, this model uses statistical methods based on the co-occurrence of terms to identify the score of a given tweet with the assumption that tweets which have similar terms will have similar sentiment score. Below is the detail explanation about how to identify the sentiment step by step.

With the given tweet t_k which is needed to be analyzed the sentiment, it is extracted into the set of terms W_k for the target of finding tweets in the Training

Table 2. The set of clusters $\mathbf{C_k}$ which is generated from $\mathbf{W_k}$

W_k	Clusters		
	3-term cluster	2-term cluster	1-term cluster
{so, happy, #sarcasm}	{so, happy, #sarcasm}	{so, happy}	{so}
		{so, #sarcasm}	{happy}
		{happy, #sarcasm}	{#sarcasm}

set which are similar with t_k. The most important notion here is that we only consider terms which belong to W_T

$$W_k = \left\{ \bigcup_{i=1}^{n_k} w_i \;\middle|\; w_i \in t_k, \; w_i \in W_T \right\} \tag{3}$$

where W_k is the set of terms which is extracted from tweet t_k, w_i is a term which belongs to tweet t_k, W_T is the set of terms in the Training set, n_k is the number of terms which is extracted from tweet t_k.

Example 1. With the given tweet t_k: "@BrianRawchester: So happy performance #sarcasm". At first, t_k is considered in lower-case mode and "@BrianRawchester" is removed. Assuming that term "performance" doesn't belong to W_T. Therefore, the set of terms W_k that we have after extracting from t_k is {so, happy, #sarcasm}.

Further, all the possible combinations of terms in W_k which are considered as clusters are generated. From W_k, we have the set of clusters as shown in Table 2. Each cluster expresses the presence of terms in a tweet (e.g., if a tweet belongs to cluster {so, happy}, the content of this tweet will include term "so" and term "happy"). As we mentioned before, our assumption is that tweets which have similar terms will have similar sentiment score. Therefore, the goal of next step is to find all the tweets in T which are similar with t_k by grouping tweets in T into clusters in C_k. To do this, each cluster in C_k is first represented as a featured vector with the dimension equals with the number of terms in W_k.

In order to group tweets into clusters in C_k, tweets in T are also represented as vectors based on W_k. To assign a tweet into a cluster, the distance from this tweet has to be minimum compared to other distance. The distance between tweet t_i and cluster c_i is calculated by using the following function with $dis(t_i, c_i)$ is the distance from tweet t_i to cluster c_i

$$dis(t_i, c_i) = 1 - \frac{t_i \cdot c_i}{\| t_i \| \| c_i \|} \tag{4}$$

Definition 1 (Cluster coefficient). *Cluster coefficient is a number to indicate how similar between tweets in a cluster and the given tweet. It is calculated based on the number of featured terms of a cluster. In this paper, we propose power function for calculating the cluster coefficient as follows*

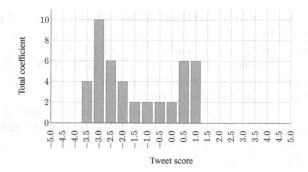

Fig. 1. Histogram of score distribution

$$\Psi_c = \alpha^{\nu_c}, \ \alpha \in \mathbb{N}^+ \tag{5}$$

where Ψ_c is the cluster coefficient, α is a constant, ν_c is the quantity of featured terms of a cluster.

A histogram is constructed to show the distribution of score through the use of tweets in clusters with their scores and cluster coefficients. We select the peak of histogram as the result which is annotated by Textual term-based module.

Example 2. Assuming that we have total 6 non-empty clusters after clustering tweets in T into C_k. We consider the situation of using Eq. 5 with $\alpha = 2$. Table 3 shows the data with its respective coefficient and Fig. 1 expresses this data as histogram. The score which is annotated by Textual term-based module in this situation is -3.0.

4.2   Sentimental Pattern-Based Module

Each term in the Training set has a score from $[-5, 5]$ to indicate the sentiment degree of a term. We express tweets in the Training set as sentimental patterns which are constructed by using term scores for the purpose of learning. Decision tree learning model is used to classify a tweet into a group with highest similarity

Table 3. Tweets and their cluster coefficients in $\alpha = 2$

Tweets	Clusters	Score	Coefficient	Tweets	Clusters	Score	Coefficient
t_1	{so, happy, #sarcasm}	-3.0	8	t_8	{so}	1.0	2
t_2	{so, happy}	0.5	4	t_9	{happy}	0.0	2
t_3	{so, happy}	1.0	4	t_{10}	{#sarcasm}	-0.5	2
t_4	{so, #sarcasm}	-2.0	4	t_{11}	{#sarcasm}	-1.0	2
t_5	{so, #sarcasm}	-2.5	4	t_{12}	{#sarcasm}	-1.5	2
t_6	{so, #sarcasm}	-3.5	4	t_{13}	{#sarcasm}	-3.0	2
t_7	{so}	0.5	2	t_{14}	{#sarcasm}	-2.5	2

pattern. We select Decision tree as the classifier because it can well perform with multiclass (classification task with more than two classes) classification model.

Term scores are calculated by using only tweets in the Training set without any dictionary. At the beginning, $P(s|w)$ is computed to show the probability that a tweet will have score s if its content includes term w. For instance, Fig. 2 shows the distribution of $P(s|\text{"}\#sarcasm\text{"})$.

Probability of a range is the sum of all its element's probability. In this step, we filter continuous ranges which have lower probability than the others by comparing their value with the threshold. Threshold value is calculated by using the following equation

$$\theta_w = \frac{\sum_{i=1}^{n} R_i}{n} \text{ with } R_i = \sum_{j=1}^{m} P(s_j|w) \tag{6}$$

where θ is the threshold for filtering continuous ranges, R is the probability value of a range, n is the number of ranges, and m is the number of range's elements.

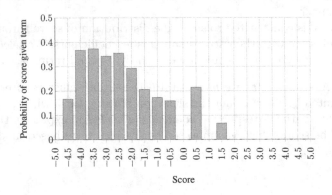

Fig. 2. The distribution of $P(s|\text{"}\#sarcasm\text{"})$

Term score is the expected value which is computed from selected ranges to show how sentimental a term is. Positive and negative terms are more important to the sentiment of a tweet rather than neutral terms. Term score is calculated by using the following formula

$$s_w = E(s_w) = \frac{\sum_{i=1}^{n} (s_i \times P(s_i|w))}{\sum_{i=1}^{n} P(s_i|w)} \tag{7}$$

where s_w is score of term w which belongs to $[-5, 5]$, n is the number of selected range's elements, s is score, and $P(s|w)$ is the probability of that score given term w.

Sentimental patterns are constructed by using the term score to express the grammar structures or the writing styles that people usually use to write figurative tweets. We build a vector space which represents extracted patterns as the input for decision tree learning model. To adapt the input's condition, every patterns have to be scaled to the same dimension by using an interpolative function as proposed by [10]. Here, vectors are scaled to the maximum possible terms that a tweet in the Training set contains (i.e., due to our Training set, the number of maximum terms is 25). We then train a decision tree-based classifier to predict from these patterns the sentiment score of the given tweet with the range belongs to $[-5, 5]$.

Table 4. Terms which are extracted from tweet t_k and their respective sentiment scores

No	0	1	2	3	4	5	6
Term	could	this	day	get	any	better	#sarcasm
Score	1.95	0.26	-0.48	-0.30	0.48	2.02	-2.27

Example 3. A tweet t_k with its content: "Could this day get any better #sarcasm". Table 4 shows the list of terms with their respective scores, and Fig. 3 is the expression of the sentimental pattern before and after length normalization.

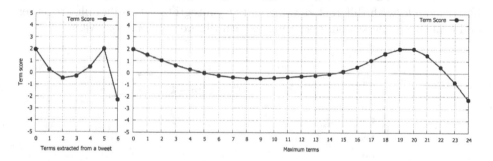

Fig. 3. Sequential pattern of tweet t_k before and after length normalization

5   Performance Measurement

Our Data set includes two parts (i.e., Training set with 8,000 tweets including 5,000 sarcastic, 1,000 ironic, and 2,000 metaphorical tweets for the purpose of training and two Test sets with about 3,800 tweets for evaluating our system) which are collected from SemEval-2015[1]. All of these tweets are written in English from 1st Jun, 2014 to 30th Jun, 2014 and are annotated by seven annotators

---

[1] http://alt.qcri.org/semeval2015/.

(i.e., three of them are native English speakers and the others are competent non-native speakers of English) on the CrowdFlower crowd-sourcing platform using 11 point sentiment scale from -5 (extreme discontent) to 5 (extreme pleasure). By its very nature, most of the sarcasm, irony, and metaphor are negative.

In sentiment analysis field, correlation measure is the most appropriate choice because human raters typically agree 79 % of the time. We selected Cosine similarity as the metric to calculate the performance of Fi-Senti model because it takes into account how closed between the predicted value and the actual value. The value which is measured by Cosine similarity has score range from 0 to 1 to express the similarity between actual results and the expected results. These two sets of results are first represented as two vectors

$$A = \{a_1, a_2, ..., a_n\} \text{ and } E = \{e_1, e_2, ..., e_n\} \tag{8}$$

where A is the actual results which is annotated by our model, E is the expected results, n is the number of tweets that needs to be evaluated.

Then, the performance of our system $sim(A, E)$ is calculated by using the following formula

$$sim(A, E) = \frac{A \cdot E}{\parallel A \parallel \parallel E \parallel} \tag{9}$$

We first conduct the experiment on the Test set 1 with 927 tweets for obtaining the value of α in Eq. 5 and W_{TT}, W_{SP} in Eq. 2 in which the system achieves highest performance through two tasks: (i) independently testing Textual term-based module with different α value and (ii) integrating two modules together and evaluating the system performance with different W_{TT}, W_{SP} value.

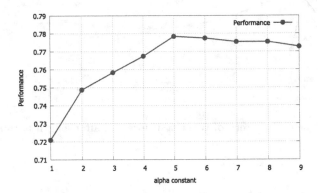

Fig. 4. The performance of Textual term-based module on the Test set 1

Regarding the first task, we evaluate the Textual term-based module by testing with 10 different α constant values which are used for calculating the cluster coefficient by using Eq. 5. From Fig. 4, it could be observed that the system performance increases to the maximum value at the point $\alpha = 5$ and then gradually diminishes. It means that we can achieve highest performance with Textual term-based by using $\alpha = 5$. With the second task, we combine Textual term-based module using $\alpha = 5$ and Sentimental pattern-based module together by using Eq. 2. Figure 5 shows the evaluation that we verify with different values of W_{TT} and W_{SP}. The highest performance is obtained with $W_{TT} = 0.55$ and $W_{SP} = 0.45$.

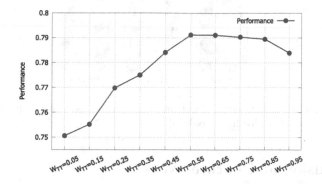

Fig. 5. The performance of Fi-Senti model on the Test set 1 ($W_{TT} + W_{SP} = 1$)

Using the coefficient and weight which are determined from the above step, we conduct the experiment on Test set 2 with 2,800 tweets and compare with other related work from SemEval-2015 Task 11. Three types of figurative language (i.e., 1,200 sarcastic, 800 ironic, and 800 metaphorical tweets) are considered. There are total of 15 teams with 35 different runs. From the results of this challenge, we identify the baseline and state of the art based on the best performance of teams for the target of demonstrating the effectiveness of Fi-Senti model as shown in Table 5.

Table 5. Baseline and State of the art of related work

Type	Baseline		State of the art	
	Performance	Team	Performance	Team
Sarcasm	0.681	SHELLFBK [3]	0.904	Elirf [4]
Irony	0.652	SHELLFBK [3]	0.918	LLT-PolyU [25]
Metaphor	0.291	RGU	0.655	ClaC [11]

Figure 6 shows that our system obtain good performance with sarcasm and irony. However, it still shows the limitation with metaphor. Hence, improving system performance with metaphorical tweets is determined as our next essential work.

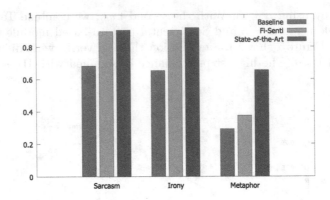

Fig. 6. The comparison between Fi-Senti model with related work from SemEval-2015 Task 11 on Test set 2 (Color figure online)

6   Conclusion and Future Work

In this paper, we proposed a language-independent model for figurative language sentiment analysis based on the statistical method. With this approach, we solve the problem without needing to understand the English meanings. However, there are other issues which are need to be solved in our next works:

- Focusing on improving the performance of metaphor. Moreover, we only consider to analyze the sentiment of figurative tweets in this work. For further research, we will add more non-figurative tweets into the Training set and the Test set to demonstrate the effect of our system for not only figurative tweets but also non-figurative tweets.
- The process of Textual term-based module is still time-consuming due to the large number of combinations. If the length of tweets increases, it can lead to the combinatorial explosion.
- Data from other SNSs (e.g., Facebook, Instagram, and Google Plus) will be combined [6] to test with our approach. This can prove the effectiveness of our system with various types of texts on different SNSs. Moreover, extending the sentiment score to smoother value by using real number is considered to develop in the future.

Acknowledgments. This work was supported by the Ministry of Education of the Republic of Korea and the National Research Foundation of Korea (NRF-2015S1A5B6037297). Also, this work was supported by the National Research Foundation of Korea (NRF) grant funded by the Korea government (MSIP) (NRF-2014R1A2A2A05007154).

References

1. Amigó, E., et al.: Overview of RepLab 2013: evaluating online reputation monitoring systems. In: Forner, P., Müller, H., Paredes, R., Rosso, P., Stein, B. (eds.) CLEF 2013. LNCS, vol. 8138, pp. 333–352. Springer, Heidelberg (2013)
2. Davidov, D., Tsur, O., Rappoport, A.: Semi-supervised recognition of sarcastic sentences in twitter and amazon. In: Proceedings of the 14th Conference on Computational Natural Language Learning, pp. 107–116 (2010)
3. Dragoni, M.: Shellfbk: an information retrieval-based system for multi-domain sentiment analysis. In: Proceedings of the 9th International Workshop on Semantic Evaluation, pp. 502–509 (2015)
4. Giménez, M., Pla, F., Hurtado, L.-F.: Elirf: a SVM approach for SA tasks in twitter at SemEval-2015. In: Proceedings of the 9th International Workshop on Semantic Evaluation, pp. 574–581 (2015)
5. Hao, Y., Veale, T.: An ironic fist in a velvet glove: creative mis-representation in the construction of ironic similes. Mind. Mach. 20(4), 635–650 (2010)
6. Long, N.H., Jung, J.J.: Privacy-aware framework for matching online social identities in multiple social networking services. Cybern. Syst. 46(1–2), 69–83 (2015)
7. Karanasou, M., Doulkeridis, C., Halkidi, M.: Dsunipi: an SVM-based approach for sentiment analysis of figurative language on twitter. In: Proceedings of the 9th International Workshop on Semantic Evaluation, pp. 709–713 (2015)
8. Kaur, A., Gupta, V.: A survey on sentiment analysis and opinion mining techniques. J. Emerg. Technol. Web Intell. 5(4), 367–371 (2013)
9. Medhat, W., Hassan, A., Korashy, H.: Sentiment analysis algorithms and applications: a survey. Ain Shams Eng. J. 5(4), 1093–1113 (2014)
10. Nguyen, H.L., Jung, J.E.: Statistical approach for figurative sentiment analysis on social networking services: a case study on twitter. Multimedia Tools Appl. (2016). doi:10.1007/s11042-016-3525-9
11. Özdemir, C., Bergler, S.: Clac-sentipipe: SemEval 2015 subtasks 10 b, e, and task 11. In: Proceedings of the 9th International Workshop on Semantic Evaluation, pp. 479–485 (2015)
12. Pang, B., Lee, L., Vaithyanathan, S.: Thumbs up? Sentiment classification using machine learning techniques. In: Proceedings of the ACL-02 Conference on Empirical Methods in Natural Language Processing, vol. 10, pp. 79–86 (2002)
13. Rajadesingan, A., Zafarani, R., Liu, H.: Sarcasm detection on twitter: a behavioral modeling approach. In: Proceedings of the 8th ACM International Conference on Web Search and Data Mining, pp. 97–106 (2015)
14. Reyes, A., Rosso, P.: Making objective decisions from subjective data: detecting irony in customers reviews. Decis. Support Syst. 53(4), 754–760 (2012)
15. Reyes, A., Rosso, P.: On the difficulty of automatically detecting irony: beyond a simple case of negation. Knowl. Inf. Syst. 40(3), 595–614 (2014)
16. Reyes, A., Rosso, P., Buscaldi, D.: From humor recognition to irony detection: the figurative language of social media. Data Knowl. Eng. 74, 1–12 (2012)
17. Reyes, A., Rosso, P., Veale, T.: A multidimensional approach for detecting irony in twitter. Lang. Resour. Eval. 47(1), 239–268 (2013)
18. Sang, E.T.K., Bos, J.: Predicting the 2011 Dutch Senate election results with twitter. In: Proceedings of the Workshop on Semantic Analysis in Social Media, pp. 53–60 (2012)
19. Shutova, E., Sun, L., Korhonen, A.: Metaphor identification using verb and noun clustering. In: Proceedings of the 23rd International Conference on Computational Linguistics, pp. 1002–1010 (2010)

20. Tsur, O., Davidov, D., Rappoport, A.: ICWSM - a great catchy name: semi-supervised recognition of sarcastic sentences in online product reviews. In: Proceedings of the 4th International AAAI Conference on Weblogs and Social Media, pp. 162–169 (2010)
21. Veale, T.: Creative language retrieval: a robust hybrid of information retrieval and linguistic creativity. In: Proceedings of the 49th Annual Meeting of the Association for Computational Linguistics: Human Language Technologies, vol. 1, pp. 278–287 (2011)
22. Veale, T.: Detecting and generating ironic comparisons: an application of creative information retrieval. In: Proceedings of the AAAI Fall Symposium: Artificial Intelligence of Humor, pp. 101–108 (2012)
23. Veale, T., Hao, Y.: Comprehending and generating apt metaphors: a web-driven, case-based approach to figurative language. In: Proceedings of the 22nd National Conference on Artificial Intelligence, vol. 2, pp. 1471–1476 (2007)
24. Veale, T., Keane, M.T.: Conceptual scaffolding: a spatially-founded meaning representation for metaphor comprehension. Comput. Intell. 8(3), 494–519 (1992)
25. Xu, H., Santus, E., Laszlo, A., Huang, C.-R.: LLT-PolyU: identifying sentiment intensity in ironic tweets. In: Proceedings of the 9th International Workshop on Semantic Evaluation, pp. 673–678 (2015)

Detection and Prediction of Users Attitude Based on Real-Time and Batch Sentiment Analysis of Facebook Comments

Hieu Tran(✉) and Maxim Shcherbakov

Computer Aided Design Department, Volgograd State Technical Univeristy,
Lenin Avenue 28, 400005 Volgograd, Russia
trantrunghieu7492@gmail.com, maxim.shcherbakov@vstu.ru
http://www.vstu.ru/en

Abstract. The most of the people have their account on social networks
(e.g. Facebook, Vkontakte) where they express their attitude to different
situations and events. Facebook provides only the positive mark as a like
button and share. However, it is important to know the position of a certain user on posts even though the opinion is negative. Positive, negative
and neutral attitude can be extracted from the comments of users. Overall
information about positive, negative and neutral opinion can bring understanding how people react in a position. Moreover, it is important to know
how attitude is changing during the time period. The contribution of the
paper is a new method based on sentiment text analysis for detection and
prediction negative and positive patterns for Facebook comments which
combines (i) real-time sentiment text analysis for pattern discovery and
(ii) batch data processing for creating opinion forecasting algorithm. To
perform forecast we propose two-steps algorithm where: (i) patterns are
clustered using unsupervised clustering techniques and (ii) trend prediction is performed based on finding the nearest pattern from the certain
cluster. Case studies show the efficiency and accuracy (Avg. MAE = 0.008)
of the proposed method and its practical applicability. Also, we discovered
three types of users attitude patterns and described them.

Keywords: Opinion mining · Sentiment analysis · Text classification ·
Social networks · Facebook comments · Real-time and batch processing ·
Clustering analysis · Forecasting techniques

1 Introduction

Sentiment analysis of textual content is used for opinion mining of people who
express their emotions and thoughts by text messages. New communication platforms such as social networks (e.g. Facebook or VKontakte) gives a new opportunity for better understanding information using natural language processing
and sentiment analysis. According to the article published by zephoria.com in
December 2015, nowadays Facebook has more than 1.55 billion monthly active

© Springer International Publishing Switzerland 2016
H.T. Nguyen and V. Snasel (Eds.): CSoNet 2016, LNCS 9795, pp. 273–284, 2016.
DOI: 10.1007/978-3-319-42345-6_24

users [21]. These users write more 510 000 comments every minute and this is a source of large information on the Internet. Usually, these textual comments are the results of the reaction of people regarding recent news or happened events. Understanding of users attitude helps to know how a certain person or groups respond to the particular topic, and it serves to draw relevant conclusions or make efficient decisions based on feedback [8,20]. For example, in the political field. Assume there is news regarding a particular decision of the government in a certain country, which published in social networks by BBC or CNN. Based on examination of the textual comments, we can understand people positions and either a certain person supports this decision of the government or not.

From the business point of view, sentiment analysis helps companies to improve customer development process, enhance business intelligence systems, and change their marketing strategies to get more profit. Moreover, using this type of text analysis, the trend of people's attitude to certain events or typical groups of events can be predicted. This foresight is valuable for proactive actions development for the future expected situation in every domain we refer to, such as politic, economic, business and so on. So, in fact, the questions is how to understand users behaviour and opinions according to processing textual comments in social network and how to predict either these opinions remain the same or will be changed in time? Opinions give the intuition about a person or customer preferences.

The main problem, which is considered in the current research, is how to understand positive or negative user's opinion about published posts and news using sentimental text analysis. Are there any laws and consistent timewise patterns in user's comments, and how to detect these patterns and predict them? The contribution of the paper is a new method based on sentiment text analysis for detection and prediction negative and positive patterns in Facebook comments.

The paper contains the following sections besides of introduction. The next section contains the literature review and analysis of the recent related works on sentiment text analysis. After it describes the main idea of the proposed method of Facebook comments sentiment analysis using a combination of the real-time and batch data processing. Results and discussion are covered in the last section.

2   Related Works

Sentiment text analysis is a large but still growing research domain. An early, and still common, approach to sentiment analysis has been to use the called 'semantic orientation' (SO) of terms as the basis for classification [11].

Turney showed that semantic orientation is useful for classifying more general product reviews [24]. The work suggests that product reviews may be classified easier than movie reviews, as product reviews tend to be shorter and be more explicitly evaluative. In [19] authors classified movie reviews used standard bag-of-words techniques with limited success. Twitter is a social network which represented as a sources of customer opinions to analyze. The early results of Twitter data sentiment analysis presented in work [2]. The authors of the paper "Sentiment Analysis on Twitter" tentatively conclude that sentiment analysis for Twitter data is not that different from sentiment analysis for other genres [1]. In [14] authors used the similar way (Twitter API) to collect training data and then to perform a sentiment investigation. An analysis of Twitter followers reveals networks of users who are related by current news tops rather than by personal interactions. Furthermore, a database of sensor data from the reality mining corpus is used for dynamic social network analysis [6]. Besides social networks, common websites have a large data needed to investigated. As practical implementation, sentiment analysis was applied to the feedback data from Global Support Services survey [7]. It helps organizations determine the quality of services. Text information from social networks provides information about entities (people and organizations) and their corresponding relations and involvement in events [26]. Up to this time, there are not too many articles about real-time text data analysis. In the research [3] proposed the theory of big data stream analytics for language modeling for sentiment analysis but did not put into practice with a certain network. A group authors from the University of Southern California described a system for real-time analysis of public sentiment toward presidential candidates in the 2012 U.S. election as expressed on Twitter [25] but this system only works with the twitters, and could not reach to users replies.

Almost all of research is concerned about status, twitter, or post to analysis sentiment [12,13,15]. There is a small number of research papers mentioned the replies and comments analysis. However, replies to Twitter and Facebook comments have values of sentiment expressions from users. In [10], authors implemented analysis of Facebook posts [27] on the 2014 Thai general election. They did not use a large information from the comments of each post. Therefore, we focus on these comments which will be performed in this paper. Also in [4] presented prediction the vote percentage that individual candidate will receive on Singapore Presidential Election 2011 using Twitter data with census correction but they met issues with fake tweeter sentiment. And the source may be related to the scenario where the voters do not truly reflect their on-line sentiments from their choice of candidate. Several case studies have found that the online information has been quite successful acting as an indicator for electoral success [27]. There are some research on the use of twitter such as [23]. There is the Vietnamese research group do analysis based on Facebook comments, but they worked in a sphere of Vietnamese language [22].

Based on literature review we conclude, that sentiment analysis of text comments on the social network as a mechanism to understand the people's attitude is still an open question.

3   A Method

3.1   General Scheme

The main problem we would like to solve is the creation of the technique, which helps understand and predict user attitude expressed in Facebook comments regarding published news or post. We propose the following method: (*1*) using collected from Facebook data we perform batch analysis to make the precise forecasting model to detect and predict negative and positive behaviour patterns; (*2*) using real-time text processing we detect a pattern and understand what is the current situation looks like; (*3*) understand how this situation will be developed based on pattern prediction; and (*4*) perform actions to change the pattern if it needed.

To be more concrete, we implement two main approaches for analysis sentiment in our method. The first approach is real-time comments analysis. The solution allows to analyze and update results right in time when data generated by users on Facebook. The second approach analyses stored data in batch mode. This is a cluster-based approach for the negative/positive attitude patterns detection and prediction. To evaluate the performance of forecasting, the median absolute deviation measure is used.

3.2   Real-Time Stream Sentiment Analysis and Event Generation

Real-time stream processing solution retrieves data from Facebook server continually and then, processes a data package in minor period of time almost real-time processing. The NLTK library is used for sentiment text analysis [16]. The results of data processing are checked by predefined user's conditions. If it satisfies conditions, the solution creates an event to update dashboard's status. Moreover, the real-time solution includes a procedure for listing to events. If a certain event occurred, the dashboard will be updated. Figure 1 presents a proposed scheme for real-time stream processing.

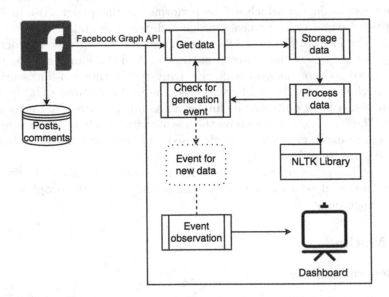

Fig. 1. A scheme for real-time stream processing

Data Retrieving. To keep a set of comments updated, a loop statement is used to receive comment data from Facebook. Time (T_u) between two loops statements is defined in advance. We set $T_u = 0.1$s as it is nearly real-time and allows to store data continually updated. However, this parameter might be changed if needed. Graph API Facebook is used in every step of the loop to obtain all comments of the selected post and to transfer data for storing and further analysis. This approach provides assurance, that collected data is fresh.

Event Generation. In the inner loop, when data is transferred for further processing, the comparison of a new segment of data, which has been obtained from Facebook recently with cached data, is made. It is crucial as it helps us find out the changes in data. If changes in data are discovered, the event needs to be generated. Observers will receive this event further.

Event Observation. To evaluate the performance of the method, the program implementation includes the procedure which allows observing any generated event. In occasion of the event reflecting changing data, the solution updates system automates re-analyze sentiment data. This involves the change of dashboard status. Furthermore, cache data is updated as well to be sure we have newest data from the server. For instance, we analyse the sentiment of comments on a post about a political topic entitled 'Obama bans solitary confinement for juveniles and low-level offenders' on CNN news channel on Facebook[1].

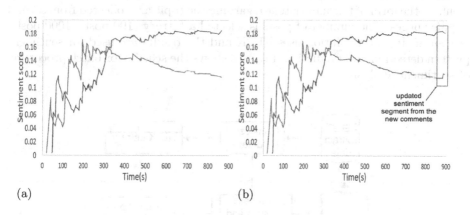

(a)   (b)

Fig. 2. Dashboard represents (a) negative/positive sentiment of the first 750 s of post's life; (b) updated results after synthetic negative comment (Color figure online)

---

[1] The title of the post: Obama bans solitary confinement for juveniles and low-level offenders, https://www.facebook.com/bbcnews/posts/10153348871732217.

Figure 2 shows the dashboard representing the negative and positive attitude in different timestamps. The red line represents values of negative sentiment. The green line is the positive's. These values are fall into the interval $[0; 1]$. Every moment t, we have a sum of negative, positive and neutral equals to 1:

$$V_t^{(p)} + V_t^{(n)} + V_t^{(u)} = 1, \tag{1}$$

where, $V^{(p)}$ - denotes the positive score, $V^{(n)}$ - denotes the negative score, $V^{(u)}$ - denotes the neutral score. In this paper, we do not place the neutral scores in the graph. The scale 'time' represents the time of comment posting as the interval from creating a post in seconds. The 'sentiment analysis' scale is the polarity value of comments.

The first figure explains the behavior for every 780 s of post life and the second one, reaction on the posted synthetic negative comment. Real-time analysis allows to detect current patterns and to compare obtained pattern with expected or required. However, finding and adjusting those references patterns depends on expert (human) intervention and due to high velocity and the variety of data, this procedure is very costly.

3.3 Batch Data Processing

As method includes pattern detection and forecasting models, another component of our system implements pattern discovery in batch mode via processing of high volumes of certain topic data collected over a long period of time. Comments in the form of textual content regarding the topic are collected from several popular pages. The number of posts is selected arbitrary (100 posts, 1000 posts, or even more). Our solution uses NLTK and the results are used for sentiment pattern detection and prediction. Figure 6 shows the scheme for batch processing of sentiment analysis.

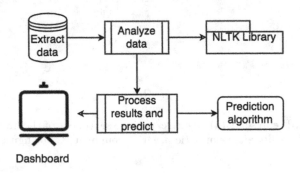

Fig. 3. Scheme for batch processing

Data Collecting. Implementation of batch data processing makes sense in the case of high volumes data. Firstly, we chose a topic, which is popular recently. For each post, using Facebook Graph API, all comments have been collected during the first 30000 s. Data is stored in flat table format (e.g. CSV file) which is easy to save in distributed file system. The header of CSV file contains the following columns: [Datetime] [Topic] [Post] [Comment] [Positive] [Negative].

The first column contains a value in seconds when a comment was appeared, which is counted from initial post's appearance. Data type of the value is the integer. The second column contains the topic's title. The third column is post's title. The fourth column is the content the text of a comment. The last two columns represent sentiment analysis scores (negative and positive) of comment on the post. The data type of scores is the float number.

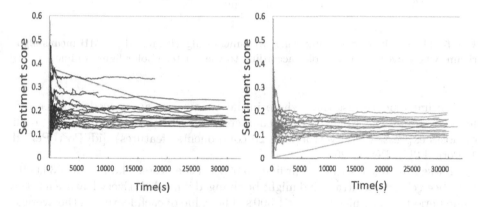

Fig. 4. Positive sentiment time series (left side) for a set of posts and negative sentiment time series (right side). Data about U.S. presidential election 2016 topics was obtained by CNN and BBC feeds

Analysis Data. As it was mentioned above, the NLTK sentiment analysis was applied for each comment for the post. As the results, we obtain time series which describes people's negative $V^{(n)}$ and positive $V^{(p)}$ scores during the time. Figure 4 expresses positive scores time series (left side) for a set of posts and the negative (right side).

Clustering Data. The output of the previous step is a set of time series and the number of time series is equal to a number of examined posts. We use the fact, that some of the time series has a similar pattern and they could be arranged into 3 or 4 different groups. It allows defining a 'typical' behaviour of the people from a sentiment point of view. For instance, the certain post could have a high negative expression in the beginning and fade negative afterwards. The well-known technique for unsupervised grouping is clustering. In this paper, two machine learning approaches: k-means and MB-means were used to cluster sentiment of posts. Needless to say, k-means has been studied and applied in a wide range of domains, e.g. transportation network analysis [9], information security [28], pattern recognition [5], text classification [19] and many others domain.

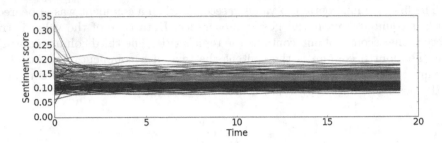

Fig. 5. The results of clustering using (a) k-means algorithm and (b) MB-means algorithm, where every plot has color according to the cluster (Color figure online)

We implement clustering, where initial time series are placed in three difference groups according to their characteristics. To perform clustering, each of posts expressed as a vector with 20 components (features): [id][Post][Period 1][Period 2]... [Period 20].

These features describe to our first 30,000 s of the posts. The values of 30,000 s have been chosen arbitrary and might be changed if needed. Every feature has the same period time, which is about 1,500 s. The value of each feature is the average value of all value in this period. All-time series (or vectors) have been divided into three clusters using k-means algorithm. The number of clusters picked up according to preliminary analysis. For comparison with k-means method, we also use the other algorithm to cluster which is named MB-means. Figure 5 shows the results of clustering, every time series has the color according to the cluster. However, in spite of difference clustering techniques, the results look quite similar.

We can describe three clusters in the following way.

– The first cluster which is red lines (based on k-means algorithm results) begins with high positive scores. It decreased quickly from the beginning to the third period, then it continuously went downward slowly. From the fifth period, it felt the lowest value. Then it remained stable in the next periods.
– With the second cluster which is green lines using k-means, we have a graph remained relatively stable from the beginning. Almost periods time, there were small increases or decreases.
– The last cluster is blue lines using k-means. It started with low positive scores. It grew up rapidly, and then it reached the peak of the score at the second period. Then there were slightly drop in the third period and leveled off. From the next period, it stayed constant and there are no more changes.

In conclusion, each cluster has characters itself obviously. It is the good signification to cluster testing data.

Prediction. Clustering allows defining typical patterns in people's behaviour. The next task is prediction the trend development of people's attitude on a post using data about the first reaction (or 5 values in vector representation). Note, we choice 5 values for the current research, but this parameter is subject to choice.

The prediction is performed in two steps. The first step including cluster detection procedure. Based on the first 5 features (or 5 values in vector representation) of given pattern which needs to be predicted, we select the appropriate cluster. The selected cluster is the cluster having the average profile for this 5 features with minimum deviation with our post (in terms of Euclidean distance). In the second step, we predict the trend development of people's attitude on this post. The nearest (in terms of distance) time series from the certain cluster is found and use as the prediction for the rest time interval. The Mean Absolute Error (MAE) is used to evaluate the performance of the forecasting technique.

$$MAE = \frac{1}{n_{test}} \sum_{i=1}^{n_{test}} \left(\frac{\sum_{j=6}^{20} |h_j - h_j^*|}{15} \right)_i , \tag{2}$$

where, h_j – denotes the real value at the j-th timestamp, h_j^* – denotes the predicted value at the j-th timestamp, n_{test} – a number of time series included in test data set.

4   Results and Discussion

To evaluate our approach we designed and implemented software solution using Python and Facebook API. For our experiments, we used the topic "the United States presidential election 2016". All data was received from two famous new channels: BBC news and CNN on Facebook. We collected 200 posts about the mentioned above topic. For every post, using Facebook Graph API, all comments have been collected during the first 30,000 s. Approximately, the total number of comments is about 100,000 comments.

The results of real-time sentiment analysis indicate user's emotions and thoughts in a real-time stream. Based on monitoring of dashboards, we can understand that the ratio of positive sentiment is higher that the negative sentiment. That means that the proportion of people who supports the news is more than the proportion who against. At the time when the intersection of the red line and the green line is observed, it's the point that the trend of attitude is changed. It might the point that perhaps people are changing the attitude to the topic. Based on these points, we can create a line which is the general trend of people's attitude. Also, it is possible to set up triggers indicates when an interested event occurs.

Using clustering techniques we are able to detect the most typical behaviour of the users and describes them. For instance, we observe that negative or positive estimations asymptotically approaching to a certain level and never exceed the threshold. Negative and positive attitude is fading during the time, and we are able to estimate time of popularity of the post and advise actions to support popularity. Also, our technique allows predicting the trend development of people's attitude. It could be a framework to detect the outliers in the comment of Facebook's community. To evaluate forecasting performance MAE has been applied as error measurement. To avoid a case where results obtained by chance, we developed cross validation (with folds $= 10$) and get average MAE $= 0.008$. Figure 6 gives the representation of results of trend forecasting. The green line is the real comment sentiment on this post. And the red line is the prediction line for the development of peoples attitude on the post.

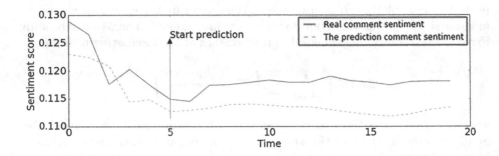

Fig. 6. Prediction trending development of positive sentiment

5   Conclusion

In this study, we perform actions to understand users preferences and attitude based on sentiment analysis of Facebook comments and application of machine learning techniques. Detection of laws and consistent patterns in users comments published in time framework allows providers of services and sales to react in real time and be more proactive using trend prediction. We propose a new method based on sentiment text analysis for detection and prediction negative and positive patterns for Facebook comments which combines (i) real-time sentiment text analysis for pattern detection and (ii) batch data processing for creating forecasting algorithms. To perform forecast we propose two-steps algorithm where: (i) patterns are clustered using unsupervised clustering techniques and (ii) trend prediction is performed based on finding the nearest pattern from a certain cluster.

Based on the results, we found three types of user behavior in their opinion expression and find that our simple forecasting technique is very accurate. Proposed method can be readily used in practice by sales companies who can use the real-time approach for learning their customer attitude about products and making the assessment of a product. Some social and political organizations use to analyze community on a certain event such as The 2016 U.S. election. Our future work will be continued by focusing on improvement the model training, improvement method of prediction using MAE. Besides, the next stage of our research will analyze a group of people such as Vietnamese and Russian [17,18].

Acknowledgments. The reported study was partially supported by RFBR, research project No. 16-37-60066 and research project MD-6964.2016.9.

References

1. Agarwal, A., Xie, B., Vovsha, I., Rambow, O., Passonneau, R.: Sentiment analysis of twitter data. In: Proceedings of the Workshop on Languages in Social Media, LSM 2011, pp. 30–38. Association for Computational Linguistics, Stroudsburg (2011)
2. Bermingham, A., Smeaton, A.: Classifying sentiment in microblogs: is brevity an advantage is brevity an advantage? In: 19th ACM International Conference on Information and Knowledge Management, pp. 1833–1836. ACM (2010)
3. Cheng, O.K.M., Lau, R.: Big data stream analytics for near real-time sentiment analysis. J. Comput. Commun. **3**, 189–195 (2015)
4. Choy, M., Cheong, M.L.F., Laik, M.N., Shung, K.P.: A sentiment analysis of Singapore Presidential Election 2011 using Twitter data with census correction. arXiv:1108.5520 (2011)
5. Dash, M., Liu, H.: Feature selection for classification. Intell. Data Anal. **1**(3), 131–156 (1997)
6. Eagle, N., Pentland, A.: Reality mining: sensing complex social systems. Pers. Ubiquit. Comput. **10**(4), 255–268 (2006)
7. Gamon, M.: Sentiment classification on customer feedback data: noisy data, large feature vectors, and the role of linguistic analysis. In: 20th International Conference on Computational Linguistics, p. 841-es. Association for Computational Linguistics, Stroudsburg (2004)
8. Godbole, N., Srinivasaiah, M., Skiena, S.: Large-scale sentiment analysis for news and blogs. In: ICWSM, vol. 7, no. 21, pp. 219–222 (2007)
9. Golubev, A., Chechetkin, I., Solnushkin, K.S., Sadovnikova, N., Parygin, D., Shcherbakov, M.V.: Strategway: web solutions for building public transportation routes using big geodata analysis. In: 17th International Conference on Information Integration and Web-Based Applications and Services (iiWAS 2015), pp. 91:1–91:4. ACM, New York (2015)
10. Grömping, M.: Echo chambers: partisan facebook groups during the 2014 Thai election. Asia Pacific Media Educ. **24**, 39–59 (2014)
11. Hatzivassiloglou, V., McKeown, K.R.: Predicting the semantic orientation of adjectives. In: 35th ACL and 8th EACL, pp. 174–181. ACL, Somerset (1997)
12. Honeycutt, C., Herring, S.C.: Beyond microblogging: conversation and collaboration via twitter. In: 42nd Hawaii International Conference on System Sciences, pp. 1–10 (2009)
13. Jansen, B.J., Zhang, M., Sobel, K., Chowdury, A.: Twitter power: tweets as electronic word of mouth. J. Am. Soc. Inform. Sci. Technol. **60**, 1–20 (2009)

14. Kwak, H., Lee, C., Park, H., Moon, S.: What is twitter, a social network or a news media? In: 19th International Conference on World Wide Web, pp. 591–600 (2010)

15. Lau, R.Y.K., Xia, Y., Ye, Y.: A probabilistic generative model for mining cyber-criminal networks from online social media. IEEE Comput. Intell. Mag. **9**, 31–43 (2014)

16. Natural Language Toolkit. http://www.nltk.org/

17. Nguyen, V.H., Nguyen, H.T., Snasel, V.: Normalization of Vietnamese tweets on twitter. In: Abraham, A., Jiang, X.H., Snášel, V., Pan, J.-S. (eds.) ECC 2015. AISC, vol. 370, pp. 179–189. Springer International Publishing, Cham (2015)

18. Nguyen, H.T., Duong, P.H., Le, T.Q.: A multifaceted approach to sentence similarity. In: Huynh, V.N., Inuiguchi, M., Demoeux, T. (eds.) IUKM 2015. LNCS, vol. 9376, pp. 303–314. Springer, Heidelberg (2015). doi:10.1007/978-3-319-25135-6_29

19. Pang, B., Lee, L., Vaithyanathan, D.: Thumbs up? Sentiment classification using machine learning techniques. In: ACL-02 Conference on Empirical Methods in Natural Language Processing (EMNLP 2002), vol. 10, pp. 79–86. Association for Computational Linguistics, Stroudsburg (2002)

20. Pang, B., Lee, L.: Opinion mining and sentiment analysis. Found. Trends Inf. Retr. **2**, 1–135 (2008)

21. The Top 20 Valuable Facebook Statistics, December 2015. https://zephoria.com/top-15-valuable-facebook-statistics/

22. Trinh, S., Nguyen, L., Vo, M., Do, P.: Lexicon-based sentiment analysis of facebook comments in Vietnamese language. In: Król, D., Madeyski, L., Nguyen, N.T. (eds.) Recent Developments in Intelligent Information and Database Systems. SCI, vol. 642, pp. 263–276. Springer, Cham (2016)

23. Tumasjan, A., Sprenger, T.O., Sandner, P.G., Welpe, I.M.: Predicting elections with twitter: what 140 characters reveal about political sentiment. In: Fourth International AAAI Conference on Weblogs and Social Media, pp. 178–185. AAAI (2010)

24. Turney, P.D.: Thumbs up or thumbs down? Semantic orientation applied to unsupervised classification of reviews. In: 40th Annual Meeting of the ACL (ACL 2002), pp. 417–424. ACL, Philadelphia (2002)

25. Wang, H., Can, D., Kazemzadeh, A., Bar, F., Narayanan, S.: A system for real-time twitter sentiment analysis of 2012 U.S. presidential election cycle. In: The ACL 2012 System Demonstrations, pp. 115–120. Association for Computational Linguistics, Stroudsburg (2012)

26. Weinstein, C., Campbell, W.M., Delaney, B.W., OLeary, G.: Modeling and detection techniques for counter-terror social network analysis and intent recognition. In IEEE Aerospace Conference, pp. 1–16. IEEE Press (2009)

27. Williams, C., Gulati, G.: What is a social network worth? Facebook and vote share in the 2008 presidential primaries. In: Annual Meeting of the American Political Science Association, Boston, MA, pp. 1–17 (2008)

28. Zhang, Y., Dang, Y., Chen, H., Thurmond, M., Larson, C.: Automatic online news monitoring and classification for syndromic surveillance. Decis. Support Syst. **47**(4), 508–517 (2009)

Aspect-Based Sentiment Analysis Using Word Embedding Restricted Boltzmann Machines

Bao-Dai Nguyen-Hoang, Quang-Vinh Ha, and Minh-Quoc Nghiem$^{(\boxtimes)}$

Faculty of Information Technology, Ho Chi Minh City University of Science,
227 Nguyen Van Cu Street, Ward 4, District 5, Ho Chi Minh City, Vietnam
{1212069,1212304}@student.hcmus.edu.vn, nqminh@fit.hcmus.edu.vn

Abstract. Recent years, many studies have addressed problems in sentiment analysis at different levels, and building aspect-based methods has become a central issue for deep opinion mining. However, previous studies need to use two separated modules in order to extract aspect-sentiment word pairs, then predict the sentiment polarity. In this paper, we use Restricted Boltzmann Machines in combination with Word Embedding model to build the joined model which not only extracts aspect terms appeared and classifies them into respective categories, but also completes the sentiment polarity prediction task. The experimental results show that the method we use in aspect-based sentiment analysis tasks is better than other state-of-the-art approaches.

Keywords: Aspect-based sentiment analysis · Opinion mining · Restricted Boltzmann Machine · Supervised learning · Word Embedding

1 Introduction

Sentiment Analysis (also known as opinion mining) is the process of determining whether a piece of writing is positive or negative. With the development of opinionated user-generated review sites, many customers can write reviews and express their opinions about the products (or services). Sentiment Analysis could help not only users to choose the right products but also companies to improve their products based on these reviews.

Aspect-based Sentiment Analysis (ABSA) has received much attention in recent years since each review might contain many aspects. For example, in a restaurant review, we may have opinions about *food, staff, ambience*, etc. Conventional ABSA systems normally have two separated modules: one for aspect extraction and another one for sentiment classification [1–3]. Recently, Wang et al. [4] introduced a joint model, called Sentiment-Aspect Extraction based on Restricted Boltzmann Machines (SERBM), that extract aspects and classify sentiments at the same time. In this model, they used unsupervised Restricted Boltzmann Machine (RBM) and three different types of hidden units to represent aspects, sentiments, and background information, respectively. Furthermore, they added prior knowledge into this model to help it acquire more accurate feature representations. The visible layer **v** of SERBM is represented as a

© Springer International Publishing Switzerland 2016
H.T. Nguyen and V. Snasel (Eds.): CSoNet 2016, LNCS 9795, pp. 285–297, 2016.
DOI: 10.1007/978-3-319-42345-6_25

$K \times D$ matrix, where K is the dictionary size and D is the document length. They showed that their model is well-suited for solving aspect-based sentiment analysis tasks.

However, there are two main problems still exist in the SERBM model. Firstly, an unsupervised method can only cluster reviews into categories and we can not know the name of the category. No information was given to determine which position of the hidden units to represent aspects, sentiments or background words during the training process. Secondly, a visible layer will be a matrix combined by high-dimensional vectors if training data has a large set of vocabulary, which requires much computational resource.

In this paper, we propose combining Restricted Boltzmann Machine with Word Embedding model to overcome the limitations of existing method. Word Embedding model has the capability of reducing the dimensionality of the input vectors. Therefore, we can use it to reduce the dimensionality of the input in visible layer while keeping the semantics of the reviews. We encode the input document as a vector, created by the Word Embedding model, instead the vector of one-hot encoding Bag Of Words model. Furthermore, we use RBM in a supervised setting. We move the output component from hidden layer to visible layer. The hidden layer now acts as the dependencies between the components in the visible layer. Doing like this, we can fix the units for desired categories.

We call our model Word Embedding Restricted Boltzmann Machine (WE-RBM). Overall, our main contributions are as follows:

- This is the first work that combines Word Embedding model and supervised RBM for the ABSA task. Compared with other state-of-the-art methods, our model can identify aspects and sentiments efficiently, yielding 1–6 % improvements in accuracy for sentiment classification task and 1.73 % to 7.06 % improvements in F1 score for aspect extraction task.
- By using Word Embedding model, we can efficiently reduce the size of input vectors up to 100 times, which in turn reduces the training time greatly.
- We also introduce a simple yet efficient way to incorporate prior knowledge into RBM model. Prior knowledge is the advantage of Word Embedding model, which can help RBM to be well-suited for solving aspect-based opinion mining tasks.

The rest of this paper is organized as follows. Section 2 introduces the related work. Section 3 overviews the background information, then describes our approach to classify reviews into aspect categories and predict sentiment polarity of the reviews. Experimental results are presented in Sect. 4. Finally, Sect. 5 concludes the paper and discusses future work.

2 Related Work

ABSA approaches may be divided into three main categories: rule-based, supervised learning, and unsupervised learning.

Rule-based approaches [1,5] can perform quite well in a large number of domains. They use a sentiment lexicon, expressions, rules of opinions, and the sentence parse tree to help classify the sentiment orientation on each aspect appeared in a review. They also consider sentiment shifter words (i.e. *not, none, nobody*, etc.). However, these rule-based methods have a shortcoming in processing complex documents where the aspect is hidden in the sentence, and failing to group extracted aspect terms into categories.

For the supervised learning approach, Wei and Gulla [6] propose a hierarchical classification model to determine the dependency and the other relevant information in the sentence. Jiang et al. [7], Boiy and Moens [8] use dependency parser to generate a set of aspect-dependent features for classification, which weighs each feature based on the position of the feature relative to the target aspect in the parse tree. Several other supervised learning models have been published, such as Hidden Markov Models [9], SVMs [10], Conditional Random Fields [11,12].

It has been demonstrated that a classifier trained from labeled data in one domain often performs poorly in another domain [13]. Hence, unsupervised methods are often adopted to avoid this issue. Several recent studies investigate statistical topic models which are unsupervised learning methods. They assume that each document consists of a mixture of topics. Specifically, Latent Dirichlet Allocation (LDA) [14], Multi-Grain LDA model [15] are used to model and extract topics from document collections. A number of authors have considered the effects of topic models on ABSA task, such as the two-step approach [16], joint sentiment/topic model [17], and topic-sentiment mixture model [18]. A recent study by Wang et al. [4] proposed the SERBM model which also jointly address these two tasks in an unsupervised setting.

3 Proposed Method

3.1 Background

Restricted Boltzmann Machine. RBM model, a generative stochastic artificial neural network, is treated as a model in the field of deep learning. RBM can be used to learn important aspects of an unknown probability distribution based on samples from this distribution [19]. Recently, researchers have applied RBM in the field of Natural Language Processing, including topic modeling and sentiment analysis [4]. One special characteristic of RBM is that it can be used in both supervised and unsupervised ways, depending on the task.

As shown in Fig. 1, RBM model is a two-layer neural network which contains one visible layer and one hidden layer. The visible layer is constituted by visible units correspond to the components of an observation (e.g., one visible unit for each word in an input document). The hidden layer composed of hidden units which model the dependencies between the components of observations (e.g., dependencies between words in the document).

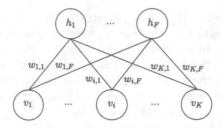

Fig. 1. The network graph of an RBM model with K visible and F hidden units

Word Embedding Model. The Word Embedding model (WEM) is a proven and powerful paradigm in Natural Language Processing, in which words are represented as vectors in a high-dimensional space [20]. The idea of this model is representing words as vectors. Therefore, the similarity of two words i and j can be calculated based on the cosine of the angle between the vectors as shown in Eq. 1.

$$\cos \theta = \frac{w_i.w_j}{||w_i||||w_j||} \tag{1}$$

where $w_i.w_j$ is the dot product of these two documents while $||w_i||$ and $||w_j||$ are the norm of vector w_i and w_j, respectively.

For document representation, every word appeared in the document is represented as a vector. We then sum all of these vectors to get a vector that represent the document [20].

3.2 Our Aspect-Based Sentiment Analysis Model

Structure. In the previous approach, Wang et al. [4] proposed an unsupervised RBM model to ABSA. As mention before, there are two main shortcomings in this SERBM model. Firstly, an unsupervised method can only cluster reviews into categories and we can not know the name of the category. This model fixes hidden units 0–6 to represent the target aspects *Food, Staff, Ambience, Price, Ambience, Miscellaneous,* and *Other Aspects,* respectively. There is no way we can determine the aspects (e.g. which unit represents *Food,* which unit represents *Staff*) based on the position of the hidden units alone.

Secondly, when training, each document is transformed into a $K \times D$ matrix **v**, where K is the dictionary size, and D is the document length. If visible unit i in **v** takes the k-th value, v_i^k is set to 1. If the training data has a large set of vocabulary, the visible layer will be a matrix combined by high-dimensional vectors. These sparse input vectors lead to not only the decrease in the model's accuracy but also the increase in computational resources.

To overcome these problems, we propose a method using supervised RBM model which is illustrated in Fig. 2. The ***Output units*** include the units represent the aspects and sentiment orientations of the reviews. We call them *Aspects identifying units* and *Sentiments identifying units,* respectively. These units are put together with **Input units** in the visible layers, instead of being placed in hidden layers. Meanwhile, the units in the hidden layer represent the relationship between the units in the visible layers. We encode input units in the visible layer as a vector, created by the Word Embedding model [21], instead the vector of one-hot encoding Bag Of Words model. This helps reduce the dimensionality of the input matrix while keeping documents' semantics.

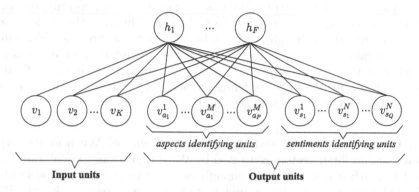

Fig. 2. The network graph of our Supervised Sentiment-Aspect Extraction RBM model

Compared to standard RBMs, the first difference of this model is the visible layer. Apart from the input units, there are also aspect and sentiment identifying units that represent the output component of the model. Suppose in the training data, there are reviews talking about P aspects and having Q sentiment orientations in total. In particular, the set of aspects is $A = \{a_1, a_2, ..., a_P\}$, and the set of sentiment orientations is $S = \{s_1, s_2, ..., s_Q\}$.

For each aspect, the model set M units to capture that aspect. For example, if the review mentioned aspect $a_i \in A$, we set the units from $v_{a_i}^1$ to $v_{a_i}^M$ to 1, and 0 otherwise. We use the same setting for sentiment units. The model set N units to capture each sentiment polarity. If the sentiment polarity is $s_j \in S$, we set the units from $v_{s_j}^1$ to $v_{s_j}^N$ to 1, and 0 otherwise. The idea of setting M and N units to capture aspects and sentiment polarity is obtained from previous work [19]. Our model has the property that input units and output units stay in the same layer. If we used only one output unit for each aspect and sentiment polarity while there are too many input units, the model would be imbalance and need more time to converge.

With this structure, our model can solve two tasks simultaneously: aspect identification and sentiment classification. This ability of our model is reflected in the aspect and sentiment identifying units which play important roles in the sampling process of RBM. Meanwhile, weight values of connecting edges contain semantic information of the review, which help identifying units to communicate with the hidden layer. In addition, the fixed dimension of the input vector does not depend on the number of vocabulary words. This capability not only helps the model prevent the occurrence of decreasing speed and accuracy, but also solves the semantic problem in the review.

For the customer restaurant reviews analyzing task, one word in a review may mention a certain aspect (e.g. "delicious" corresponds to *food* aspect), or a certain opinion (e.g. "good" is about positive sentiment, while "bad" is about negative sentiment). Furthermore, there are other words do not mention about aspect or sentiment, they can be removed during the preprocessing phase. These latent topics in a review are considered as the factors which generate aspect and sentiment words within that review. To technically illustrate this, hidden layer contains information of the latent topics. In the generation process, hidden units produce the values of visible units, which are also the information of the aspect and sentiment words in reviews. Moreover, we can increase the number of hidden units in order to increase the modeling capacity of the RBM, which make the model powerful enough to represent complicated distributions.

Word Embedding Restricted Boltzmann Machine. When we use supervised RBM with input vectors generated by Word Embedding model, the results showed that RBM does not have enough capacity to regenerate visible units. Much of the instability in this approach stems from two reasons. Firstly, RBM has only hidden and visible layers to capture the diversity of semantic of documents. Secondly, in visible and hidden units, the values vary continuously by each epoch in the training process. This leads to loss of semantic information of documents. One way to solve this problem is increasing the number of hidden units for RBM to accommodate more information. However, increasing the number of hidden units causes the model to increase the number of visible units. This leads to the problem encountered in the previous approach [4], which is insufficient computational resources.

Therefore, we propose Word Embedding Restricted Boltzmann Machine (WE-RBM) to take full advantage of WEM's structure and supervised RBM. Beside input units, WE-RBM also uses prior knowledge obtained from the vector comparing process of WEM. Hence, the model would have more information to make up for the loss in the training process. Our WE-RBM model is expressed in Fig. 3.

Assuming that WEM has been trained independently before, we put every document into WEM to generate the corresponding vector. The dimension of each vector is equal to the number of input units of RBM. After document vectors generation process, we make the training phase for the RBM model in a supervised setting. This phase helps the model be able to understand the vectors' pattern generated by WEM.

In the testing phase, each document is also converted into a vector by WEM. Unlike previous work, the aspect identification and sentiment classification task are firstly performed by WEM instead of RBM model. Particularly, WE-RBM computes the cosine similarity score (shown in Eq. 1) between the document vector and each of all aspect and sentiment vectors, which are also generated by WEM. The prior knowledge is determined by the highest similar pair of vectors.

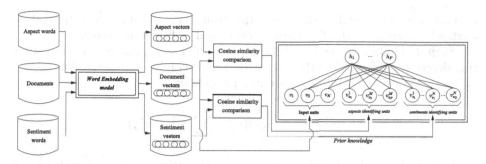

Fig. 3. WE-RBM model overview. Double boxed items are main components in WE-RBM model.

For instance, a document d can be categorized into one of these aspects $S = a_1, a_2, ..., a_P$. Let us call the vector form of document or aspect x $vec(x)$. Prior knowledge would be labeled i if cosine similarity between $vec(d)$ and $vec(a_i)$ is the highest one. The model performs similarly with sentiment classification task. This prior knowledge can help improve the ability of the model to extract aspects and identify sentiments.

One characteristic of WEM is two vectors which represent two different words would have high cosine similarity if these two words have close meaning to each other [21,22] (e.g. "strong" may have close meaning with "powerful", whereas "strong" and "London" are more distant). Inspired by this, we use WEM to categorize aspect and classify sentiment before feeding the input vector into RBM model. Hence, each document can have its own prior knowledge before RBM's classification. The final classification result is determined by both WEM and RBM. Therefore, we call our novel combination model is WE-RBM.

There are three main reasons we propose this novel WE-RBM model to solve the ABSA's tasks. Firstly, when a document is converted into vector form, it can be compared with other documents by cosine similarity while semantic information is conserved. This is the advantage of word representation technique that WE-RBM use in the classification process. Secondly, WE-RBM can save computational resources and processing time. The model does not need to adapt input units for huge dictionary size thanks to fixed dimension of the input vector. Using prior knowledge instead of increasing hidden units, WE-RBM can handle big training set in reduced processing time. Last but not least, our WE-RBM model also has the capability of jointly modeling aspect and sentiment information together.

Training and Testing. In the training process, Contrastive Divergence (CD), also called Approximate Gradient Descent, is the way that helps RBM learn the connection weights in the network. CD has two phases, which are Positive phase and Negative phase. In each phase, we compute the positive and negative value by the multiplication of visible and hidden units. Then, we update the connection weight based on the subtraction from positive value of negative value. Particularly, each epoch of CD can be expressed in five steps below:

Step 1. Update the states of the hidden units using the logistic activation rule described in Eq. 2. For the j-th hidden unit, compute its activation energy and set its state to 1 with corresponding probability.

$$P(h_j = 1|\mathbf{v}) = sigm(a_j + \sum_{i=1}^{D} \sum_{k=1}^{K} v_i^k W_{ij}^k) \tag{2}$$

Step 2. For each connection edge e_{ij}, get the value from positive phrase by Eq. 3.

$$pos(e_{ij}) = v_i h_j \tag{3}$$

Step 3. Reconstruct the visible units in a similar manner by using the logistic activation rule described in Eq. 4. For the i-th visible unit, compute its activation energy and set its state to 1 with corresponding probability. Then do **Step 1** to update the hidden units again.

$$P(v_i^k = 1|h) = sigm(b_i^k + \sum_{j=1}^{F} h_j W_{ij}^k) \tag{4}$$

Step 4. For each connection edge e_{ij}, get the value from negative phrase by by Eq. 5.

$$neg(e_{ij}) = v_i h_j \tag{5}$$

Step 5. Update the connection weights W_{ij} by Eq. 6.

$$W_{ij} = W_{ij} + \mathbf{lr}(pos(e_{ij}) - neg(e_{ij})) \tag{6}$$

where $sigm(x) = \frac{1}{(1+e^{-x})}$ is the logistic function, and **lr** is a learning rate.

After m epochs of transfer between visible and hidden layers in a CD-m run of the above steps, values of the hidden units reflect the relationship between the visible units in the model. The connection weight matrix between the two layers helps hidden layer generate visible units which include input units, aspect identifying units and also sentiment identifying units.

In the testing process, we convert all documents into vector form using WEM. Then, each document vector is compared with each aspect vector by cosine similarity score. The most similar pair of vectors gives the model prior knowledge about the aspect label of this document. Final aspect label of the document is determined based on the result of aspects identifying units after WE-RBM generation process.

4   Experimental Results

In this section, we present two experiments to evaluate the performance of our model on the aspect identification and sentiment classification tasks.

4.1 Data

To evaluate our model performance, we used Restaurant Review Dataset [23]. This data is also used widely in previous work [16,23,24]. Data contain reviews about the restaurant with 1,644,923 tokens and 52,574 documents in total. Documents in this dataset are annotated with one or more labels from a gold standard label set $S = \{Food, Staff, Ambience, Price, Anecdote, Miscellaneous\}$.

4.2 Aspect Extraction

Experimental Setup. Following the previous studies (Brody and Elhadad [16] and Zhao et al. [24]), reviews with less than 50 sentences are chosen. From that, we only use sentences with a single label for evaluation to avoid ambiguity. These sentences are selected from reviews with three major aspects chosen from the gold standard labels $S' = \{Food, Staff, Ambience\}$. After choosing suitable sentences from the data, we have 50303, 18437, 10018 sentences which labeled Food, Staff, Ambience. Then, we change the case of any document to lower case and remove stop words.

To convert each word in the sentence into vector form, we use Word Embedding technique [21] combined with Google pre-trained model[1]. This model has been trained on part of Google News dataset (about 100 billion words). It contains 300-dimensional vectors for 3 million words and phrases. Each sentence's representation is a vector which is a sum of vectors that represent for words appeared in that sentence [20].

We use 300 visible units in our WE-RBM as aspects identifying units, where units 1–100 capture *Food* aspect, units 101–200 capture *Staff* aspect and units 201–300 capture *Ambience* aspect. For sentiment classification, we also use 200 visible units, where units 1–100 capture positive information, units 101–200 represent negative information. Initially, these units are set to 0. After Gibbs sampling process, we create a sum for each group of 100 units to determine the aspect and sentiment polarity appeared in the document.

Evaluation. To evaluate the model's performance, we use Precision, Recall and F_1 scores for each aspect identification on Restaurant Review Dataset.

As a baseline, we implement *Prior knowledge only* method, which only uses cosine similarity of the document vector and aspect vector to identify aspect. For detail, we generate the vectors of "food", "staff" and "ambience" words using WEM. Then, we compute the cosine similarity between the vector of document d_i and each of those 3 aspect vectors. The document will be put into aspect a_i category if the cosine similarity between $vec(d_i)$ and $vec(a_i)$ is the highest one.

We also use SVM, a well-known machine learning technique [25], to compare with our WE-RBM model. We implement SVM by LIBSVM[2] which is an integrated tool for support vector classification. With linear kernel, SVM model

---

[1] https://code.google.com/archive/p/word2vec/.
[2] https://www.csie.ntu.edu.tw/~cjlin/libsvm/.

is trained with feature vectors generated by bag-of-words model. Without prior knowledge, standard RBM using Word Embeddings is also re-implemented. This method processes the same Restaurant Review Dataset and identifies aspects for every document in this dataset under the same experimental conditions. Evaluation results for aspect identification are given in Table 1.

Table 1. Aspect identification results in terms of precision, recall, and F_1 scores on the restaurant reviews dataset

Aspect	Method	Precision	Recall	F_1
Food	Prior knowledge only	88.09	87.60	87.84
	SVM	**94.10**	75.74	83.93
	RBM + Word Embeddings	74.58	14.79	24.68
	WE-RBM	84.04	**94.43**	**88.93**
Staff	Prior knowledge only	**95.96**	45.39	61.63
	SVM	90.25	**78.14**	**83.76**
	RBM + Word Embeddings	27.73	53.32	36.48
	WE-RBM	88.27	65.28	75.06
Ambience	Prior knowledge only	16.02	**86.93**	27.06
	SVM	29.15	82.70	43.11
	RBM + Word Embeddings	14.85	60.12	23.81
	WE-RBM	**69.22**	59.64	**64.07**

Discussion. Considering the results from Table 1, we find that WE-RBM performs better than other methods. Specifically, it is evident that our WE-RBM model outperforms previous methods' F_1 scores on *Food* and *Ambience* aspects. Compared with Prior knowledge only, the F_1 scores improve by 1.09 %, 13.43 % and 37.01 % respectively, for the *Food*, *Staff*, and *Ambience* aspects. This result proves that our WE-RBM model is not entirely based on Prior knowledge obtained from WEM to have the better performance. Inheriting RBM's ability in modeling latent topics and WEM's capability in identifying aspects, WE-RBM model can achieve higher Precision and Recall scores for the imbalanced dataset. Compared with RBM using Word Embeddings, the F_1 scores yield relative improvements by 63.25 %, 38.58 %, and 40.26 % respectively, on the same aspects. This result reveals that RBM model performs badly without the prior knowledge obtained from WEM.

Comparing with SVM performance, precision scores in food and staff domains of SVM are higher than WE-RBM's. But SVM can not address the imbalanced data problem, which results in reduced precision in *Ambience* domain. Moreover, WE-RBM is a joint model which has ability to identify aspects and classify sentiment polarities simultaneously, while SVM is a classification model and we have to train two separated models to solve these two tasks.

In our model's performance evaluation process, we can not re-implement SERBM due to insufficient computational resources. However, we compare WE-RBM's result with SERBM's result which was presented by Wang et al. [4]. With the same settings in aspect identification task on Restaurant Review Dataset, WE-RBM outperforms SERBM with improvement in F_1 scores by 1.73 % and 7.06 % on *Food* and *Staff* aspects, respectively.

4.3   Sentiment Classification

Sentiment classification task is modified similar to aspect identification task in our WE-RBM model. We assign a sentiment score to every document in the Restaurant Review Dataset based on the output of WE-RBM's sentiment identifying units in the visible layer. Then we use SentiWordNet[3], a famous lexical resource for opinion mining [26], and adapt SVM to compare the result with our WE-RBM model.

Following the previous study [4], we consult SentiWordNet to obtain a sentiment label for every word and aggregate these to judge the sentiment information of an entire review document in terms of the sum of word-specific scores. For SVM, we use the linear kernel to train the model and the other setting is the same as WE-RBM's. Table 2 shows the comparison between SentiWordNet, SVM and WE-RBM with Accuracy as the evaluation metric.

As we can observe in Table 2, the best sentiment classification accuracy result is 79.79 % achieved by WE-RBM. Compared with two baselines, our WE-RBM yields a relative improvement in the overall accuracy by 6.43 % over SentiWordNet and by 1.53 % over SVM. Comparing the result of WE-RBM with SERBM's which was presented by Wang et al. [4], WE-RBM increases the classification accuracy by 0.99 %.

The reason for better performance of WE-RBM compared with other methods is the combination of prior knowledge and generating process of RBM. Prior knowledge can be considered as the first classification phase of WE-RBM. This knowledge boosts WE-RBM classifier in both aspect and sentiment identification.

Table 2. Accuracy of SentiWordNet, SVM and WE-RBM on sentiment classification task

Method	Accuracy
SentiWordNet	73.36
SVM	78.26
WE-RBM	**79.79**

---

[3] http://sentiwordnet.isti.cnr.it.

5   Conclusion and Future Work

In this paper, we have proposed Word Embedding Restricted Boltzmann Machines model to jointly identify aspect categories and classify sentiment polarities in the supervised setting. Our approach modifies standard RBM model and combine it with WEM, which not only helps reduce dimension of the input vectors but also gives the RBM model prior knowledge for more accurate classification. Our experimental results show that this model can outperform state-of-the-art models.

In the future, we plan to collect Vietnamese review dataset with diverse domains and combine our WE-RBM with stacked RBMs to form Deep Belief Networks. We hope that it will solve the aspect identification and sentiment classification tasks as well as RBM family methods.

Acknowledgments. This research is supported by research funding from Honors Program, University of Science, Vietnam National University - Ho Chi Minh City.

References

1. Hu, M., Liu, B.: Mining and summarizing customer reviews. In: Proceedings of KDD, pp. 168–177 (2004)
2. Blair-goldensohn, S., Neylon, T., Hannan, K., Reis, G.A., Mcdonald, R., Reynar, J.: Building a sentiment summarizer for local service reviews. In: NLP in the Information Explosion Era (2008)
3. Kumar, R.V., Raghuveer, K.: Web user opinion analysis for product features extraction and opinion summarization. Int. J. Web Semant. Technol. **3**, 69–82 (2012)
4. Wang, L., Liu, K., Cao, Z., Zhao, J., de Melo, G.: Sentiment-aspect extraction based on Restricted Boltzmann Machines. In: Proceedings of the 53rd Annual Meeting of the ACL, pp. 616–625 (2015)
5. Ding, X., Liu, B., Yu, P.S.: A holistic lexicon-based approach to opinion mining. In: Proceedings of the Conference on Web Search and Web Data mining, pp. 231–240 (2008)
6. Wei, W., Gulla, J.A.: Sentiment learning on product reviews via sentiment ontology tree. In: Proceedings of the 48th Annual Meeting of the ACL, pp. 404–413 (2010)
7. Jiang, L., Yu, M., Zhou, M., Liu, X., Zhao, T.: Target-dependent twitter sentiment classification. In: Proceedings of the 49th Annual Meeting of the ACL, pp. 151–160 (2011)
8. Erik, B., Moens, M.F.: A machine learning approach to sentiment analysis in multilingual web texts. Inf. Retrieval **12**, 526–558 (2009)
9. Jin, W., Ho, H.H.: A novel lexicalized hmm-based learning framework for web opinion mining. In: Proceedings of the International Conference on Machine Learning, pp. 465–472 (2009)
10. Varghese, R., Jayasree, M.: Aspect based sentiment analysis using support vector machine classifier. In: Advances in Computing, Communications and Informatics (ICACCI), pp. 1581–1586 (2013)

11. Yejin, C., Cardieo, C.: Hierarchical sequential learning for extracting opinions and their attributes. In: Proceedings of the Annual Meeting of the ACL, pp. 269–274 (2010)
12. Jakob, N., Gurevych, I.: Extracting opinion targets in a single-and cross-domain setting with conditional random fields. In: Proceedings of the Conference on Empirical Methods in NLP, pp. 1035–1045 (2010)
13. Liu, B.: Sentiment Analysis and Opinion Mining. Morgan and Claypool, San Rafael (2012)
14. Blei, D.M., Ng, A.Y., Jordan, M.I.: Latent Dirichlet Allocation. J. Mach. Learn. Res. **3**, 993–1022 (2003)
15. Titov, I., McDonald, R.: A joint model of text and aspect ratings for sentiment summarization. In: Proceedings of the Annual Meeting of the ACL, pp. 308–316 (2008)
16. Brody, S., Elhadad, N.: An unsupervised aspect-sentiment model for online reviews. In: Proceedings of NAACL-HLT 2010, pp. 804–812 (2010)
17. Lin, C., He, Y.: Joint sentiment/topic model for sentiment analysis. In: Proceedings of the ACM International Conference on Information and Knowledge Management, pp. 375–384 (2009)
18. Mei, Q., Ling, X., Wondra, M., Su, H., Zhai, C.: Topic sentiment mixture: modeling facets and opinions in weblogs. In: Proceedings of the International Conference on World Wide Web, pp. 171–180 (2007)
19. Fischer, A., Igel, C.: Training Restricted Boltzmann Machines: an introduction. In: Progress in Pattern Recognition, Image Analysis, Computer Vision, and Applications, pp. 14–36 (2012)
20. Le, Q.V., Mikolov, T.: Distributed representations of sentences and documents. CoRR abs/1405.4053, pp. 1188–1196 (2014)
21. Řehůřek, R., Sojka, P.: Software framework for topic modelling with large corpora. In: Proceedings of the LREC 2010 Workshop on New Challenges for NLP Frameworks, ELRA, pp. 45–50 (2010)
22. Mikolov, T., Sutskever, I., Chen, K., Corrado, G.S., Dean, J.: Distributed representations of words and phrases and their compositionality. In: Burges, C.J.C., Bottou, L., Welling, M., Ghahramani, Z., Weinberger, K.Q. (eds.) Advances in Neural Information Processing Systems, vol. 26, pp. 3111–3119. Curran Associates, Inc., Red Hook (2013)
23. Ganu, G., Elhadad, N., Marian, A.: Beyond the stars: Improving rating predictions using review text content. In: Proceedings of WebDB 2009, pp. 1–6 (2009)
24. Zhao, W.X., Jiang, J., Yan, H., Li, X.: Jointly modeling aspects and opinions with a MaxEnt-LDA hybrid. In: Proceedings of EMNLP 2010, pp. 56–65 (2010)
25. Chih-Chung, C., Chih-Jen, L.: LIBSVM: a library for support vector machines. ACM Trans. Intell. Syst. Technol. **2**, 27:1–27:27 (2011). Software, http://www.csie.ntu.edu.tw/~cjlin/libsvm
26. Baccianella, S., Esuli, A., Sebastiani, F.: Sentiwordnet 3.0: an enhanced lexical resource for sentiment analysis and opinion mining. In: Proceedings of the Seventh Conference on International Language Resources and Evaluation (LREC 2010), pp. 2200–2204 (2010)

Lifelong Learning for Cross-Domain Vietnamese Sentiment Classification

Quang-Vinh Ha$^{(\boxtimes)}$, Bao-Dai Nguyen-Hoang, and Minh-Quoc Nghiem

Faculty of Information Technology, Ho Chi Minh City University of Science,
227 Nguyen Van Cu Street, Ward 4, District 5, Ho Chi Minh City, Vietnam
{1212304,1212069}@student.hcmus.edu.vn, nqminh@fit.hcmus.edu.vn

Abstract. This paper proposes an improvement to lifelong learning for cross-domain sentiment classification. Lifelong learning is to retain knowledge from past learning tasks to improve the learning task on a new domain. In this paper, we will discuss how bigram and bag-of-bigram features integrated into a lifelong learning system can help improve the performance of sentiment classification on both Vietnamese and English. Also, pre-processing techniques specifically for our cross-domain, Vietnamese dataset will be discussed. Experimental results show that our method achieves improvements over prior systems and its potential for cross-domain sentiment classification.

Keywords: Sentiment classification · Vietnamese · Supervised learning · Lifelong learning

1 Introduction

The rapid growth of e-commerce and the Web age quickly makes the sentiment knowledge become an advantage to contribute more values to market predictions. Sentiment analysis remains a popular topic for research and developing sentiment-aware applications [1]. Sentiment classification, which is a subproblem of sentiment analysis task, is the task of classifying whether an evaluative text is expressing a positive, negative or neutral sentiment. In this paper, we focus on document-level binary sentiment classification, in which the sentiment is either positive or negative.

In recent years, most studies on sentiment classification adopt machine learning and statistical approaches [2]. Such approaches hardly perform well on real-life data, which contains opinionated documents from domains different from the domain used to train the classifier. To overcome this limitation, lifelong learning [3], transfer learning [4], self-taught learning [5] and other domain adaptation techniques [4] were proposed. All mentioned methods is to transfer the knowledge gained from source domains to improve the learning task on the target domain.

Chen et al. [3] proposed a novel approach of lifelong learning for sentiment classification, which is based on Naïve Bayesian framework and stochastic gradient descent. Although this approach could deal with cross-domain sentiment

© Springer International Publishing Switzerland 2016
H.T. Nguyen and V. Snasel (Eds.): CSoNet 2016, LNCS 9795, pp. 298–308, 2016.
DOI: 10.1007/978-3-319-42345-6_26

classification, it used the "bag-of-words" model and faces difficulties when represent the relationship between words. For example, the phrase "have to", which is a common phrase in the negative text (but much less important in positive text), cannot be taken advantage of with bag-of-words feature. This is especially true in isolated languages, such as Vietnamese, where words are not separated by white spaces.

As a resource-poor language, Vietnamese has quite a few accomplishments in the field of sentiment classification. To the best of our knowledge, there is no study on Vietnamese cross-domain sentiment classification. There is also no suitable dataset with a reasonable amount of reviews and variance of products to apply lifelong learning on Vietnamese.

In this paper, we propose the use of bigram feature to lifelong learning approach on sentiment classification. Wang and Manning [6] proved that adding bigrams improves sentiment classification performance because they can capture modified verbs and nouns. We also created a dataset for Vietnamese cross-domain sentiment classification by collecting more than 15,000 reviews from the e-commerce website Tiki.vn[1] with 17 distinctive domains. We proposed combining the bigram feature with the Nave Bayesian optimization framework. The proposed method has leveraged the phrases that contain sentiment better than that of Chen et al. [3] and outperforms other methods in both Vietnamese and English datasets.

The remainder of this paper is organized as follows. Section 2 provides a brief overview of the background and related work. Section 3 presents our method including how we add bigram and bag-of-bigram features to the lifelong learning, and how we processed the raw reviews of the Vietnamese dataset to improve the performance. Section 4 describes the experimental setup and results. Section 5 concludes the paper and points to avenues for future work.

2   Related Work

Our work is related to lifelong learning, multi-task learning, transfer learning and domain adaptation. Chen and Liu have exploited different types of knowledge for lifelong learning on mining topics in documents and topic modeling [7,8]. Chen and Liu [3], in their other work, also proposed the first lifelong learning approach for sentiment classification. Likewise, Ruvolo and Eaton [9] developed a method for online multi-task learning in the lifelong learning setting, which maintains a sparsely shared basis for all task models. About domain adaptation, most of the work can be divided into two groups: supervised (Finkel and Manning [10], Chen et al. [11]) and semi-supervised (Kumar et al. [12], Huang and Yates [13]).

There are also many previous works on transfer learning and domain adaptation for sentiment classification. Yang et al. [14] proposed an approach based on feature-selection for cross-domain sentence-level classification. Other approaches include structural correspondence learning (Blitzer et al. [15]), spectral feature

---

[1] http://tiki.vn/.

alignment algorithm (Pan et al. [16]), CLF (Li and Zong [17]). Similar methods can be found in the work of Liu [2].

In the field of sentiment analysis for Vietnamese, Duyen et al. [18] has published an empirical study which compared the use of Nave Bayes, MEM and SVM with hotel reviews. Also, using the corpus from Duyen, Bach et al. [19] proposed the use of user-ratings for the task. Term feature selection approach was investigated by Zhang and Tran [20], while Kieu and Pham [21] investigated a rule-based system for Vietnamese sentiment classification. As that being said, to the best of our knowledge, there is no previous work on domain adaptation or lifelong learning as well as a appropriate dataset for Vietnamese (with a reasonable amount of reviews and variance of products).

3   Our Proposed Method

In this section, we describe our system for sentiment classification in a lifelong learning setting, which is a combination of components to analyze reviews from many domains. The system takes customer reviews, from multiple types of products, as source domains. Each review can contain multiple sentences and it is labeled positive, negative or neutral based on how users rated them. From the source domains mentioned above, the system gains knowledge valuable to the learning task on the target domain. Such knowledge is used to optimize the classifier on the target domain using stochastic gradient descent (SGD).

3.1   Overview of Lifelong Learning for Sentiment Classification

As described in Fig. 1, the system contains three main modules: knowledge storing, optimization, and sentiment classification.

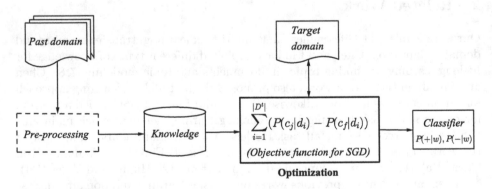

Fig. 1. Lifelong learning for sentiment classification

Knowledge Storing. The system extracts knowledge from the past domains, which is used to optimize the classifier on the target domain. There are three types of knowledge, including:

- The Prior probability $P_+^t(w|c)$ and $P_-^t(w|c)$ of each word, where t is a past learning task.
- Number of times a word appears in positive or negative in learning task: $N_{+,w}^t$, $N_{-,w}^t$. Similarly, the number of occurrences of w in the positive and negative documents are respectively $N_{+,w}^{KB} = \sum N_+^t$ and $N_{-,w}^{KB} = \sum N_-^t$.
- Number of past tasks in which $P_{w|+} > P_{w|-}$ or vice versa: $M_{+,w}^{KB}$, $M_{-,w}^{KB}$. The two figures are used to leverage domain knowledge via a penalty term to penalize the words that appear in just a few domains.

Optimization. With the help of all three types of knowledge mentioned above, this component is used to optimize the objective function on the training set of the target domain. The objective function is $\sum_{i=1}^{|D_i|} P(c_j|d_i) - P(c_f|d_i)$, in which c_j is the actual labeled class, c_f is the wrong class of the document d_i. We follow the SGD with similar regularization techniques proposed by Chen et al. [3]. Our optimized variables are $X_{+,w}$ and $X_{-,w}$, which are the occurrences of a word w in a positive and negative class, respectively. The objective function is optimized on each document of the target domain until convergence. After SGD, we use Bayes formula (see Eqs. 1 and 2) to create a classifier optimized for the target domain. Note that Laplace smoothing is applied in both cases.

$$P(+|w) = \frac{\lambda + X_{+,w}}{\lambda|V| + \sum_{v=1}^{V} X_{+,v}} \tag{1}$$

$$P(-|w) = \frac{\lambda + X_{-,w}}{\lambda|V| + \sum_{v=1}^{V} X_{-,v}} \tag{2}$$

Sentiment Classification. With the classifier optimized for the target domain, the system does sentiment classification task on each document of the test domain. Although the approach still follows Nave Bayes framework, the way we classify differentiates between unigrams, bigrams, and bag-of-bigrams.

3.2   Bigrams

We propose the use of bigram feature, instead of unigram, on this type of sentiment classification. Wang and Manning [6] has proved that using bigram always improve the performance on sentiment classification. For instance, phrases such as "have to" in English or "không thích" (dislike) in Vietnamese can express sentiment well in the documents. These noun phrases and verb phrases cannot be captured by using unigram feature alone.

The way we integrate bigram feature into Nave Bayesian framework for lifelong learning is described below:

- In **Knowledge storing** step, beside $P_+^t(w|c)$ and $P_-^t(w|c)$, we also store $P_+^t(w_i|w_{i-1})$ and $P_-^t(w_i|w_{i-1})$ whereas $P_+^t(w_{i+1}|w_i) = \frac{\lambda+N_{+,w_iw_{i+1}}}{\lambda|V|+N_{+,w_i}}$ and $P_-^t(w_{i+1}|w_i) = \frac{\lambda+N_{-,w_iw_{i+1}}}{\lambda|V|+N_{-,w_i}}$). The number of occurrences of each bigram on each class ($N_{+,w_iw_{i+1}}^t$ and $N_{-,w_iw_{i+1}}^t$) and the domain-level knowledge ($M_{+,w_iw_{i+1}}^{KB}$, and $M_{-,w_iw_{i+1}}^{KB}$) are also stored.
- In **Optimization** step, due to the use of bigram, the probability for each document is modified as Eqs. 3 and 4:

$$P(+|d) = \frac{P_+}{P_-}.P_+(w_0).P_+(w_1|w_0).P_+(w_2|w_1)P_+(w_n|w_{n-1}) \qquad (3)$$

$$P(-|d) = \frac{P_-}{P_+}.P_-(w_0).P_-(w_1|w_0).P_-(w_2|w_1)P_-(w_n|w_{n-1}) \qquad (4)$$

- The positive and negative probabilities for each document on the test data also have to follow the Eqs. 3 and 4 for the **Sentiment Classification** step.

3.3   Bag of Bigrams

Although using bigram help taking advantage of the phrases that express sentiments, using the standard Bayes formula still relies on the probabilities and number of occurrences of unigrams on all the documents. Our alternative way to leverage bigram is to treat each bigram as a unigram and apply the normally used Bayes formula ($P_{+|d} = \frac{P_+}{P_-}.P_+(w_0w_1).P_+(w_1w_2).P_+(w_2w_3)\ldots P_+(w_{n-1}w_n)$) to create the classifier. Such formula is applied to **Optimization** and **Sentiment classification** steps. We will compare how the two solutions improve the classification performance on both Vietnamese and English dataset.

3.4   Pre-processing on Vietnamese Dataset

Different to the dataset from Chen et al. [3] on English, the Tiki.vn dataset contains many emoticons. Therefore, we need to pre-process the data before **Knowledge storing** step to leverage all lexical resources in the dataset. In most online forums or discussion groups, users often use emoticons such as ":)", ":(" or punctuations such as "!!!!!" to express their opinions. However, during the task, we standardize the emoticons used by users, e.g. changing ":((((((" to ":(". We treat each emoticon or punctuation as a unigram and follow the other steps as normal. In this pre-processing step, we also perform word segmentation by following the maximum entropy approach of Dinh and Thuy [22]. Word segmentation can model the sentiment adjectives which often contain two or more morphemes, hence, provide a better vocabulary set for classification on Vietnamese using unigram feature.

4 Experimental Results

4.1 Dataset

In this study, we used two datasets for sentiment classification, one is Vietnamese and the other is English. The English one has been used by Chen et al. [3] for lifelong learning, in which there are 20,000 product reviews from Amazon divided into 20 domains. The Vietnamese dataset was also crawled from an e-commerce website, Tiki.vn. The two datasets can contribute great values to different tasks of cross-domain sentiment analysis on both languages.

Labeled Vietnamese Reviews. For this study, we crawled the reviews from Tiki.vn, which is a large e-commerce website with quality reviews from the customers. It is a large corpus of 17 diverse domains or products and a total of 15,394 product reviews, but we selected a group of 10 with a fair amount of negative reviews for experiments (including 13,652 reviews), which we name "A Community Resource for sentiment analysis on Vietnamese" (CRSAVi). This selection not only helped reduce the imbalanced distribution, but also committed enough lexical resources for creating a classifier. We followed the previous works [23,24] to treat reviews with more than 3 star as positive reviews, equal to 3 star as neutral and fewer than 3 star as negative ones. The number of positive, neutral and negative reviews are shown as in the Table 1:

Table 1. Names of 10 domains and the number of positive, neutral and negative reviews

Product	Positive	Neutral	Negative
TrangDiem (Cosmetics)	3,629	792	154
Dungcuhocsinh (Tools for students)	1,803	164	37
Sanphamvegiay (Papers)	1,778	144	34
Butviet (Pens and pencils)	1,044	125	28
Dodungnhabep (Kitchen)	987	100	24
DauGoi (Shampoo)	347	59	18
Tainghe (Headphones)	698	90	18
DoDungChoBe (Baby)	658	61	14
Filehosobiahoso (Files)	157	47	14
Phukiendienthoaimaytinhbang (Accessories)	583	32	13
Total (13,652 reviews)	11,684	1,614	354

It is noted that the all product reviews from Tiki was checked by the website administrators before publishing, which helps guarantee low rate of low quality reviews from online users. In fact, all of them contain Vietnamese tone marks, some contain emoticons. On our dataset, the average unigram per document on each domain varies from 66 to just above 75 unigrams. The information packed in a single review in our dataset consists of the product name, author name, rating, headline, bought-already, time of review and details. From the Table 1, it can be seen that the proportion of negative class among the dataset is only around 2.6 %. As that being said, to experiment lifelong learning, a mass of reviews among multiple product types are required, although there is no Vietnamese sentiment dataset that can meet the requirements. Although different types of products are crawled for the task and Tiki has a great deal of book reviews, CRSAVi does not include books because most of the book reviews mention the book content, not the overall quality like other products.

Because the difference between the number of reviews across domains might result in the efficiency of the system, for each experiment, we selected randomly a maximum amount of 100 reviews each class on each domain to conduct the experiments.

Labeled English Reviews. The corpus from Chen et al. [3] was utilized to compare directly to their lifelong learning approach in English sentiment classification. The corpus contains reviews of 20 different products crawled from Amazon. The experiments were on a dataset which has a reasonable proportion of negative reviews across domains, varies from 11.97 to 30.51 %.

4.2 Evaluation Metrics

The evaluation method used is 5-fold cross validation. While dividing a domain into groups, we tried to keep the class distribution to avoid the case of no negative review on a segment due to the small proportion of negative class mentioned above. F1-measure on negative and positive class in types of Micro-average and Macro-average are applied.

4.3 Baseline

Our method is compared to VietSentiWordnet by Vu et al. [25]. The approach uses a dictionary which contains a list of segmented words or phrases in Vietnamese that express sentiment. For each word or phrase, the dictionary provides corresponding positive and negative score. For each document, the score is evaluated by summing up all (positive score - negative score) of all sentiment words or phrases that are available in the dictionary. if the score is positive, the document is labeled as positive and vice versa. It is noted that VietSentiWordnet can only work on a single domain data.

In English, we compare our proposed method to Chen et al. [3] to illustrate the benefits of our approach on lifelong learning.

4.4 Bigram Feature Improves the Classification on English Dataset

We compare our result to the original lifelong learning approach of Chen et al. [3] (LSC) on the balanced class distribution. We created a balance dataset of

200 reviews (100 positive and 100 negative) in each domain dataset for this experiment. On balanced class distribution, how the accuracy is improved is expressed as in Table 2

Table 2. Accuracies on English balanced distribution over 20 domains

LSC	LSC-bag-of-bigram	LSC-bigram
83.34	**85.92**	85.44

Our method exceeds LSC to get to a high of 85.92 %. This improvement confirms the results of Wang and Manning [6] and proves that the use of bigram and bag-of-bigram features also improve the performance on cross-domain sentiment classification.

4.5 Vietnamese Cross-Domain Sentiment Classification

We compare our proposed method in different settings to the baseline method on the Vietnamese dataset. The average F1-score for the positive class is not shown because being the majority class makes the classifiers perform well and do not show much difference between multiple settings, although they all perform better than VietSentiWordnet. Table 3 compares VietSentiWordnet (VSWN) to our proposed method (lifelong learning for Vietnamese sentiment classification - now called LLVi) using unigram feature with segmentation (LLVi-uni) and without segmentation (LLVi-uniWS).

The LLVi with unigram feature (no segmentation) which also counts emoticons (LLVi-e) is also compared.

Table 3. Macro, micro average F1-score of the negative class on CRSAVi

	VSWN	LLVi-uni	LLVi-e	LLVi-uniWS
Macro F1	33.21	47.19	**51.20**	50.87
Micro F1	40.85	61.33	**62.12**	61.93

The Table 3 has obviously shown that while segmentation task helps improving the performance on lifelong learning with unigram feature. For example, the word "tuy_nhin" (however) can classify well in our dataset, but cannot be leveraged effectively without segmentation. However, lifelong learning with emoticons still performs slightly better. The two emoticons ":(" and ":)" provides significantly biased probability thus become good classifiers. The Table 3 also confirms that the lifelong learning approach has a huge advantage over VietSentiWordnet, which can only work on the target domain.

Table 4 compares the performance is a collection of lifelong learning approaches with different features applied. The group includes LLVi-uni, LLVi with bigram feature (LLVi-bi), LLVi with bag-of-bigram feature (LLVi-bb) in two settings, segmentation and without segmentation.

Table 4. Macro, micro average F1-score on negative class with Vietnamese dataset, unigram vs. bigram vs. bag-of-bigram. Unit: %

		LLVi-uni	LLVi-bi	LLVi-bb
Macro F1	No segmentation	47.19	56.10	60.56
	With segmentation	50.87	64.37	**65.85**
Micro F1	No segmentation	61.33	56.52	**66.27**
	With segmentation	61.93	61.53	60.59

Similar to English, using bigram and bag-of-bigram features make a huge improvement to the performance compared to using unigram feature only. With segmentation combined, the lifelong learning using unigram feature improves significantly, while it does not have clear impact on lifelong learning with bigram and bag-of-bigram features. 'cuc'_kỳ tôt' (extremely good), 'khó chiu' (frustrating), 'rát da' (burning skin sensation), 'chẳng mê' (cannot love) are examples of how using bigram performs better.

5   Conclusion

In this paper, we have presented our method that uses lifelong learning for cross-domain sentiment classification on English and Vietnamese. Experimental results on both corpus showed that:

- Lifelong learning approach is effective for cross-domain sentiment classification in Vietnamese as well as in English.
- Incorporating bigram and bag-of-bigram features into lifelong learning improved the performance of the system.
- Emoticons and word segmentation made a slight improvement in sentiment classification on the Vietnamese dataset.

There is abundant room for further progress of our work. We would like to further exploit the sentiments from emoticons due to the high rate of occurrences of these in our dataset. Besides, future work could be focused on another collection of reviews with different qualities and different types of products to verify our proposed method.

Acknowledgments. This research is supported by research funding from Honors Program, University of Science, Vietnam National University - Ho Chi Minh City.

References

1. Pang, B., Lee, L.: Opinion mining and sentiment analysis. Found. Trends Inf. Retr. **2**, 1–135 (2008)
2. Liu, B.: Sentiment analysis and opinion mining. Synth. Lect. Hum. Lang. Technol. **5**, 1–167 (2012)
3. Chen, Z., Ma, N., Liu, B.: Lifelong learning for sentiment classification. In: Proceedings of the 53rd Annual Meeting of the Association for Computational Linguistics and the 7th International Joint Conference on Natural Language Processing: Short Papers, Beijing, China, vol. 2, pp. 750–756. Association for Computational Linguistics (2015)
4. Pan, S.J., Yang, Q.: A survey on transfer learning. IEEE Trans. Knowl. Data Eng. **22**, 1345–1359 (2010)
5. Raina, R., Battle, A., Lee, H., Packer, B., Ng, A.Y.: Self-taught learning: transfer learning from unlabeled data. In: Proceedings of the 24th International Conference on Machine Learning, ICML 2007, pp. 759–766. ACM, New York (2007)
6. Wang, S., Manning, C.: Baselines and bigrams: simple, good sentiment and topic classification. In: Proceedings of the 50th Annual Meeting of the Association for Computational Linguistics: Short Papers, Jeju Island, Korea, vol. 2, pp. 90–94. Association for Computational Linguistics (2012)
7. Chen, Z., Liu, B.: Mining topics in documents: standing on the shoulders of big data. In: Proceedings of the 20th ACM SIGKDD International Conference on Knowledge Discovery and Data Mining, pp. 1116–1125. ACM (2014)
8. Chen, Z., Liu, B.: Topic modeling using topics from many domains, lifelong learning and big data. In: Jebara, T., Xing, E.P. (eds.) Proceedings of the 31st International Conference on Machine Learning (ICML 2014), pp. 703–711. JMLR Workshop and Conference Proceedings (2014)
9. Ruvolo, P., Eaton, E.: Scalable lifelong learning with active task selection. In: AAAI Spring Symposium: Lifelong Machine Learning (2013)
10. Finkel, J.R., Manning, C.D.: Hierarchical Bayesian domain adaptation. In: Proceedings of the Human Language Technologies: The 2009 Annual Conference of the North American Chapter of the Association for Computational Linguistics, NAACL 2009, Stroudsburg, PA, USA, pp. 602–610. Association for Computational Linguistics (2009)
11. Chen, M., Weinberger, K.Q., Blitzer, J.: Co-training for domain adaptation. In: Advances in Neural Information Processing Systems, pp. 2456–2464 (2011)
12. Kumar, A., Saha, A., Daume, H.: Co-regularization based semi-supervised domain adaptation. In: Advances in Neural Information Processing Systems, pp. 478–486 (2010)
13. Huang, F., Yates, A.: Exploring representation-learning approaches to domain adaptation. In: Proceedings of the 2010 Workshop on Domain Adaptation for Natural Language Processing, pp. 23–30. Association for Computational Linguistics (2010)
14. Yang, H., Callan, J., Si, L.: Knowledge transfer and opinion detection in the TREC 2006 blog track. In: TREC (2006)
15. Blitzer, J., Dredze, M., Pereira, F., et al.: Biographies, bollywood, boom-boxes and blenders: domain adaptation for sentiment classification. In: ACL, vol. 7, pp. 440–447 (2007)
16. Pan, S.J., Ni, X., Sun, J.T., Yang, Q., Chen, Z.: Cross-domain sentiment classification via spectral feature alignment. In: Proceedings of the 19th International Conference on World Wide Web, pp. 751–760. ACM (2010)

17. Li, S., Zong, C.: Multi-domain sentiment classification. In: Proceedings of the 46th Annual Meeting of the Association for Computational Linguistics on Human Language Technologies: Short Papers, pp. 257–260. Association for Computational Linguistics (2008)
18. Duyen, N.T., Bach, N.X., Phuong, T.M.: An empirical study on sentiment analysis for Vietnamese. In: Advanced Technologies for Communications (ATC), pp. 309–314 (2014)
19. Bach, N.X., Phuong, T.M.: Leveraging user ratings for resource-poor sentiment classification. Procedia Comput. Sci. **60**, 322–331 (2015). Proceedings of the 19th Annual Conference on Knowledge-Based and Intelligent Information Engineering Systems, KES 2015, Singapore, September 2015
20. Zhang, R., Tran, T.: An information gain-based approach for recommending useful product reviews. Knowl. Inf. Syst. **26**, 419–434 (2011)
21. Kieu, B.T., Pham, S.B.: Sentiment analysis for Vietnamese. In: 2010 Second International Conference on Knowledge and Systems Engineering (KSE), pp. 152–157. IEEE (2010)
22. Dien, D., Thuy, V.: A maximum entropy approach for Vietnamese word segmentation. In: 2006 International Conference on Research, Innovation and Vision for the Future, pp. 248–253 (2006)
23. Blitzer, J., Dredze, M., Pereira, F.: Biographies, bollywood, boom-boxes and blenders: domain adaptation for sentiment classification. In: Proceedings of the 45th Annual Meeting of the Association of Computational Linguistics, Prague, Czech Republic, pp. 440–447. Association for Computational Linguistics (2007)
24. Pang, B., Lee, L., Vaithyanathan, S.: Thumbs up? Sentiment classification using machine learning techniques. In: Proceedings of the 2002 Conference on Empirical Methods in Natural Language Processing, pp. 79–86. Association for Computational Linguistics (2002)
25. Vu, X.-S., Song, H.-J., Park, S.-B.: Building a Vietnamese SentiWordNet using Vietnamese electronic dictionary and string kernel. In: Kim, Y.S., Kang, B.H., Richards, D. (eds.) PKAW 2014. LNCS, vol. 8863, pp. 223–235. Springer, Heidelberg (2014)

Determing Aspect Ratings and Aspect Weights from Textual Reviews by Using Neural Network with Paragraph Vector Model

Duc-Hong Pham[1], Anh-Cuong Le[2(✉)], and Thi-Thanh-Tan Nguyen[3]

[1] University of Engineering and Technology, VNU, Hanoi, Vietnam
hongpd@epu.edu.vn
[2] Faculty of Information Technology, Ton Duc Thang University,
Ho Chi Minh City, Vietnam
leanhcuong@tdt.edu.vn
[3] Electric Power University, Hanoi, Vietnam
tanntt@epu.edu.vn

Abstract. Aspect-based analysis currently becomes a hot topic in opinion mining and sentiment analysis. The major task here is how to detect rating and weighting for each aspect based on an input of a collection of users' reviews in which only the overall ratings are given. Previous studies usually use a bag-of-word model for representing aspects thus may fail to capture semantic relations between words and cause an inaccuracy of aspect ratings prediction. To overcome this drawback, in this paper we will propose a model for aspect analysis, in which we first use a new deep learning technique from [8] for representing paragraphs and then integrate these representations into a neural network model to infer aspect ratings and aspect weights. The experiments are carried out on the data collected from hotel services with the aspects including "cleanliness", "location", "service", "room", and "value". Experimental results show that our proposed method outperforms the well known studies for the same problem.

1 Introduction

In recent years, opinion mining and sentiment analysis has been one of the attracting topics of knowledge mining and natural language processing. It is the task of detecting, extracting and classifying opinions and sentiments concerning different topics, as mentioned in textual input. Some works have been done to this task such as rating the overall sentiment of a sentence/paragraph, or a textual review regardless of the entities (e.g., movies) from reviews, [13,15], opinion extraction and sentiment classification [3,4], detect comparative sentences from reviews [6,7], extracting information and summarization from reviews [9,12,23]. However these works fail to capture the sentiments over the aspects on which an entity can be reviewed. For example, the entity is a hotel which can contains some aspects as "cleanliness", "location" and "service".

© Springer International Publishing Switzerland 2016
H.T. Nguyen and V. Snasel (Eds.): CSoNet 2016, LNCS 9795, pp. 309–320, 2016.
DOI: 10.1007/978-3-319-42345-6_27

Sentiment for each aspect is the important information, therefore there are now more studies working on aspect based sentiment analysis. Hu and Liu [5] focused on the taks of determining aspects in given textual reviews. They assumed that product aspects are expressed by nouns and noun phrases and their frequencies are used for identifying aspects. Wu et al. [20] used a language model and a phrase dependency parser to detect product aspects, expression of opinion and relations between them. Several other works focused on rating aspects such as Snyder and Barzilay [16] proposed the good grief algorithm for modeling the dependencies among aspects and Titov and McDonald [17] used a topic based model and a regression model for extracting aspect terms as well as detecting ratings. However these studies based on the assumption that the aspect ratings are explicitly provided in the training data.

Different from [16,17] some other studies consider aspect ratings as latent factors and develop models for determining them. Wang et al. [18] proposed a probabilistic rating regression model to infer aspect ratings and aspect weights for each review. An extension of this model was provided by Wang et al. [19] which is a unified generative model (called Latent Aspect Rating Analysis). Note that this model does not need to predefine the aspect seed words. Xu et al. [21] proposed a model called Sparse Aspect Coding Model (SACM) by considering the user and item side information of textual reviews. They used two latent variables, namely, user intrinsic aspect interest and item intrinsic aspect to identify aspects. After obtaining aspects, they predict the rating on each aspect for each review. However these works represent aspect features based on bag of word model to fail to capture semantic relations between different words and it is the cause to lead inaccuracies on the result of aspect rating predictions.

Recently, deep learning models can capture semantic relations between different words and learn feature representation such as (Bengio et al. [1], Mikolov et al. [11], Yih et al. [22]) learning word vector representations through neural language models, Collobert et al. [2] applying technical convolution to extract higher level features from word vectors, Le and Mikolov [8] learning sentence/paragraph or document vector representations. To improve the result of aspect rating predictions, in this paper we propose a new model based on neural network to discover both aspect ratings and aspect weights for each review, in which we use the learned aspect features from paragraph vector model as input for our neural network model.

We evaluate the proposed model on the data collected from Tripadvisor[1]. We will focus on the five aspects including "cleanliness", "location", "service", "room", and "value". Experimental results show that our model can obtain better results in comparison with the model proposed in Wang et al. [18].

The rest of this paper is organized as follows: Sect. 2 presents paragraph vector model; Sect. 3 presents our proposed model, we first give the problem definition, next aspect feature representation for each review is presented, and then we present our model; Sect. 4 describes our experiments and results. Some conclusions are presented in the last section.

---

[1] www.tripadvisor.com.

2   Paragraph Vector

Le and Mikolov [8] proposed the paragraph vector model for learning representations of sentences and documents. It is an unsupervised framework that learns
continuous distributed vector representations for pieces of texts with variable-
length of the texts are ranging from sentences to documents. Specifically, in
paragraph vector framework, every paragraph is mapped to a unique vector and
represented by a column in matrix D. Every word is mapped to a unique vector and word vectors are concatenated or averaged to predict the context, i.e.
the next word. The contexts are fixed-length and sampled from a sliding window
over the paragraph. The paragraph vector is shared across all contexts generated
from the same paragraph but not across paragraphs. The word vector matrix
W, however, is shared across paragraphs. i.e. the vector for "better" is the same
for all paragraphs.

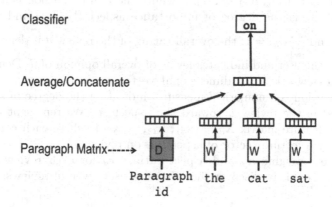

Fig. 1. Framework for learning paragraph vectors [8]

In Fig. 1, the paragraph is mapped to a vector via matrix D, the concate-
nation or average of this vector with a context of three words "the", "cat" and
"sat" is used to predict the fourth word "on". The paragraph vector represents
the missing information from the current context and can act as a memory of
the topic of the paragraph. The advantages of paragraph vector model are that
they inherit the property of word vectors, i.e., the semantics of the words. In
addition, they also take into consideration a small context around each word
which is in close resemblance to the n-gram model with a large n. This property
is crucial because the n-gram model preserves a lot of information of the sen-
tence/paragraph, which includes the word orderalso. This model also performs
better than the bag of word model which wouldcreate a very high-dimensional
representation that has very poor generalization.

3 Proposed Method

In this section, we first present the problem definition and then we present learning aspect feature representation. Finally, we propose a new model based on neural networks for discovering aspect weights and aspect ratings for each review.

3.1 Problem Definition

Given a collection of textual reviews $D = \{d_1, d_2, ..., d_{|D|}\}$ discussing about an entity or topic, each textual review is assigned a numerical overall rating. We assume that the set of reviews D have k-aspects and are denoted by $\{A_1, A_2, ..., A_k\}$, where an aspect A_i is a set of terms that characterize a rating factor in the reviews.

For each $d \in D$, we denote the aspect weights for the review d is a k - dimensional vector $\alpha_d = (\alpha_{d1}, \alpha_{d2}, ..., \alpha_{dk})$, where the i-th dimension is a numerical measure, indicating the degree of importance aspect A_i on d and we require $0 \leq \alpha_{di} \leq 1$ and $\sum_{i=1}^{k} \alpha_{di} = 1$, the overall rating of the review d is denoted by O_d, it is given by the user and indicates levels of overall opinion of d. Denote aspect ratings for review d is a k - dimensional vector $r_d = (r_{d1}, r_{d2}, ..., r_{dk})$, where the i-th dimension is a numerical measure, indicating the degree of satisfaction demonstrated in the review d toward the aspect A_i. We represent aspects for review d is a feature matrix $X_d = (x_{d1}, x_{d2}, ..., x_{dk})$, where each column x_{di} is represented by a feature vector of aspect A_i on review d.

Both aspect weights α_d and aspect ratings r_d for each review $d \in D$ are unknown, our goal is to discover them from the set given of reviews.

3.2 Learning Aspect Feature Representation

Unlike previous works that represent aspect features by a bag of word model which fail to capture semantic relations between different words or between words and aspects, we apply paragraph vector model [8] to learn semantic representations for aspects (i.e. we also know as learn feature representations). Specifically, given a collection of textual reviews $D = \{d_1, d_2, ..., d_{|D|}\}$, we first apply the aspect segmentation algorithm [18] to identify aspects and segment reviews. Next, for each review $d \in D$, we use concatenation of all sentences with the same aspect category as a paragraph of a specific aspect and we will be obtained k-paragraphs corresponding to k-aspects. Then we apply paragraph vector model to learn the aspect feature matrix $X_d = (x_{d1}, x_{d2}, ..., x_{dk})$ for each review d. According to X_d, the aspect feature representations of review d can be identified.

3.3 Latent Rating Neural Network Model

This section presents a new model based on neural networks for discovering aspect weights and aspect ratings for each review. We assume that both aspect weights and aspect ratings are latent in the model and we call this model as Latent Rating Neural Network Model (LRNN). In Fig. 2, we show the architecture of the LRNN model.

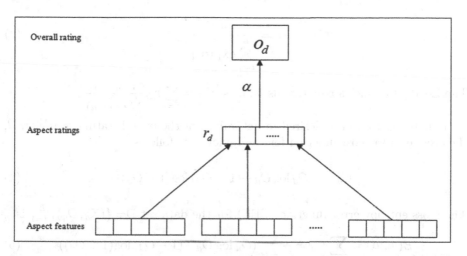

Fig. 2. An illustration of the LRNN model discovers aspect ratings r_d and aspect weights a_d for review d, in which r_d are k units at the hidden layer, a_d are the weights between the hidden layer and the output layer. The input are aspect feature vectors x_{d1}, x_{d2}, ..., x_{dk} which learned from paragraph vector model

After learning aspect feature representations, we use the learned aspect vectors as the input for our model LRNN. We denote $w_i = (w_{i1}, w_{i2}, ..., w_{in})$ as a weight vector of aspect feature A_i. Then, the aspect rating r_{di} of the review d generated based on a linear combination of aspect feature vector and the weight vector as $r_{di} \sim \sum_{l=1}^{n} x_{dil}.w_{il}$ [18]. Specifically, we assume that the aspect rating r_{di} is generated at the hidden layer of the neural network and it is computed as follows:

$$r_{di} = \text{sigm}(\sum_{l=1}^{n} x_{dil}w_{il} + w_{i0}) \qquad (1)$$

where $\text{sigm}(y) = 1/(1 + e^{-y})$, w_{i0} is a bias.

The aspect weights of the review d are assumed as the weights between the hidden layer and the output layer. The overall rating is generated at the output layer of the neural network and it is computed based on the weighted sum of a_d and r_d as follows:

$$\hat{O}_d = \sum_{i=1}^{k} r_{di}\alpha_{di} \qquad (2)$$

subject to $\sum_{i=1}^{k} \alpha_{di} = 1, 0 \leq \alpha_{di} \leq 1$ for $i = 1, 2, \ldots, k$.

In order to support $\sum_{i=1}^{k} \alpha_{di} = 1$ and $0 \leq \alpha_{di} \leq 1$, we use the auxiliary aspect weight $\hat{\alpha}_{di}$ instead of the aspect weight α_{di} as follows:

$$a_{di} = \frac{\exp(\hat{\alpha}_{di})}{\sum_{l=1}^{k} \exp(\hat{\alpha}_{dl})} \tag{3}$$

The Eq. (2) becomes a equation as follows: $\hat{O}_d = \sum_{i=1}^{k} r_{di} \frac{\exp(\hat{\alpha}_{di})}{\sum_{l=1}^{k} \exp(\hat{\alpha}_{dl})}$

Denote O_d to be the desired target values of the overall rating of review d, the cross entropy cost function for the review d is follows:

$$C_d = -O_d \log \hat{O}_d - (1 - O_d) \log(1 - \hat{O}_d) \tag{4}$$

The cross entropy error function (CEE) for the data set D= $\{(X_d, O_d)\}_{d=1}^{|D|}$ is:

$$E(w, \hat{\alpha}) = \sum_{d \in D} C_d = -\sum_{d \in D} (O_d \log \hat{O}_d + (1 - O_d) \log(1 - \hat{O}_d)) \tag{5}$$

Next, to avoid over-fitting and no loss of generality, we add regularizer to the function $E(w, \hat{\alpha})$. Regularization terms are ubiquitous. They typically appear as an additional term in an optimization problem.

$$E(w, \hat{\alpha}) = -\sum_{d \in D} (O_d \log \hat{O}_d + (1 - O_d) \log(1 - \hat{O}_d)) + \frac{1}{2} \lambda \sum_{i=1}^{k} |w_i|^2 \tag{6}$$

where $|w_i|^2 = \sum_{l=1}^{n} (w_{il})^2$, λ is the regularization parameter.

We denote $W = [w]_{kxn}$ is a matrix which each row is a weight vector of a aspect feature; $w_0 = (w_{01}, w_{02}, ..., w_{0k})$ is a bias vector, where w_{0i} is bias of aspect A_i; $\hat{\alpha} = \left[\hat{\alpha}\right]_{|D|xk}$ is a auxiliary aspect weight matrix which each row is a auxiliary aspect weight vector of a review; $\alpha = [\alpha]_{|D|xk}$ is a aspect weight matrix which each row is a aspect weight vector of a review; $R = [r]_{|D|xk}$ is a aspect rating matrix which each row is a aspect rating vector of a review.

Our goal is to determine W, w_0 and $\hat{\alpha}$ to the function $E(w, \hat{\alpha})$ reaches the minimum value, this is the problem of nonlinear square optimization, it has no closed-form solution and is solved by iterative algorithm.

The gradient of $E(w, \hat{\alpha})$ with respect to $\hat{O}_d$ is,

$$\frac{\partial E(w, \hat{\alpha})}{\partial \hat{O}_d} = -\left(\frac{O_d}{\hat{O}_d} - \frac{1 - O_d}{1 - \hat{O}_d}\right) \tag{7}$$

The gradient of $E(\text{w}, \overset{\wedge}{\alpha})$ with respect to $\overset{\wedge}{\alpha}_{di}$ is, $\dfrac{\partial E(\text{w}, \overset{\wedge}{\alpha})}{\partial \overset{\wedge}{\alpha}_{di}} = \dfrac{\partial E(\text{w}, \overset{\wedge}{\alpha})}{\partial \hat{O}_d} \cdot \dfrac{\partial \hat{O}_d}{\partial \overset{\wedge}{\alpha}_{di}}$

$$= \frac{\partial E(\text{w}, \overset{\wedge}{\alpha})}{\partial \hat{O}_d}\left(\sum_{l=1}^{k}\delta(\text{i}=\text{l})\alpha_{di}(1-\alpha_{di})r_i-\sum_{l=1}^{k}\delta(\text{i}\neq\text{l})\alpha_{di}\alpha_{dl}r_{dl}\right) \qquad (8)$$

where $\delta(y) = \begin{cases} 1; & \text{if } y = true \\ 0; & \text{if } y = false \end{cases}$

The gradient of $E(\text{w}, \overset{\wedge}{\alpha})$ with respect to w_{il} is, $\dfrac{\partial E(\text{w}, \overset{\wedge}{\alpha})}{\partial \text{w}_{il}} = \dfrac{\partial E(\text{w}, \overset{\wedge}{\alpha})}{\partial \hat{O}_d} \cdot \dfrac{\partial \hat{O}_d}{\partial \text{w}_{il}}$

$$= \frac{\partial E(\text{w}, \overset{\wedge}{\alpha})}{\partial \hat{O}_d}\cdot\alpha_{di}\cdot r_{di}(1-r_{di})\cdot\left\{\begin{matrix}\sum\limits_{d=1}^{|D|} x_{dil};(1\leq i\leq k)\\ 1;(i=0)\end{matrix}\right\} + \lambda\text{w}_{il} \qquad (9)$$

At time $t = 0$, initialize the weight matrix W and the auxiliary aspect weight matrix $\overset{\wedge}{\alpha}$.

Two phases: propagation and weight update are as follows:

Phases 1: propagation, the rating r_{di} of aspect A_i in review d at time t at the hidden layer is given by the formula:

$$r_{di}(t) = \text{sigm}\left(\sum_{l=1}^{n} x_{dil}\text{w}_{il}(t) + \text{w}_{i0}(t)\right) \qquad (10)$$

The overall rating $\hat{O}_d$ of review d at time t at the output layer is given by the formula:

$$\hat{O}_d(t) = \sum_{i=1}^{k}\alpha_{di}(t)\cdot r_{di}(t) \qquad (11)$$

Phases 2: weight update, each element of the weight vector w_i and bias w_{i0} is updated at time $t + 1$ according to the formula:

$$\text{w}_{il}(t+1) = \text{w}_{il}(t) + \Delta\text{w}_{il}(t) \qquad (12)$$

where $\Delta\text{w}_{il}(t) = -\eta\frac{\partial E(\text{w}, \overset{\wedge}{\alpha})(t)}{\partial \text{w}_{il}(t)}$, $\eta \in (0,1)$ is the learning rate.

Each element of the weight vector $\overset{\wedge}{\alpha}_d$ is updated at time $t + 1$ according to the formula:

$$\overset{\wedge}{\alpha}_{di}(t+1) = \overset{\wedge}{\alpha}_{di}(t) + \Delta\overset{\wedge}{\alpha}_{di}(t) \qquad (13)$$

where $\Delta\overset{\wedge}{\alpha}_{di}(t) = -\eta\frac{\partial E(\text{w}, \overset{\wedge}{\alpha})(t)}{\partial \overset{\wedge}{\alpha}_{di}(t)}$

The Algorithm 1 presents the process for discovering aspect ratings and aspect weights.

Input: A collection of textual reviews $D = \{d_1, d_2, ..., d_{|D|}\}$, each textual review $d \in D$ is given an overall rating O_d, the learning rate η, the error threshold ε, the iterative threshold I and the regularization parameter λ

Step 0: $t=0$; initialize W, w_0, $\hat{\alpha}$

Step 1: for $iter=0$ to I do

for each textual review $d \in$ D do

    1.1. Calculate α_{di} According to Eq. (3);

    1.2. Calculate r_{di} at time t at hidden layer According to Eq. (10);

    1.3. Calculate $\hat{O}_d$ at time t at output layer According to Eq. (11);

    1.4. Update weights w_i and bias w_{i0} at time $t+1$ According to Eq. (12);

    1.5. Update auxiliary aspect weight $\hat{\alpha}_{di}$ at time $t+1$ according to Eq. (13);

Step 2: For offline learning, the step 1 may be repeated until the iteration error $\frac{1}{|D|} \sum_{d \in D} \left| O_d - \hat{O}_d(t) \right|$ is less than the error threshold or the number of iterations have been completed.

Output: W, w_0, $\hat{\alpha}$, R

After obtaining W, w_0, R and $\hat{\alpha}$, for review d, we compute each aspect weight $\alpha_{di} \in \alpha_d$ according to Eq. (3).

4 Experiment

4.1 Experimental Data

We use the data including 157214 reviews of 1105 hotels collected from the very famous tourist website[2]. This dataset is a part of the data used in the work in [18, 19] and it is downloaded from[3]. We choose five aspects to do with including "cleanliness", "location", "service", "room", and "value" and the ratings are in the range from 1 star to 5 stars. In summary, Table 1 shows the statistics on the data in our experiments.

Table 1. Evaluation Data Statistics

Number of reviews	157214
Number of hotels	1105
Number of sentence	1962888
Number of aspects	5

---

[2] www.tripadvistor.com.

[3] http://times.cs.uiuc.edu/~wang296/Data/.

Algorithm 1. The algorithm discover aspect ratings and aspect weights for each review

4.2   Experimental Result

We first apply the aspect segmentation algorithm [18] for obtaining aspect segmentations of each review. After for each review, we mix the sentences/segments of the same aspect into an unifined text which is considered as a paragraph. It means that for each aspect we have a corresponding paragraph saying sentiments about it. Then we apply paragraph vector model Doc2Vec[4] to learn the paragraph vectors for each paragraph corresponding with the aspect mentioned by this paragraph. These paragraph vectors are aspect feature representations of reviews and they are used as input for Algorithm 1. We perform the Algorithm 1 to determine aspect ratings and aspect weights for each review with the learning rate $\eta = 0.015$, the error threshold $\varepsilon = 10^{-4}$, the iterative threshold $I=1500$ and the regularization $\lambda = 10^{-3}$. In Table 2, we show the aspect rating determining for five hotels with the same mean (average) overall rating as 3.5 which we randomly select from our results achieved, note that the ground-truth aspect ratings in parenthesis. We can see that the result of aspect rating prediction is very close to the value rating in the ground-truth aspects.   In Table 3, we show

Table 2. The aspect ratings determined for the five hotels

Hotel name	Values	Rooms	Location	Cleanliness	Service
Barcelo Punta Cana	3.5(3.3)	2.9(3.0)	3.4(4.0)	3.2(3.2)	3.2(3.1)
The Condado Plaza Hilton	3.2(3.3)	3.5(3.8)	3.1(4.0)	3.5(3.7)	3.4(3.5)
King George Hotel	3.6(3.6)	3.0(3.1)	3.6(4.3)	4.2(3.7)	3.8(3.8)
Astoria Hotel	3.1(3.6)	2.7(2.6)	3.2(4.5)	3.7(3.3)	3.4(3.1)
Radisson Ambassador Plaza Hotel	3.2(3.2)	3.2(3.6)	3.2(3.6)	3.8(3.7)	3.6(3.5)

the results of aspect weights detection for five hotels. We can see that the aspect weights of the hotel: *King George Hotel* and *Astoria Hotel* have high values of aspect weights for the aspect "Values", it means this aspect is important.

Table 3. Determining aspect weights for the five hotels

Hotel name	Values	Rooms	Location	Cleanliness	Service
Barcelo Punta Cana	0.157	0.004	0.819	0.004	0.015
The Condado Plaza Hilton	0.001	0.008	0.373	0.002	0.616
King George Hotel	0.939	0.040	0.011	0.008	0.002
Astoria Hotel	0.685	0.004	0.050	0.260	0.002
Radisson Ambassador Plaza Hotel	0.236	0.448	0.300	0.010	0.006

---

[4] https://github.com/piskvorky/gensim/.

4.3   Evaluation

We represent aspect features in four following cases:

Bag of Words: We use the dictionary with 3987 word sentiments which is created in the process of applying the boot-strapping algorithm [18] to aspect segmentation for each review. We represent aspect features according to this dictionary.

Word Vector Averaging: We first apply the Word2Vec[5] with the window size of context is 7, the word frequency threshold is 7 (note that ignore all words with total frequency lower than this) and the size of word vector is 400 to learn word vector for each word. Then for each aspect on a review, we represent it by averaging word vectors of words appear in the text assigned it.

Sentence Vector Averaging: We apply the Sentence2Vec[6] with the window size of context is 7, the word frequency threshold is 7 and the size of sentence vector is 400 to learn sentence vectors for each sentence. Then for each aspect on a review, we represent each aspect by averaging sentence vectors.

Paragraph Vector: We apply paragraph vector model Doc2Vec[7] with the window size of context is 7, the word frequency threshold is 7 and the size of paragraph vector is 400 to learn aspect feature representations (note that we mentioned that learning aspect feature representations in Sect. 3.2).

To compare our proposed method with other methods, we use Latent Rating Regression model (LRR) [18] to compare with our LRNN model, the LRR is a novel probabilistic rating regression model and solve the same tasks as the model LRNN (i.e., discover aspect ratings and aspect weights). We evaluate two methods on four cases of aspect feature representations. In each case, the models use the same data set, we perform 5 times for training and testing, and report the mean value of metrics. In each time, we randomly select 75 % of given reviews to train, the remaining 25 % of given reviews to test.

We use the three metrics for evaluating aspect rating prediction including: (1) root mean square error on aspect rating prediction (Δ_{aspect}, lower is better), (2) aspect correlation inside reviews [18] (ρ_{aspect}, higher is better), (3) aspect correlation across reviews prediction [18] (ρ_{review}, higher is better). In Table 4, we show the mean value of three metrics for each method in each case of aspect features.

We can see that when using bag of words for representing aspect features, our model LRNN performs better than LRR on P_{aspect} but on two metrics Δ_{aspect}, P_{review} the LRR perform better than it. For aspect features represented by word vector averaging or sentence vector averaging, our model LRNN perform better than LRR on P_{aspect} and P_{review}. For aspect features represented by paragraph vector, our model LRNN perform slightly better than LRR on all metrics. In all cases of aspect feature representations, we see that both LRNN model and LRR

---

[5] https://github.com/piskvorky/gensim/.

[6] https://github.com/klb3713/sentence2vec.

[7] https://github.com/piskvorky/gensim/.

Table 4. Comparison with other models

Aspect feature	Method	Δ_{aspect}	P_{aspect}	P_{review}
Bag of words	LRR	**0.711**	0.307	**0.661**
	Our LRNN	0.787	**0.422**	0.557
Word vector averaging	LRR	0.583	0.314	**0.697**
	Our LRNN	**0.534**	**0.423**	0.677
Sentence vector averaging	LRR	0.586	0.372	**0.700**
	Our LRNN	**0.512**	**0.474**	0.686
Paragraph vector	LRR	0.485	0.367	0.708
	Our LRNN	**0.461**	**0.388**	**0.711**

model perform best on Δ_{aspect} and only slightly better on P_{review} when they use paragraph vector.

5 Conclusion

In this paper, we have proposed a new model based on neural network using aspect feature representations which learned from a paragraph vector model to discover aspect ratings and aspect weights for each review. Through experimental results, we have demonstrated that using paragraph vector model gives better results in comparison with using bag-of-word representation or using word vector. In addition, our LRNN model also shows its stronger than the LRR model with the same input representations.

Acknowledgments. This paper is partly funded by The Vietnam National Foundation for Science and Technology Development (NAFOSTED) under grant number 102.01-2014.22.

References

1. Bengio, Y., Ducharme, R., Vincent, P., Jauvin, C.: Neural probabilitistic language model. J. Mach. Learn. Res. **3**, 1137–1155 (2003)
2. Collobert, R., Weston, J., Bottou, L., Karlen, M., Kavukcuglu, K., Kuksa, P.: Natural language processing (almost) from scratch. J. Mach. Learn. Res. **12**, 2493–2537 (2011)
3. Dave, K., Lawrence, S., Pennock, D.M.: Mining the peanut gallery: opinion extraction and semantic classification of product reviews. In: Proceedings of WWW, pp. 519–528 (2003)
4. Devitt, A., Ahmad, K.: Sentiment polarity identification in financial news: a cohesion-based approach. In: Proceedings of ACL, pp. 984–991 (2007)
5. Hu, M., Liu, B.: Mining and summarizing customer reviews. In: Proceedings of SIGKDD, pp. 168–177 (2004)

6. Jindal, N., Liu, B.: Identifying comparative sentences in text documents. In: Proceedings of SIGIR 2006, pp. 244–251 (2006)
7. Kim, H., Zhai, C.: Generating comparative summaries of contradictory opinions in text. In: Proceedings of CIKM 2009, pp. 385–394 (2009)
8. Le, Q.V., Mikolov, T.: Distributed representations of sentences and documents. In: Proceedings of ICML, pp. 1188–1196 (2014)
9. Liu, B., Hu, M., Cheng, J.: Opinion observer: analyzing and comparing opinions on the web. In: Proceedings of WWW, pp. 342–351 (2005)
10. Lu, Y., Zhai, C., Sundaresan, N.: Rated aspect summarization of short comments. In: Proceedings of WWW, pp. 131–140 (2009)
11. Mikolov, T., Sutskever, I., Chen, K., Corrado, G., Dean, J.: Distributed representations of words and phrases and their compositionality. In: Proceedings of NIPS, pp. 1–9 (2013)
12. Morinaga, S., Yamanishi, K., Tateishi, K., Fukushima, T.: Mining product reputations on the web. In: Proceedings of KDD, pp. 341–349 (2002)
13. Pang, B., Lee, L.: Seeing stars: exploiting class relationships for sentiment categorization with respect to rating scales. In: Proceedings of ACL, pp. 115–124 (2005)
14. Pang, B., Lee, L.: Opinion mining and sentiment analysis. Found. Trends Inf. Retrieval 2(1–2), 1–135 (2008)
15. Pang, B., Lee, L., Vaithyanathan, S.: Thumbs up?: sentiment classification using machine learning techniques. In: Proceedings of EMNLP, pp. 79–86 (2002)
16. Snyder, B., Barzilay, R.: Multiple aspect ranking using the good grief algorithm. In: Proceedings of NAACL HLT, pp. 300–307 (2007)
17. Titov, I., McDonald, R.: A joint model of text and aspect ratings for sentiment summarization. In: Proceedings of ACL, pp. 308–316 (2008)
18. Wang, H., Lu, Y., Zhai, C.: Latent aspect rating analysis on review text data: a rating regression approach. In: Proceedings of SIGKDD, pp. 168–176 (2010)
19. Wang, H., Lu, Y., Zhai, C.: Latent aspect rating analysis without aspect keyword supervision. In: Proceedings of SIGKDD, pp. 618–626 (2011)
20. Wu, Y., Zhang, Q., Huang, X., Wu, L.: Phrase dependency parsing for opinion mining. In: Proceedings of ACL, pp. 1533–1541 (2009)
21. Xu, Y., Lin, T., Lam, W.: Latent aspect mining via exploring sparsity and intrinsic information. In: Proceedings of CIKM, pp. 879–888 (2014)
22. Yih, W., Toutanova, K., Platt, J., Meek, C.: Learning discriminative projections for text similarity measures. In: Proceedings of the Fifteenth Conference on Computational Natural Language Learning, pp. 247–256 (2011)
23. Zhuang, L., Jing, F., Zhu, X.Y.: Movie review mining and summarization. In: Proceedings of CIKM, pp. 43–50 (2006)

Stance Analysis for Debates on Traditional Chinese Medicine at Tianya Forum

Can Wang[✉] and Xijin Tang

Institute of Systems Science, Academy of Mathematics and Systems Science,
Chinese Academy of Sciences, Beijing 100190, People's Republic of China
wangcan@amss.ac.cn, xjtang@iss.ac.cn

Abstract. Internet and social media devices have created a new public space for debates on societal topics. This paper applies text mining methods to conduct stance analysis of on-line debates with the illustration of debates on traditional Chinese medicine (TCM) at one famous Chinese BBS Tianya Froum. After crawling and preprocessing data, logistic regression is adopted to get a domain lexicon. Words in the lexicon are taken as features to automatically distinguish stances. Furthermore a topic model latent Dirichlet allocation (LDA) is utilized to discover shared topics of different camps. Then further analysis is conducted to detect the focused technical terms of TCM and human names referred during the debates. The classification results reveal that using domain discriminating words as features of classifier outperforms taking nouns, verbs, adjectives and adverbs as features. The results of topic modeling and further analysis enable us to see how the different camps express their stances.

Keywords: Stance analysis · Opinion mining · Latent Dirichlet allocation · Traditional Chinese medicine

1 Introduction

With the development of Internet, people can easily express and exchange their opinions through on-line forums or social media. It is widely recognized that mining public opinion from on-line discussions is an important task, which is related to a wide range of applications. There exist two streams of literature in this domain. One is distinguishing subjective expressions from factual information [1, 2]. The other is detecting the text polarity, positive or negative. The bulk of such works have focused on feature selection [3–5], classifiers optimization [6], and finally improving the precision of classifiers.

Despite the fair amount of studies in the opinion mining domain, there are several limitations of the existing literature. Firstly, opinion mining and sentiment analysis are usually used as synonyms, for both fields apply data mining and natural language processing (NLP) techniques to deal with textual information [7]. However, sentiments cannot truly represent stances [8]. Secondly, corpora are important for opinion mining. Many of previous studies used users' comments[1] or news[2] as corpora. Unlike those

---

[1] http://www.cs.cornell.edu/people/pabo/movie-review-data/.
[2] http://mpqa.cs.pitt.edu/.

© Springer International Publishing Switzerland 2016
H.T. Nguyen and V. Snasel (Eds.): CSoNet 2016, LNCS 9795, pp. 321–332, 2016.
DOI: 10.1007/978-3-319-42345-6_28

corpora, the debates on societal problems on Internet are more diverse and conversational. They are highly contextualized, depending on rich background of shared knowledge and assumptions. Thirdly, previous researches on opinion mining mostly depended on existing lexicons, or generated lexicons by seed words [9]. The lexicons or the seed words came from people's experiences. While one word may have opposite meanings within different contexts. Fourthly, some previous studies focused on automatically determining the stance of a debate participant [10–13]. There are limited researches on how people express their different perspectives towards an issue.

In this paper we focus on stance analysis of debates rather than sentiment analysis. There are two camps of people by their attitudes towards traditional Chinese medicine (TCM). Some people take the "abolishing TCM" stance. In their opinion TCM should be abolished from the national health system. Other people take the "preserving TCM" stance and insist that TCM should be preserved. The debate started since the modern medicine entered into China. The discussion on TCM is always 2-paralyzation that is correlated to culture, philosophy, history and economy. Now Internet provides a public space for people to voice and exchange their opinions on societal hot spots and the livelihood issues. We select on-line discussion on TCM as our corpus since it enables us to understand different perspectives of debates on TCM directly from the public. Considering the context of the debate, we use logistic regression to generate discriminating words relevant to TCM. Latent Dirichlet allocation (LDA) is utilized to generate topics of the two camps. We try different ways to capture how people from different camps express their viewpoints.

The rest of the paper is organized as follows. Section 2 describes related work. Section 3 discusses our corpus in more details and describes the preprocessing of data. Section 4 presents our stance classification experiments, including two policies of feature words selection. Section 5 describes topics of the two different camps. Section 6 describes further analysis to detect the focused technical terms of TCM and human names referred during the debate. Conclusions are presented in Sect. 7.

2 Literature Review

To some extent, stance analysis is related to arguing or debate. Somasundaran and Wiebe [10] from University of Pittsburgh explained that "arguing is a type of linguistic subjectivity, where a person is arguing for or against something or expressing a belief about what is true, should be true or should be done in his or her view of the world". They focused on automatically determining the stances of debate participants with respect to a particular issue. In their research, they used the MPQA (Multiple-Perspective Question Answering) corpus to get arguing lexicon for debate. They combined the arguing lexicon and sentiment lexicons as opinion features to discriminate the debate stances and improved the precision of the classifier. Anand et al. [12, 13] from University of California Santa Cruz, taking debates from open debating websites "ConvinceMe.net" and "4forums.com" as corpora, tried a variety of features to get one's stance within debate, such as repeated punctuation, syntactic dependency, posts per author, words per sentence, etc. Their research illustrated that subjective expressions varied across debates.

Stance classification, by recognizing politically oriented polarity in texts, can be widely applied in political domain. Tikves et al. [14, 15] from Arizona State University research on profiling Islamic organizations' ideology and activity patterns along a hypothesized radical/counter-radical scale. They utilized ranked perspectives to map Islamic organizations in UK and Indonesia on a set of socio-cultural, political and behavioral scales based on their web corpus. Gryc and Moilanen [16] focused on modeling blogosphere sentiments centered around Barack Obama during the 2008 U.S. presidential election. Lin et al. [17] used statistical models to identify perspectives about "Palestinian" or "Israeli" at the document and sentence levels.

3 Data Collection and Preprocessing

3.1 Debate on TCM at BBS

Because of anonymity, bulletin board systems (BBS, in this paper as "forum") are good platforms for Internet users to freely express their opinions. Tianya Forum is one of the most popular Chinese BBS sites and there are many hot posts on TCM at Tianya Forum. Some of these posts are listed in Table 1 [8]. In this paper we take the hottest post "2822432" as our corpus.

Table 1. Hot posts about TCM at Tianya Forum

Post-ID	Replies	Participants	Start time	End time
2822432	117318	4890	2012-10-16	2013-11-29
2121178	36592	5522	2011-03-21	2015-01-24
2317943	33547	6067	2011-11-12	2015-01-24

3.2 Preprocessing of Data

Firstly, we label the replies by user IDs' stances. There are 4890 authors (user IDs) who participate the debate. 267 authors who have replied more than 5 times are chosen and their stances are manually labeled. There are 84 authors who hold "abolishing TCM" stance and 183 authors who hold "preserving TCM" stance.

Secondly, we preprocess the labeled replies as follows:

(1) Remove replies with no texts.
(2) Filter out urls.
(3) Segment words with the ICTCLAS tool[3], keep the user ID names and technical terms of TCM as reserved words. We use a TCM terminology dictionary from Sougou Cell dictionary[4] which contains 28428 TCM technical terms.
(4) Remove stop words (such as "oh") from the bag of words and words with only one character.

---

[3] http://ictclas.nlpir.org/.

[4] http://pinyin.sogou.com/dict/detail/index/20664.

4 Stance Classification

4.1 Features and the Classifier

Lin et al. [17] observed that people from different perspectives seemed to use words with different frequencies. For example, a participant who talks about "child" and "life" at an abortion debate is more likely from an against-abortion side, while someone who talks about "woman", "rape" and "choice" is more likely from a for-abortion side. To automatically distinguish the stances of the participants, either support or oppose, in this paper we use logistic regression to get the stance feature words. The process is as follows.

(1) Calculate the frequencies of the words appeared within a reply;
(2) Create a term-document matrix of frequencies. In our research terms mean words, documents mean replies;
(3) Label the replies' stances with "1" and "−1", "1" means "preserving TCM" and "−1" means "abolishing TCM";
(4) Use the MATLAB implementation of the SLEP package[5] to run the logistic regression. The vector of labeled stances and the term-document matrix are inputs, and the vector of words' coefficients is the output;
(5) Filter words with a threshold of absolute coefficient 0. Words with positive coefficients are chosen as "preserving TCM" feature words, and words with negative coefficients are taken as "abolishing TCM" feature words;
(6) Take the selected words as features, use the "e1071" package[6] in R to train a support vector machine (SVM) model to predict replies' stances.

Adjectives words were employed as features in opinion mining, as many researches on subjectivity detection revealed a high correlation between adjectives and sentences subjectivity [18]. Benamara et al. [19] demonstrated that features with both adjectives and adverbs outperformed features with only adjectives. Subrahmanian and Reforgiato [20] added verbs to feature words besides adjectives and adverbs. Turney and Littman [9] proposed a new method to get the semantic orientation of words by using adjectives, adverbs, verbs and nouns. In this paper, we select words including all the nouns, adjectives, adverbs and verbs in the corpus as a baseline.

Pang et al. [6] employed three machine learning methods to determine whether a review was positive or negative. The results showed that SVM model outperformed Naive Bayes and maximum entropy classifier. So we approach the classification work by using SVM. Figure 1 shows the experimental process of the paper.

4.2 Results and Discussions

After preprocessing, 44940 replies are labeled "preserving TCM" and 28646 replies are labeled "abolishing TCM". To avoid the imbalance problem, we randomly sample

---

[5] http://www.yelab.net/software/SLEP/.
[6] http://cran.r-project.org/web/packages/e1071/.

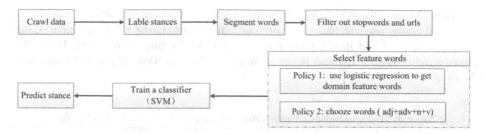

Fig. 1. The experimental process with different policies of feature selection

10000 "abolishing TCM" replies and 10000 "preserving TCM" replies. To guarantee enough text information in the replies, we select replies with more than 15 characters. We randomly split the data into training set and predicting set, each set respectively contains half of the sample data.

By logistic regression, each word has a coefficient contributing to stance towards TCM. With the threshold of absolute coefficient 0, we get 2879 discriminating words from 23441 words, including 1288 words with positive coefficients (related to "preserving TCM" stance) and 1491 words with negative coefficients (related to "abolishing TCM" stance). Table 2 lists top 15 discriminating words from both stances.

Table 2. Top 15 discriminating words in each camp

Preserving TCM			Abolishing TCM		
No.	Original Chinese words	English translation	No.	Original Chinese words	English translation
p1	反中医	Opposition to TCM	a1	郎中	Quack doctor
p2	正义	Justice	a2	中医粉	TCM fans
p3	儿童	Children	a3	博大精深	Vast
p4	废除	Abolish	a4	虫草	Cordyceps
p5	先生	Doctor	a5	巫医	Witch doctor
p6	小学	Primary school	a6	粪坑	Cesspit
p7	西医	Western medicine	a7	愚昧	Ignorance
p8	不要脸	Shameless	a8	禽流感	Avian influenza
p9	天理	Justice	a9	网友	Internet users
p10	根本	Fundamental	a10	特别	Special
p11	而已	Only/nothing more	a11	国人	Compatriots
p12	中西	Chinese and Western	a12	人中黄	Pulvis glycyrrhizae Praeparatus
p13	救死扶伤	Heal the wounded and rescue the dying	a13	质疑	Doubt
p14	一直	Always	a14	养生	Health preservation
p15	无效	Ineffectiveness	a15	郎中	Quack

Table 3 shows the experiments' results. 15669 words including adverbs, adjectives, verbs and nouns are selected. Using our domain discriminating words as features, the precision of the SVM model predicting the stances of replies is 63.13 %. Using the adverbs, adjectives, verbs and nouns as features, the precision of the SVM model is 51.18 %.

Table 3. The comparison of two feature selection policies for SVM classifier

Selection policies of feature words	Feature words	Precision
Adverbs, adjectives, verbs and nouns	15669	51.18 %
Domain discriminating words	2879	63.13 %

Shen et al. [21] attempted to identify perspectives of TCM. They collected Sina Weibo users whose tags contain their given TCM related words, crawled down these users' tweets and labeled the tweets "supporting TCM" and "opposing TCM". The differences between their research and ours are as follows. Firstly, their corpus is selected from Weibo posts and the length of the posts are limited in 140 characters. Our corpus is selected from Tianya Forum and there is no limitation of the length of the replies. So authors can fully express themselves. Secondly, their data are imbalanced, including 40,888 "supporting TCM" posts and 6,975 "opposing TCM" posts due to their biased data collection policy. We sample our data unbiased from the replies. Thirdly, there is no interaction between their subjects from Weibo while our subjects from Tianya Forum reply to the seed post or others' replies. Our corpus from Tianya Forum is more "discussion" oriented.

5 Topic Analysis Based on Camps

Latent Dirichlet allocation (LDA) is a topic model to generate topics of a group of documents based on the words of the documents [22]. We utilize LDA on the "preserving TCM" replies and the "abolishing TCM" replies to get more details to see how the opposite camps express their stances.

Table 4 shows the ten topics from the "preserving TCM" replies. We list the top 15 words of each topic. We label the topics by words distributed in the topic. There are mainly five groups of topics from the "preserving TCM" stance holders. Firstly, people in this camp doubt the motivation of the "abolishing TCM" stance holders. In their standpoint, the "abolishing TCM" stance holders are traitors of the traditional culture (e.g., Topics p1 & p3). Secondly, they mention TCM which can actually treat some diseases (e.g., Topics p7 & p10). Thirdly, they list some health preserving theories in TCM (e.g., Topics p4, p5 & p9). Fourthly, they use the national policy and the curriculum setup in colleges and universities to demonstrate the scientific nature of TCM (e.g., Topics p2 & p8). Additionally, rude Internet behaviors appear during the debate (e.g., Topic p6).

Table 5 shows the ten topics from the "abolishing TCM" replies. There are mainly five groups of topics from the "abolishing TCM" stance holders. Firstly, people in this camp doubt the rationality of TCM, especially the theories of *yin-yang* and five elements, feeling the pulse and acupuncture points (e.g., Topics a6, a7 & a9). Secondly, they emphasize that some TCM contain abnormal materials even materials with toxicity (e.g., Topics a3 & a8). Thirdly, they list some pseudo TCM experts or some people related to illegal practice of medicine (Topic a1), in their opinion we should discard the dross of traditional things (e.g., Topic a2). Fourthly, they introduce the modern medicine, for example the virus theory (e.g., Topic a4). Additionally, rude Internet behaviors also appear in this camp. Some of the "abolishing TCM" stance holders write doggerel to express their opinions (e.g., Topics a5 & a10).

Table 4. Topics from "preserving TCM" replies

No.	Topics	Words related to topics
p1	Motivation of "abolishing TCM"	科学 西医 没有 人类 真理 生命 创新 先生 实验 技术 癌症 能够 转基因 神经 理论
p2	Curriculum setup in colleges	西医 知道 没有 骗子 治病 医学 废除 西药 骗人 东西 理论 大學 否定 反对 中醫藥
p3	Water army	人士 问题 理解 没有 智商 证明 不能 中药 回答 事实 认为 逻辑 青蒿 实践 天涯
p4	Health preserving theory	人體 方法 疾病 患者 没有 可能 科盲 请问 血氣 网友 中醫 能力 血液 水平 能量
p5	TCM theory	理论 实践 科学 医学 中医 经络 研究 方法 人体 物质 发展 疾病 存在 系统 认识
p6	Rude Internet behaviors	中药 中医 智商 知道 证明 儿童 出来 砒霜 逻辑 东西 板蓝根 告诉 孙子 承认 无效
p7	TCM with good effectiveness	不能 治愈 患者 高血压 血压 大气 甘草 山药 饮食 人参 医学 大便 疗程 知母 下陷
p8	National policy	中医 中医药 国家 医学 中药 医药 医疗 发展 结合 文化 临床 全国 研究 我国 社会
p9	Health preserving theory	治疗 药物 西药 西医 抗生素 疾病 使用 人体 病人 病毒 作用 引起 出现 没有 导致
p10	Specific examples	医院 医生 病人 没有 患者 治疗 时间 检查 手术 知道 时候 问题 结果 情况 认为

6 Further Analysis Based on Camps

In Sect. 4, discriminating words are generated by logistic regression. In Sect. 5, topics from respective camps are generated by topic modeling. For the context of TCM, we focus on the TCM technical terms in this section. Taking the TCM terminology dictionary from Sougou Cell dictionary[7] with 28428 TCM technical terms as reserved words, and filtering out other words, we do logistic regression to get more details of the debate.

---

[7] http://pinyin.sogou.com/dict/detail/index/20664.

Table 5. Topics from "abolishing TCM" replies

No.	Topics	Words related to topics
a1	Pseudo TCM experts	医院 没有 国家 医疗 行医 中医药 医生 工作 大学 患者 记者 部门 误诊 医学 国际
a2	Reject the dross	问题 世界 东西 祖先 历史 病人 人类 作为 体系 废除 作用 选择 接受 结合 五行
a3	Toxicity of TCM	方法 实验 服用 问题 植物 临床 副作用 毒性 研究 标准 朱砂 含有 动物 毒副作用
a4	Virus theory	研究 原因 禽流感 药物 引起 预防 结果 细菌 医学 方法 检查 死亡 治愈 临床 抗生素
a5	Doggerel	天涯 问题 治病 地方 老子 教养 看到 喜欢 水平 说明 世界 缺乏 相信 支持 浪费
a6	Theory of *yin-yang* and five elements	解释 认为 方法 知识 思想 文化 发展 真理 基础 不能 存在 检验 事物 社会 错误
a7	Feeling the pulse and acupuncture points	告诉 逻辑 养生 证据 傻子 时候 人们 能够 算命 事实 是不是 可能 承认 请问 不能
a8	Strange prescriptions	阴阳 垃圾 方子 回答 月经 明白 人中黄 解释 乾坤 听说 试试 五行 狗屁 不止 看到
a9	Scientific nature of TCM	认为 臆想 概念 五脏 心脏 循环 研究 致病 脏腑 解剖 六淫 组成 胃肠 阴阳 作用
a10	Rude Internet behaviors	医院 网友 手术 粉丝 结合 针灸 没有 疗效 遇到 寿命 治愈 仪器 重病 医治 神医

With the threshold of absolute coefficient 0.2, we get 562 discriminating words from 1049 words, including 305 words with positive coefficients (related to "preserving TCM" stance) and 257 words with negative coefficients (related to "abolishing TCM" stance).

Table 6 lists 12 TCM technical terms with high absolute coefficients from "preserving TCM" perspective. "Preserving TCM" stance holders always mention the TCM theories and philosophies (e.g., Nos. tp1, tp4 & tp7) and the specific medicines which are well known can actually treat some disease (e.g., Nos. tp6 & tp12).

Table 7 lists 12 TCM technical terms with high absolute coefficients from "abolishing TCM" perspective. "Abolishing TCM" stance holders have mainly four groups of technical terms. Group one (e.g., Nos. ta1, ta7, ta8 & ta11) contains those specific abstract conceptions which are difficult to be explained and understood. Group two refers to medical prescriptions contains abnormal materials (e.g., Nos. ta6, ta12) or materials with toxicity (e.g., No. ta4). Group three (e.g., No. ta4) explains that patients may recover themselves. Group four (e.g., Nos. ta3 & ta9) mentions acute diseases which cannot be cured by TCM.

Similarly, we use human names appeared in the corpus to do logistic regression because people usually quote others' sayings or list some human names related to famous events to support their stance in debates.

With the absolute threshold of absolute coefficient 0.2, we get 100 discriminating words from 7459 names, including 48 human names with positive coefficients (related to "preserving TCM" stance) and 52 human names with negative coefficients (related to "abolishing TCM" stance).

Table 6. 12 discriminating technical terms from "preserving TCM" stance

No.	Original Chinese words	Note
tp1	辩证施治	TCM philosophy
tp2	卒中	Illness
tp3	风热	Illness
tp4	脏腑学说	TCM theory
tp5	体外	TCM conceptions
tp6	桂枝汤 (Guizhi Decoction)	TCM prescription
tp7	奇经八脉	TCM theory
tp8	猪脑	TCM prescription
tp9	胃气	Illness
tp10	球后	A specific acupuncture point
tp11	实热	Illness
tp12	金匮要略	An ancient book about TCM

Tables 8 and 9 show 8 human names from each side of the debate by decreasing rank of their absolute coefficients. These human names (e.g., Nos. na1, na2 & na6) are well known to the pubic because they are pseudo experts or related to illegal practice of medicine. The historical figures (e.g., Nos. na4 & na5) are famous TCM practitioners in ancient China. The man (No. na7) is a western medicine doctor who made contribution for conducting epidemic prevention work in the 1910s. In the "preserving TCM" camp, there are mainly two groups of human whose names are referred. Group one (e.g., Nos. np1, np2, np3 & np7) are government administers who support TCM. These people in group 2 (e.g., Nos. na5 & na6) are doctors. Some journalists' names (e.g., Nos. na3 & np8) outperform as their newspaper articles supporting opposite stance are mentioned for many times. Some user IDs of Tianya Forum are referred because they are active participants during the debate (e.g., No. np4).

Table 7. 12 discriminating technical terms from "abolishing TCM" stance

No.	Original Chinese words	Note
ta1	六淫	TCM conception
ta2	康复医学	Rehabilitation medicine
ta3	肠痛	A kind of acute disease
ta4	缓解期	Remission stage
ta5	阅读障碍	Illness
ta6	齿垢 (Denticola)	TCM prescription
ta7	清热解毒	TCM conception
ta8	肝肾阴虚	Illness
ta9	脑出血 (hemorrhage)	A kind of acute disease
ta10	轻粉	A specific TCM
ta11	口舌生疮	Illness
ta12	人中黄	TCM prescription

Table 8. 8 discriminating human names from "preserving TCM" stance

No.	Human names	Note
np1	王国强	One government administer
np2	钱信忠	One government administer
np3	李斌	One government administer
np4	施正义	One active User ID during the TCM debate
np5	倪建俐	One doctor
np6	王拥军	One doctor
np7	温家宝	Former Prime Minister
np8	魏敏	One journalist

Table 9. 8 discriminating human names from "abolishing TCM" stance

No.	Human names	Note
na1	张悟本	One pseudo health expert
na2	闫芳	One pseudo Tai Chi and Kung Fu expert
na3	孙国根	One journalist
na4	华佗	One famous TCM practitioner in ancient China
na5	孙思邈	One famous TCM practitioner in ancient China
na6	胡万林	One pseudo health expert
na7	伍连德	One western medicine doctor
na8	刘海若	One journalist of a television station

7 Conclusions

This study explores a stance mining problem about a debate on societal issue TCM. We select one hot post on TCM from one of the most influential Chinese BBS, Tianya Forum, and automatically determine the replies' stances about TCM. Our results show that logistic regression can effectively select domain feature words and identify replies' stance with precision of 63.13 %, outperforming the SVM model using adjectives, adverbs, verbs and nouns as features.

Secondly, our topic modeling by LDA reveal that the emphases of the two camps are different during the debate. The "preserving TCM" stance holders concern the motivations of the other camp, the effectiveness of the TCM, etc. The "abolishing TCM" stance holders doubt the scientific nature and the rationality of TCM, introduce the modern medicine, and condemn the illegal medical practice relevant to TCM.

Thirdly, our further analysis verifies meanings of specific discriminating words present during the debate by logistic regression. The details of the concerned technical terms and human names in the different camps let us see how people express their viewpoints and perspectives during the TCM debate.

This paper provides an example for future research designed to explore stances on societal issues. In the future, we will do more study on identifying stance by interactions within debate and how opposing perspectives and arguments are put forward during debates.

Acknowledgments. This work was supported by the National Natural Science Foundation of China (Nos. 61473284 and 71371107).

References

1. Riloff, E.: Automatically generating extraction patterns from untagged text. In: 13th National Conference on Artificial Intelligence, Portland, pp. 1044–1049 (1996)
2. Riloff, E., Wiebe, J.: Learning extraction patterns for subjective expressions. In: Conference on Empirical Methods in Natural Language Processing, Sapporo, pp. 105–112 (2003)
3. Cui, H., Mittal, V., Datar, M.: Comparative experiments on sentiment classification for online product reviews. In: 21st National Conference on Artificial Intelligence, Boston, pp. 61–80 (2006)
4. Ng, V., Dasgupta, S., Arifin, S.M.N.: Examining the role of linguistic knowledge sources in the automatic identification and classification of reviews. In: International Conference on Computational Linguistics and Meeting of the Association for Computational Linguistics, Sydney, pp. 381–393 (2006)
5. Gamon, M.: Sentiment classification on customer feedback data: noisy data, large feature vectors, and the role of linguistic analysis. In: 23rd International Conference on Computational Linguistics, Beijing, pp. 841–847 (2010)
6. Pang, B., Lee, L., Vaithyanathan, S.: Thumbs up? sentiment classification using machine learning techniques. In: Conference on Empirical Methods in Natural Language Processing, Philadelphia, pp. 79–86 (2009)
7. Liu, B.: Opinion mining and sentiment analysis. Found. Trends Inf. Retrieval **2**(1–2), 1–135 (2008)
8. Zhao, Y.L., Tang, X.J.: In-depth analysis of online hot discussion about TCM. In: 15th International Symposium on Knowledge and Systems Science, pp. 275–283. JAIST Press, Sapporo (2014)
9. Turney, P.D., Littman, M.L.: Measuring praise and criticism: inference of semantic orientation from association. ACM Trans. Inf. Syst. **21**(4), 315–346 (2003)
10. Somasundaran, S., Wiebe, J.: Recognizing stances in ideological on-line debates. In: NAACL HLT 2010 Workshop on Computational Approaches to Analysis and Generation of Emotion in Text, Los Angeles, pp. 116–124 (2010)
11. Abbott, R., Walker, M., Anand, P., et al.: How can you say such things?!?: recognizing disagreement in informal political argument. In: Workshop on Languages in Social Media, Portland, pp. 2–11 (2011)
12. Anand, P., Walker, M., Abbott, R., et al.: Cats rule and dogs drool!: classifying stance in online debate. In: 2nd Workshop on Computational Approaches to Subjectivity and Sentiment Analysis, Portland, pp. 1–9 (2011)
13. Walker, M.A., Anand, P., Abbott, R., et al.: That is your evidence? Classifying stance in online political debate. Decis. Support Syst. **53**(4), 719–729 (2012)
14. Tikves, S., Banerjee, S., Temkit, H., et al.: A system for ranking organizations using social scale analysis. Soc. Netw. Anal. Min. **3**(3), 313–328 (2013)
15. Tikves, S., Gokalp, S., Temkit, M., et al.: Perspective analysis for online debates. In: International Conference on Advances in Social Networks Analysis and Mining, Istanbul, pp. 898–905 (2012)
16. Gryc, W., Moilanen, K.: Leveraging textual sentiment analysis with social network modeling: sentiment analysis of political blogs in the 2008 U.S. Presidential Election. In: Workshop on from Text to Political Positions, Amsterdam (2010)

17. Lin, W.H., Wilson, T., Wiebe, J.: Which side are you on? Identifying perspectives at the document and sentence levels. In: 10th Conference on Computational Natural Language Learning, New York, pp. 109–116 (2006)
18. Hatzivassiloglou, V., Wiebe, J.M.: Effects of adjective orientation and gradability on sentence subjectivity. In: International Conference on Computational Linguistics, Mexico, pp. 299–305 (2003)
19. Benamara, F., Cesarano, C., Picariello, A., et al.: Sentiment analysis: adjectives and adverbs are better than adjectives alone. In: Veselovská, K., Hajic, J., Šindlerová Bojar, O., Žabokrtský, Z. (eds.) International Conference on Weblogs and Social Media, Boulder (2007)
20. Subrahmanian, V.S., Reforgiato, D.: AVA: Adjective-verb-adverb combinations for sentiment analysis. IEEE Intell. Syst. 23(4), 43–50 (2008)
21. Shen, J., Zhu, P., Fan, R., Tan, W., Zhan, X.: Sentiment analysis based on user tags for traditional Chinese medicine in Weibo. In: Li, J., et al. (eds.) NLPCC 2015. LNCS, vol. 9362, pp. 134–145. Springer, Heidelberg (2015). doi:10.1007/978-3-319-25207-0_12
22. Blei, D.M., Ng, A.Y., Jordan, M.I.: Latent Dirichlet allocation. J. Mach. Learn. Res. 3, 993–1022 (2003)

Architecting Crowd-Sourced Language Revitalisation Systems: Generalisation and Evaluation to Te Reo Māori and Vietnamese

Asfahaan Mirza[✉] and David Sundaram

Department of Information Systems and Operations Management, University of Auckland,
Auckland, New Zealand
{a.mirza,d.sundaram}@auckland.ac.nz

Abstract. Many linguists claim as many as half of the world's nearly 7,105 languages spoken today could disappear by the end of this century. When a language becomes extinct, communities lose their cultural identity and practices tied to a language, and intellectual wealth. Preservation of endangered languages is a critical but challenging effort. A language is not preserved and revitalized by just documenting, archiving and developing shared resources. The revitalisation is highly dependent on the learning and usage of the language. Most current systems and approaches do one or the other. There are few systems or approaches that interweave preservation with learning. The purpose of our research is to architect a language revitalisation system that (a) leverages and integrates crowd-sourced collective intelligence approaches with knowledge management approaches to (b) capture, curate, discover, and learn endangered languages. We propose and implement an generalisable architecture that can support any language revitalisation effort in terms of capture, curate, discover, and learn. The validity of the research was tested by implementing the system to support Te Reo Maori and Vietnamese. Furthermore, we evaluate the concepts, processes, architecture, and implementation using a number of mechanisms.

Keywords: Language revitalisation · Crowd sourced · Social media · Knowledge management · Endangered languages · Collective intelligence · Mobile apps

1 Introduction

Many researchers predict that 90 % of the world's 7000 plus languages will become extinct or endangered within the next hundred years [1]. The decline of languages is due to many factors such as the globalization of culture, increase in development of web and communication technologies, and global commerce. These factors are influencing the movement towards dominance of a limited number of languages.

There has been a lot of research and development to support language documentation and revitalisation efforts. The graph in Fig. 1 illustrates current availability of systems and the problem and research gap. The y-axis shows systems that cater for language documentation and x-axis refers to systems that are available for language learning. The z-axis presents the platform of systems – desktop or mobile. There are many desktop

© Springer International Publishing Switzerland 2016
H.T. Nguyen and V. Snasel (Eds.): CSoNet 2016, LNCS 9795, pp. 333–344, 2016.
DOI: 10.1007/978-3-319-42345-6_29

applications focused towards language documentation such as capture and curation of languages (green zone in Fig. 1) but barely available on mobile applications. Furthermore, there are many applications for language learning both in desktop and mobile platforms (green zone in Fig. 1). The problem and research gap (yellow zone in Fig. 1) is that there is no integrated system that holistically integrates key language revitalisation components - capture, curate, retrieve and learn on a mobile platform.

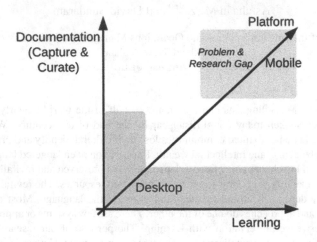

Fig. 1. Problem and research gap (Color figure online)

This research (yellow zone) tries to address the practical and research problems by exploring language revitalisation, collective intelligence and knowledge management approaches on mobile platform to capture, curate, discover and learn endangered languages anytime anywhere. Moreover, we postulate that contributors will help learners learn the language through capturing and curating data and learners will eventually become contributors themselves. In the following section we will briefly discuss the current state of endangered languages and how collective intelligence can be leverage to revitalize endangered languages.

1.1 Endangered Languages and Language Revitalisation

During the past 30 years extensive literature has been published on Language Revitalisation [2–7]. The languages are disappearing at a frightening rate. Crystal [8] estimates that an average of one language every 2 weeks may disappear over the next 100 years. Moreover, only approximately 600 languages may survive that have more than 100,000 speakers [2]. Hence, language revitalisation efforts need to be made to save languages and the culture and intellectual wealth embedded within.

Language revitalisation is to reverse the decline of a language or to revive an extinct language. Language revitalisation is also referred to as Reversing Language Shift (RLS) [3]. Language shift is the process when an individual or community language shifts from one language (generally their indigenous language) to another language. Taking essential measures to counter language shift is known as fostering language maintenance and/or language revitalisation [9].

Language revitalisation of endangered languages is essential to preserve linguistic and cultural diversity in world. In 1991, Joshua Fishman proposed an eight-staged system of reversing language shift which involves assessing the degree to endangerment of the language to identify the most effective method of assisting and revitalizing the language [3]. Therefore, the goals of a language revitalisation program should be set according to the current vitality of the language.

There are many proposed models for language revitalisation. To revitalize a language, just documentation is not sufficient, but we need to adopt or develop techniques of disseminating it to the community [10]. Hinton and Hale [11] in the "The Green Book" identify the five main categories for revitalisation as (1) school-based programs; (2) children's programs outside the school (after-school programs, summer programs); (3) adult language programs; (4) documentation and materials development; and (5) home-based programs [7, 11–13]. In the subsequent section, we briefly describe how collective intelligence and knowledge management fundamentals can be used to revitalize languages.

1.2   Language Revitalisation Using Collective Intelligence

Collective intelligence is the ability for different individuals to come together and form a group with the intention to share a common line of thought [14]. Social media is a key enabler towards collective intelligence. Social media refers to an interaction among individuals and groups where the participants are involved in the creation, sharing and exchange of information, ideas and data in a virtual community as well as networks [15]. The role played by individuals communicating in the past was not dynamic; the consumer audience and communicator were distinct groups. At present the consumers actively create, publish, produce and broadcast in the autonomous form that a platform facilitates [16]. Hence, social media concepts are optimal to apply in modern communications and interactions.

Collective intelligence dwells on three key principles that include cooperation, coordination and cognition [17]. Using these key principles, harnessing collective intelligence enables individuals and groups to solve practical problems. Most of the indigenous languages currently do not have presence in digital and social spaces. Currently there are limited indigenous language revitalisation oriented applications available for ubiquitous devices. Such applications do not allow user to holistically carry out key language documentation and learning function such as capture, curate, share, access, learn and multiple user collaboration. They are mainly focused as (1) Reference tools such as dictionaries, phrases, and stories; (2) Language learning based on already documented information (e.g. Go Vocab, Hika Explorer, Duo Lingo); (3) Capture limited data such as audio but do not allow sharing (e.g. Ma Iwaidja, Duo Lingo); (4) No curation process of user documented information; and (5) Do not facilitate collaborative user engagement towards language revitalisation.

In this research we explore how to leverage concepts of collective intelligence to save endangered languages. If a crowd sourced approach is employed using ubiquitous devices among the indigenous population, then the workload of language revitalisation is distributed. Moreover, design, develop and implement systems that can be used to mitigate the loss of languages, preserve and revitalize the affected languages.

2   Research Methodology

The primary aim of this research is to design and implement a system. The word "Design" means "to create, fashion, execute, or construct according to plan" [18]. Therefore, it is best to discover through design and adapt a multi-methodological approach to conduct this design science research [19]. For this study, Nunamaker's [20] multi-methodological approach for information systems research (ISR) will be adapted to propose and develop various artefacts. Moreover, the criteria for the design science artefacts proposed by Nunamaker et al. [20] and Hevner et al. [21] will be followed throughout the study.

The adapted multi-methodological approach is a practical way of designing and implementing a system. It consists of four research strategies/phases - observation, theory building, systems development and experimentation as illustrated in Fig. 2. The phases are not in any particular order but they are all mutually connected to support creation and validation of a system with multiple iterations. As this research focuses mainly on design and implementation of a system, the proposed approach will follow the sequence of observation, theory building, system development, and experimentation. As research progresses through each phase, the artefacts will be refined and generalised as depicted in Fig. 2. Generalisation of the artefacts is the centre focus of this research.

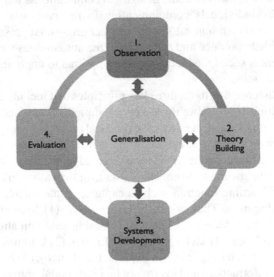

Fig. 2. Research methodology

Observation: The observation of existing literature and systems helps bring clarity to the research domain. We examined existing academic literature on language revitalisation and review existing applications that are available for indigenous languages. The outcome was comparison of existing applications available for language revitalisation [22].

Theory Building: This consists of adapting and developing ideas and concepts, creation of conceptual models, processes and frameworks. The proposed theories will help conceptualize a generic system that supports a crowd sourced approach towards language revitalisation including Te Reo Māori, Vietnamese and non-Roman languages. The outcomes were: conceptual concepts, models, processes, frameworks and architectures for crowd sourced knowledge management driven approach towards language revitalisation [22, 23].

Systems Development: The proposed concepts, models, processes, frameworks will enable us to design and implement a holistic crowd sourced knowledge management system to capture, curate, discover and learn Te Reo Māori which supports dialect variations and media such as words, phrases, imagery, poetry, proverbs and idioms that are common as well as specific to a particular tribe or family [22, 23]. This system is described in Sect. 3. The development of the Te Reo Māori revitalisation system will help demonstrate feasibility of the system for other endangered languages. The outcomes include Save Lingo – a crowd sourced knowledge management system to revitalize Te Reo Māori, Learn Lingo games (Flash cards and hangman) and a refined architecture and implementation.

Evaluation: Once the system is developed, we will adopt various evaluation mechanisms to validate and refine purposed theories (concepts, models, processes, frameworks and architectures) and to enhance and generalise our systems namely Save Lingo and Learn Lingo. Development is an iterative process and the issues identified during experimentation will lead to further refinement or creation of design artefacts. The evaluation plan is described in Sect. 5.

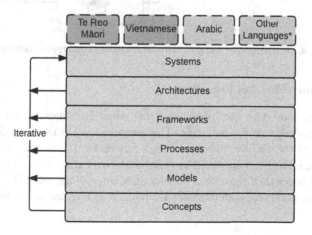

Fig. 3. Framework of common design elements

Generalisation: The generalisation of concepts, models, processes, frameworks, archi-tectures and systems is an ongoing process to make each of the artefacts applicable to all languages. The Framework of Common Design Elements as shown in Fig. 3 will be language independent. Initial implementation will be for Te Reo Māori, followed by Vietnamese, and then a non-Roman language such as Arabic. This will help us generalise our artefacts to a level that they are adaptable for majority of the languages.

3   Design and Implementation of Language Revitalisation for Te Reo Māori – Save Lingo and Learn Lingo

The synthesis of academic literature on collective intelligence, social media, knowledge management, ubiquitous systems and language revitalisation and analysis of current language systems led to the creation of the holistic crowd sourced model to revitalize endangered languages as illustrated in Fig. 4 [1, 11, 24–27]. It has five states are capture, curate, discover, learn and share. The *capture stage* allows contributors to capture words, phrases, idioms, stories and songs as text, audio, image or video in different dialects. The *curate stage* allows nominated experts approve, reject or modify captured records by wider audience. The *discover stage* allows wider community to access curated words, phrases, idioms, stories and songs. The *learn stage* facilitates using the curated data for the creation of interactive games such as Flash cards, hangman and more. While discovering and learning the language, users can share knowledge through social networks to promote the use of language.

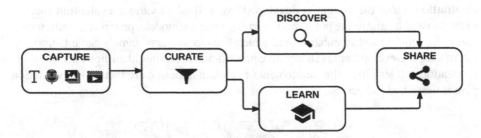

Fig. 4. Key concepts and processes to revitalize endangered languages

3.1   Governance Model of Records

The maintenance of data quality is essential when it comes to documenting and preserving knowledge of the language. The proposed governance model depicted in Fig. 5 illustrates the various stages a record can be in. Initial state of the record is *captured*. Once it has been approved by more than two expert users, it is considered to be *curated* and becomes *discoverable* to wider audience. If the record has been rejected more than being accepted, then its status is changed to rejected and is hidden to wider public.

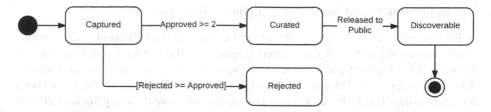

Fig. 5. Governance model of records within Save Lingo

3.2 Activity Diagram – Capture Function

In order to implement key use cases of the system [22], we need to better understand the flow of activities associated with the function. The high level activity diagram of capture functionality is illustrated in Fig. 6. It shows the interaction between the user and system to add a new record or add to an existing record available in the database.

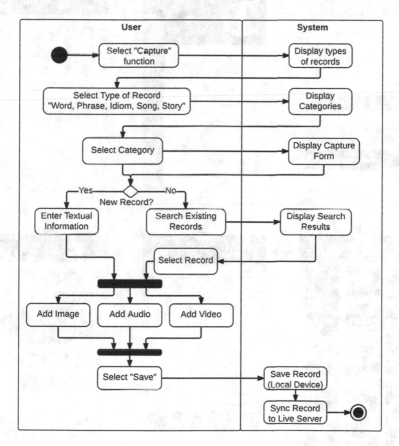

Fig. 6. Activity diagram of capture function

3.3 Implementation of Save Lingo App for Te Reo Maori

Te Reo Māori (Māori language) is the native language of New Zealand's native population. It is considered to be endangered language with decreasing number of active
speakers [28]. A prototypical implementation of the language revitalisation system
(Save Lingo) for te reo Māori in displayed in Fig. 7. The Save Lingo system has been
described in detail in the book chapter - *Design and Implementation of Socially Driven
Knowledge Management Systems for Revitalizing Endangered Languages* [22]. The key
features include (1) User personalization and registration, (2) Ability to capture content
in various formats, (3) Curation function for content governance, (4) Discover content
that has been curated, (5) Sharing via social media to encourage use of the language,

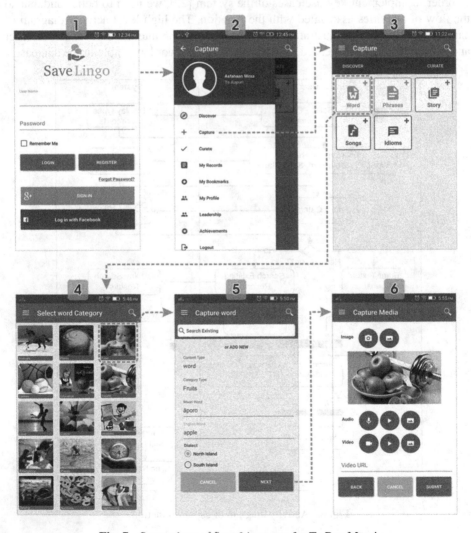

Fig. 7. Screenshots of Save Lingo app for Te Reo Māori

and lastly (6) Social Gamification via Google Play to enhance user experience and competition to help promote the language within the community.

4 Generalisation of Save Lingo to Other Languages Namely Vietnamese

In order to test the generalisability of the Save Lingo architecture, framework and system, we implemented the app for Vietnamese. Initially the system was developed specifically for Te Reo Māori which catered for language documentation functions such as capture and curate. Version 2.0 of the application was enhanced to include the ability to access/discover records that have been captured and curated. Furthermore, we incorporated social media integration, leader boards and gamification concepts. Once the application was fully functional we wanted to generalise the system so that it can be used to preserve and revitalize other endangered languages.

To facilitate multiple languages in future, major code refactoring of the app codebase as well as web services was required. Initially the web services were developed in PHP and database was in MySQL. During the refactoring phase, the web services/APIs were rewritten in ASP.NET and database was migrated to Microsoft SQL Server. To ensure the scalability of storing images, audio and video files, the system was integrated with Dropbox infrastructure. The files are stored on Dropbox, and only the reference to the file is stored in the Microsoft SQL database.

After making the necessary changes, the app was implemented to support English-Vietnamese. The app was presented at Ton Duc Thang University in Ho Chi Minh City, Vietnam on 20 March 2016. The audience were native speakers of Vietnamese, students, designers, academics and architects of information systems. The feedback from them was positive and constructive. The feedback was taken on board and has been incorporated into Save Lingo version 3.0 app displayed in Fig. 8.

Fig. 8. Screenshots of the generalised Save Lingo 3.0 app for Vietnamese language

5 Evaluation

Our research on language revitalisation systems has produced a number of artefacts. These artefacts include conceptual model, processes, frameworks, high level and detailed level architectures, and save lingo and learn lingo systems. To validate them, we adopted a variety of evaluation methods in our research methodology design. A summary of these evaluation methods is provided in the Table 1.

Table 1. Evaluation methods

Evaluation method	Description
Architecture analysis	Assessing the fit of Save Lingo frameworks and architectures into technical IS architectures
Prototyping	Validating the proposed theory of holistic crowd source approach for language revitalisation through system implementations
Generalisability	Implement Save Lingo into multiple languages – Te Reo Māori, Hawaiian, and Non-roman script (Arabic)
Static analysis	Examining the prototype at unit, technology, and system levels
Functional testing	Testing how the key functionalities of the prototype - capture, curate, discover, learn, and share
Structural testing	Testing the internal structure of the prototype
Computer simulations	Testing and executing the prototype and key functionalities with artificial data
Case study	Apply proposed artefacts to various languages, namely Te Reo Māori and Vietnamese to examine the application of the holistic crowd sourced processes and the prototype
Illustrative scenarios	Demonstrating how the prototype supports the key language revitalisation processes – capture, curate, discover, learn and share
Expert evaluation	Presenting, publishing and validating our research artefacts through peer (system, domain and academic experts) reviewed sessions, seminars, conferences and journals

In order to assess the validity of the research artefacts, one or more evaluation methods were employed according to the nature and evaluation requirements of the research artefact. Table 2 presents the summary of our research artefacts and their selected evaluation methods.

Table 2. Research artefacts and evaluation methods

Artefacts	Definition	Model	Processes	Features	Frameworks	Architecture	Prototype
Architecture analysis					✔	✔	
Prototyping			✔	✔	✔	✔	
Generalisability	✔	✔			✔	✔	✔
Static analysis							✔
Functional testing							✔
Structural testing							✔
Computer simulations							✔
Case study			✔	✔			✔
Illustrative scenarios			✔	✔			✔
Expert evaluation	✔	✔	✔	✔	✔	✔	✔

Significant mechanisms were used to evaluate the generalisability of this research. Whether our concepts, processes, framework, architecture, and implementation would be applicable to many languages. We have successfully implemented the system to support Te Reo Maori and Vietnamese. Since both these languages are based on the Roman script the challenges of generalisation were moderate. Implementation of Save Lingo and Learn Lingo to non-Roman scripts will help further generalise our concepts.

6 Conclusion

The rapid disappearance of vital knowledge and culture embedded within languages, as well as the limitation of current systems and approaches motivates this research to design and implement a holistic crowd sourced knowledge management approach to revitalise endangered languages. The primary contributions of this research are towards endangered language revitalisation. We have so far implemented the crowd-sourced language revitalisation system to save and learn Te Reo Māori and Vietnamese. The Save Lingo system has further been generalised to support non-Roman script languages including Arabic, Chinese, Hindi, Urdu and more. We have also evaluated the system using ten mechanisms that range from architectural analysis to functional testing to expert testing. The validated concepts, models, processes, framework, architecture, and implementation could potentially contribute to closely related disciplines such as education, linguistics, computer science and information systems provided in the Table 1.

References

1. Romaine, S.: Preserving endangered languages. Lang. Linguist. Compass **1**, 115–132 (2007)
2. Krauss, M.: The world's languages in crisis. Language (Baltim) **68**, 4–10 (1992)
3. Fishman, J.: Reversing language shift: theoretical and empirical foundations of assistance to threatened language. Multilingual Matters (1991)
4. Nettle, D., Romaine, S.: Vanishing Voices: The Extinction of the World's Languages. Oxford University Press, Oxford (2000)
5. Gibbs, W.W.: Saving dying languages. Sci. Am. **287**, 78–85 (2002)

6. Grenoble, L.A., Whaley, L.J.: Saving Languages: An Introduction to Language Revitalization. Cambridge University Press, Cambridge (2005)
7. Brenzinger, M., Yamamoto, A., Aikawa, N., Koundiouba, D., Minasyan, A., Dwyer, A., Grinevald, C., Krauss, M., Miyaoka, O., Sakiyama, O.: Language vitality and endangerment. In: Paris: UNESCO Intangible Cultural Unit, Safeguarding Endangered Languages, 1 July 2010. UNESCO (2003)
8. Crystal, D.: Language Death. Cambridge University Press, Cambridge (2002)
9. Dwyer, A.M.: Tools and techniques for endangered-language assessment and revitalization. Minor. Lang.Today's Glob. Soc. (2012)
10. Goodfellow, A.M.: Speaking of Endangered Languages: Issues in Revitalization. Cambridge Scholars Publishing, Newcastle upon Tyne (2009)
11. Hinton, L., Hale, K.L.: The Green Book of Language Revitalization in Practice. Academic Press, San Diego (2001)
12. Amery, R.M.: Warrabarna Kaurna! Reclaiming an Australian Language. Swets & Zeitlinger BV, Lisse (2000)
13. King, J.: Te Kohanga reo: Maori language revitalization. In: The Green Book of Language Revitalization in Practice, pp. 119–128 (2001)
14. Kubátová, J.: Growth of collective intelligence by linking knowledge workers through social media. Lex ET Sci. Int. J. 1, 135–145 (2012)
15. Kaplan, A.M., Haenlein, M.: Users of the world, unite! The challenges and opportunities of Social Media. Bus. Horiz. 53, 59–68 (2010)
16. Lewis, B.K.: Social media and strategic communications: attitudes and perceptions among college students. Public Relat. J. 4(3), 1–23 (2010)
17. Engel, D., Woolley, A.W., Jing, L.X., Chabris, C.F., Malone, T.W.: Reading the mind in the eyes or reading between the lines? Theory of mind predicts collective intelligence equally well online and face-to-face. PLoS One 9, e115212 (2014)
18. Merriam-Webster Dictionary: "Design" Meaning. http://www.merriam-webster.com/dictionary/design
19. Baskerville, R.: What design science is not. Eur. J. Inf. Syst. 17, 441–443 (2008)
20. Nunamaker Jr., J.F., Chen, M., Purdin, T.D.M.: Systems development in information systems research. J. Manag. Inf. Syst. 1(3), 89–106 (1990)
21. Hevner, A.R., March, S.T., Park, J., Ram, S.: Design science in information systems research. MIS Q. 28, 75–105 (2004)
22. Mirza, A., Sundaram, D.: Design and Implementation of socially driven knowledge management systems for revitalizing endangered languages. In: Helms, R., Cranefield, J., van Reijsen, J. (eds.) Social Knowledge Management in Action, in the Knowledge Management and Organizational Learning Series. Springer, Berlin (2016)
23. Mirza, A., Sundaram, D.: Harnessing collective intelligence to preserve and learn endangered languages. In: Proceedings of 2nd EAI International Conference on Nature of Computation and Communication, Rach Gia, Vietnam (2016)
24. Nonaka, I., Takeuchi, H., Umemoto, K.: A theory of organizational knowledge creation. Int. J. Technol. Manag. 11, 833–845 (1996)
25. Alavi, M., Leidner, D.E.: Review: knowledge management and knowledge management systems: conceptual foundations and research issues. MIS Q. JSTOR 25(1), 107–136 (2001)
26. Lévy, P.: Collective Intelligence. Plenum/Harper Collins, New York (1997)
27. Malone, T.W., Laubacher, R., Dellarocas, C.: Harnessing Crowds: Mapping the Genome of Collective Intelligence. MIT Sloan Research Paper No. 4732-09 (3 Feb 2009)
28. Statistics New Zealand: 2013 Census QuickStats about Māori - Statistics New Zealand (2013)

Collective Online Clicking Pattern on BBS as Geometric Brown Motion

Zhenpeng Li[1](✉) and Xijin Tang[2](✉)

[1] Department of Applied Statistics, Dali University, Dali 671003, China
lizhenpeng@amss.ac.cn
[2] Academy of Mathematics and Systems Sciences, Chinese Academy of Sciences,
Beijing 100190, China
xjtang@amss.ac.cn

Abstract. In this paper, we focus on massive clicking pattern on BBS. We find that the frequency of clicking volumes on BBS satisfies log-normal distribution, and both the lower-tail and upper-tail demonstrate power-law pattern. According to the empirical statistical results, we find the collective attention on BBS is subject to exponential law instead of inversely proportional to time as suggested for Twitter [4]. Furthermore we link the dynamical clicking pattern to Geometric Brown Motion (GBM), rigorously prove that GBM observed after an exponentially distributed attention time will exhibit power law. Our endeavors in this study provide rigorous proof that log-normal, Pareto distributions, power-law pattern are unified, most importantly this result suggests that dynamic collective online clicking pattern might be governed by Geometric Brown Motion, embodied through log-normal distribution, even caused by different collective attention mechanisms.

Keywords: Geometric Brown Motion · Log-normal distribution · Power-law · Collective behaviors over BBS

1 Introduction

Humans complex social behavior patterns are displayed through the cumulative effects of individual behaviors. One of the most common strategies in studying the social behaviors is to investigate and interpret whether any "pattern" is presented by fitting observed statistical regularities via data analysis. If the observed pattern can be described by a model characterized by related social psychological factors, that means we are close to the mechanisms that generate the collective regularity. As the main communication and information transmission tools in Web 1.0 era, bulletin board systems (BBS) and online communities were the main platforms for online activities in the whole Chinese cybersphere before 2005. BBS such as Tianya Forum expose digital traces of social discourse with an unprecedented degree of resolution of individual behaviors, and are characterized quantitatively through countless number of clicks, comments, replies and updates. Thanks to the different working functional designs, comparing with

© Springer International Publishing Switzerland 2016
H.T. Nguyen and V. Snasel (Eds.): CSoNet 2016, LNCS 9795, pp. 345–353, 2016.
DOI: 10.1007/978-3-319-42345-6_30

micro-blogging systems such as Twitter, long-time dynamics of human collective patterns on BBS are more stably showed out. Here we focus on massive clicking pattern on BBS. We analyze a large-scale record of Tianya Forum activity and find that the frequency of clicking volumes satisfies log-normal distribution, and both the lower-tail and upper-tail demonstrate power-law behavior. Furthermore we prove that the power-law behavior is caused by collective attention exponential decay. According to the empirical statistical results, we link the dynamical clicking pattern to Geometric Brown Motion (GBM), and rigorously provide a quantitative interpretation for the collective clicking phenomenon on BBS.

2   Data Source

Tianya Forum, as one of the most popular Internet forums in China, was founded on March, 1, 1999[1]. Till 2015, it was ranked by Alexa[2] as the 11th most visited site in the People's Republic of China and 60th overall. It provides BBS, blog, microblog and photo album services. With more than 85 million registered users, it covers more than 200 million users every month [1]. Tianya BBS, composed of many different boards, such as Tianya Zatan, entertainment gossip, emotional world, Media Jianghu, etc. is a leading focused online platform for important social events and highlights in China. We obtain the data by using automatic web mining tool - gooSeeker[3] and collect 22,760 posts from the Media Jianghu Board (MJB) of Tianya Forum during the replying time span from 13 June, 2003 to 16 September, 2015. The layout of MJB is shown in Fig. 1. Each post can be described by a 5-tuple: <title, author, clicking volumes, replying volumes, and replying time>. The 5-tuple dynamic is the feedback of user community behavior, and reflects collective online patterns. For example, posting represents that users release posts and want to be concerned, posting volumes reflect the active level of MJB, clicking means that visitors are interested in the posts or reflects the posts attraction level, while replying activities represent that users have intention to join the collective action compared with simple browsing (clicking), since replying behaviors indicate joiners have more in-deep thinking and enthusiasm towards the forum topics.

As for certain title (i.e. topic), the ratio between clicking volume and replying volume reflects the attention rate of the post and public participation degree. These cumulative micro individual behaviors (such as the number of posts, clicks and replies, the ratio between clicking volume and relying volume for each post) contribute to the global collective patterns, which could be measured by quantitative data analysis and modeling methods. Based on the above ideas, in this study, we take the replying and clicking volumes as the quantitative indexes to describe online group behaviors in the forum.

---

[1] http://bbs.tianya.cn/.

[2] Alexa Internet, Inc. is a California-based company that provides commercial web traffic data and analytics. https://en.wikipedia.org/wiki/Alexa-Internet.

[3] http://www.gooseeker.com/.

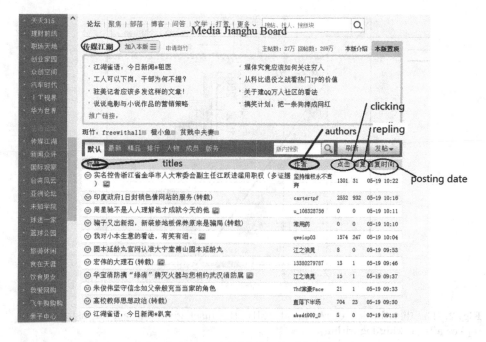

Fig. 1. Layout of Media Jianghu Board

3   The Distribution of Replies

Replying behaviors indicate visitors have more deep thinking and enthusiasm towards the forum topics. In order to study the pattern of replies, for 22,760 posts, we count each post replying volume. Around 34 % of the total posts, or 6,828 posts have no replies. After removing the no-reply records, we investigate the posts replying pattern. The statistics result is as shown in Fig. 2.

The inset in Fig. 2 suggests that the replies after taking logarithm follow exponential distribution, and log-log scale plot demonstrates power law pattern (take the logarithm for both replies and the corresponding number of posts). Next we fit the power-law distribution $f(x) \propto x^\alpha, x > x_{min}$.

We estimate the lower bound of the power-law behavior x_{min}, and scaling exponent based on the method described in [2]. We find that when ln(replying) > 3.4340, or replying volume > 31 (the estimate of the lower bound of the power-law behavior), the distribution of replies at MJB demonstrates power-law pattern, and maximum likelihood estimate of the scaling exponent $\alpha = -1.51$.

More replying activities represent the users have more active intention to join the collective action, meanwhile replying volumes show the topics' attraction or novelty levels, which means collective attention on MJB can be described by exponential distribution of replies. It is worth to note that as a function of time t, based on exponential form novelty decay, we will unify log-normal, Power law, and Pareto distribution by Geometric Brown Motion (GBM), and provide rigorous mathematic proofs in next section.

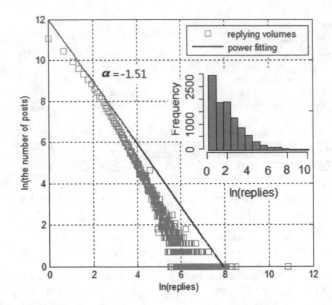

Fig. 2. The distribution of replies at MJB (the inset gives the actual histograms of replies after taking logarithm).

We use the replying number of a new post within 24 h as an index to measure collective attentions for the post. We randomly select 1000 samples from the total 22,760 posts, and average the number of replies in the first 24 h. We plot the average density distribution of replying volumes in Fig. 3. Different topics attract different users and show different attention characteristics. However the average results confirm that the collective attention is subject to exponential law.

According to the observation as shown in Fig. 3, we introduce d_t as a function of time t to account for the novelty decay [3], where d_t is defined as $d_t = \lambda e^{-\lambda t}$, $t > 0, \lambda > 0$. Here we set novelty decay d_t as exponential form instead of inversely proportional to time t in Ref. [4], where $d_t \propto 1/t$ is used to describe novelty decay of collective attentions on Twitter, as Twitter has the properties of instant arriving and fast transmission. On the contrary, usually with clearly defined title and no limitation of post length, BBS allow registered visitors to drop comments on the posts, thus generate interaction and discussion about the topics at hand. We estimate $\hat{\lambda} = 0.4888 (R^2 = 0.965)$ by nonlinear least squares method according to the selected 1000 samples. The empirical observation is not consistent with [4]. In contrast to BBS, we see Twitter has higher degree of attention as the new themes released, but the attention level declines more rapidly with the comparison as shown in Fig. 3(b). At any time of the first 24 h, Twitter users' attention level decay rate (the derivatives of the curves) is faster than that of BBS users. The characteristic is more dominant in the first 5 h, the slope $k_{Twitter}$ is obviously larger than k_{BBS} This result suggests that BBS and Twitter might have some different collective attention features also demonstrates the focusing

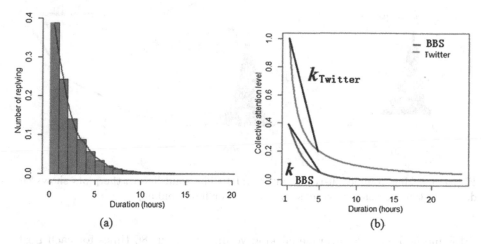

(a)   (b)

Fig. 3. Average density distribution of replying on 1000 samples in the first 24 h. (The curve line is the kernel density estimation)(Color figure Online)

patterns of behaviors emerged from the platforms between Web 1.0 and Web 2.0 are different.

4   The Distribution of Clicking Volume

We measure all the 22,760 posts on MJB with replying time span from 13 June, 2003 to 16 September, 2015. Replying and clicking time accurate to the second. We count C_t^q the clicking volumes for each post q on the Board at its corresponding replying time stamp t. The replying time stamp t is continuous, C_t^q describes the collective users' browsing pattern. At first we analyze all the 22,760 posts clicking volumes distribution in the given replying time span.

Figure 4(a) immediately suggests that the clicking volumes for the total $N = 22,760$ posts are distributed according to log-normal distribution. Since the horizontal axis is logarithmically rescaled, the histograms appear to be Gaussian function. A Kolmogorov-Smirnov normality test of $ln(N)$ with mean 4.94826 and standard deviation 1.4427 yields a p-value of 0.0536 and testing statistic $D = 0.0895$, suggests that the frequency of clicking volumes follows a log-normal distribution. Since p-value is at the critical point of rejection region, we need to check normal distribution significance further with Quantile-Quantile (Q-Q) plots. If the random variable of the data is a linear transformation of normal variate, the points will line up on the straight lines shown in the plots. Consider Fig. 4(c), it is obvious that the empirical distributions are apparently more skewed than in the normal case. However, we observe that the (logarithmically rescaled) empirical distributions exhibit normality with the exception of the high and low end of the distributions. These tail outliers occur more frequently than could be expected for a normal distribution. We estimate $ln(N) = 4.4486$ by

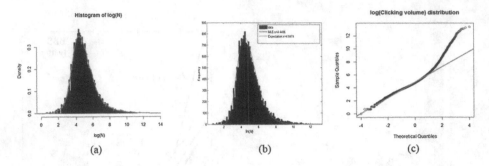

Fig. 4. Clicking volumes distribution on MJB (The solid line in the plots shows the density estimates using a kernel smoother)(Color figure online)

MLE method, e.g. the average clicking volume is about 86 times for each post, the result is as shown in Fig. 4(b).

About the tails distributions, we compute both lower tail (clicking volumes cumulative frequency below a given level) and upper tail (clicking volumes cumulative frequency above a given level) distributions. Figure 5 shows the cumulative frequency (in logarithmic scale) above (a) and below (b) a given level (in logarithmic scale), and demonstrates the upper-tail power-law behaviors, long recognized in the laws of Pareto and Zipf.

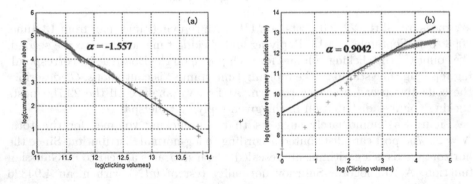

Fig. 5. Clicking volumes distribution on MJB (The "+" symbol refers to real data, and solid line in the plots is real data fitting line)

5   The Unified Stochastic Modeling on Users Clicking Pattern

As observed in Sect. 4, the frequencies of users clicking volumes satisfy log-normal distribution, in addition, both the lower-tail and upper-tail demonstrate power-law behaviors. From stochastic processes (Geometric Brown Motion) perspective, this paper contains a quantitative interpretation for this collective phenomenon. Our endeavors will focus on mathematic rigorous proofs why log-normal, Pareto distributions, lower and upper tails power-law pattern are unified. With the analytic results of visitors' clicking records on MJB, our aim is to bridge the results of theoretical modeling and empirical data analysis. In statistics, the generalized Pareto distribution (GPD) is a family of continuous probability distributions. It is often used to model the tails of another distribution. The power law distribution and Pareto distribution (sometimes called Zipf's law) are unified based on the fact that Cumulative Distributions Function (CDF) of Probability Density Function (PDF) with a power-law form follows Pareto distribution (Zipf's law) [5].

Next we model visitors' clicking patterns per unit time as Geometric Brown Motion and prove that under the condition of visitors' clicking volumes $C_{r(t)}$ as the function of attention level r a random variable subject to exponential distribution, double tails power-law characteristic is obtained for visitors' clicking volumes evolving as GBM.

The temporal evolution of many phenomena exhibiting power-law characteristic is often considered to involve a varying but size independent proportional growth rate, which mathematically can be modelled by Geometric Brownian Motion (GBM). We set the clicking volumes fluctuation for all posts on MJB as a function of random variable that is subject to exponential distribution instead of directly as a function of fixed time stamp. The general explanations root in the fact that new topics will compete with old interesting ones, due to the limited attention of visitors or novelty decay of new topics, new posts can usurp the positions of earlier topics of interest, and soon older contents attentions are replaced by newer ones, but all these are random. We use the empirical result in Sect. 4 that the novelty decay or attention level is defined as $d_t = \lambda e^{-\lambda t}$. In other words, if we look d_t as a probability density function, then collective attention time can be seen as a random variable that satisfies exponential distribution. It seems more reasonable that the attention time to one post would be considered as a random variable, and hence, attention time is assumed as an exponential distributed random variable might be more accurate for browsing scenarios of BBS post. That is why we define stochastic fluctuation of clicking volume on the BBS post as a function of exponential distributed random variable T. According to the above analysis, firstly, we define stochastic fluctuation of clicking volumes on the forum as GBM

$$dC_T = \mu C_r dr + \sigma C_r dW_r \tag{1}$$

where W_r is the standard Wiener process with $W_0 = 0, W_t - W_s \sim N(0, t-s)$ (for $0 \leqslant s < t$) and $N(\mu, \sigma^2)$ denotes the normal distribution with expected value μ and variance σ^2. With the initial state C_{r_0} after some fixed time r, by using Ito integral to (1), we have

$$C_r = C_{r_0} e^{(\mu - \frac{\sigma^2}{2})r + \sigma W_r} \tag{2}$$

Taking logarithmic on both sides of Eq. (2), we have the following logarithmic form

$$log(C_r) = log(C_{r_0}) + (\mu - \frac{\sigma^2}{2})r + \sigma W_r \tag{3}$$

Equation (3) shows that given initial state C_{r_0} and fixed r since W_r is subject to normal distribution, $log(C_{r_0}) + (\mu - \frac{\sigma^2}{2})r$ is constant, $log(C_r)$ is subject to normal distribution, with $E(log(C_r)) = log(C_{r_0}) + (\mu - \frac{\sigma^2}{2})r$ and $var(log(C_r)) = \sigma^2 r$. Hence, we rigorously prove that C_r subject to log-normal distribution, but we could not confirm if it exhibits power-law behavior.

If we regard C_r as a function of an exponential distributed random variable instead of fixed r, we prove that GBM will exhibit power law characteristic as following. Without losing generality, for the computation simplicity, we set $C_{r_0} = 1, \sigma_2 = 1, \mu = \frac{1}{2}$, i.e. $log(C_r) \sim N(0, r)$. Since

$$f(C_r) = \int_0^\infty f(C_r, r)dr = \int_0^\infty f(C_r|r)f(r)dr, \tag{4}$$

then if we stop the process at an exponentially distributed time with mean $\frac{1}{\lambda}$, i.e. $f(r) = \lambda e^{-\lambda r}, r > 0$, the density function of C_r is

$$f(C_r) = \int_0^\infty f(C_r, r)dr = \int_0^\infty \lambda e^{-\lambda r} \frac{1}{\sqrt{2\pi r}C_r} e^{\frac{-(lnC_r)^2}{2r}} dr. \tag{5}$$

Using the substitution $r = u^2$, gives

$$f(C_r) = \frac{2\lambda}{\sqrt{2\pi}C_r} \int_0^\infty e^{-\lambda u^2 - \frac{(lnC_r)^2}{2u^2}} du. \tag{6}$$

we have the integral result for $C_r \geq 1$

$$f(C_r) = \frac{\lambda}{\sqrt{2\pi}C_r} \sqrt{\frac{\pi}{\lambda}} e^{-2\sqrt{\frac{\lambda(lnC_r)^2}{2}}} = \sqrt{\frac{\lambda}{2}} C_r^{-1-\sqrt{2\lambda}} \tag{7}$$

which is named Pareto distribution, and exhibits power-law behavior in both tails as observed in Fig. 5.

With this we end the proof. Interestingly, the result shows that clicking dynamics on the forum yields power law behavior. The above results also suggest a generic conclusion that although the GBM is used to generate log-normal distributions, only a small change from the lognormal generative process might yield a different distributed pattern.

6   Conclusions

To study the dynamics of collective attention in social media, in this paper we conduct a study on the cumulative micro individual behaviors, such as clicking volume and relying volume for each post on Media Jianghu Board of Tianya

Forum. Data analysis result shows that the frequency of clicking volumes follows a log-normal distribution. In order to explain the phenomenon, we use Geometric Brownian Motion to model the collective clicking fluctuation and the model is well matched with our empirical result. Moreover we rigorously prove that the emergence of users' collective clicking volumes double tails power-law pattern is caused by the collective attention exponential decay. This result suggests that dynamic collective online clicking pattern on BBS posts might be governed by Geometric Brown Motion, embodied through log-normal distribution, and rooted in collective attention exponential decay mechanism.

Acknowledgments. This research was supported by National Natural Science Foundation of China under Grant Nos. 71171187 and 61473284, 71462001, the Scientific Research Foundation of Yunnan Provincial Education Department under Grant No. 2015Y386, and No. 2014Z137, and the Open Project Program of State Key Laboratory of Theoretical Physics, Institute of Theoretical Physics, Chinese Academy of Sciences.

References

1. Tianya Forum. http://help.tianya.cn/about/history/2011/06/02/166666.shtml
2. Clauset, A., Shalizi, C.R., Newman, M.E.J.: Power-law distributions in empirical data. SIAM Rev. **51**(4), 661–703 (2009)
3. Wu, F., Huberman, B.A.: Novelty and collective attention. Proc. Natl. Acad. Sci. U.S.A. **104**(45), 17599–17601 (2007)
4. Asur, S., et al.: Trends in social media: persistence and decay. SSRN 1755748 (2011)
5. A Ranking Tutorial. http://www.hpl.hp.com/research/idl/papers/ranking/ranking.html

Author Index

Printed in the United States
By Bookmasters